ARISTOTLE ON FEMALE ANIMALS

Aristotle's account of female nature has received mostly negative treatment, emphasizing what he says females cannot do. Building on recent research, this book comprehensively revises such readings, setting out the complex and positive role played by the female in Aristotle's thought with a particular focus on the longest surviving treatise on reproduction in the ancient corpus, the *Generation of Animals*. It provides new interpretations of the nature of Aristotle's sexism, his theory of male and female interaction in generation and his account of inherited features. It also discusses a range of more general issues which can and should be re-examined in light of Aristotle's account of female animals: his methodology, hylomorphism, teleology and psychology. *Aristotle on Female Animals* will be valuable to all those interested in Aristotle's philosophy and the history of gender.

SOPHIA M. CONNELL is an Affiliated Lecturer in the Faculty of Philosophy, University of Cambridge, a Fellow of Selwyn College and a Bye Fellow of Newnham College. She has published articles on Aristotle's biology and Galen and Aristotle on women's bodies.

CAMBRIDGE CLASSICAL STUDIES

General editors

ARISTOTLE ON FEMALE ANIMALS

A Study of the *Generation of Animals*

SOPHIA M. CONNELL
University of Cambridge

CAMBRIDGE
UNIVERSITY PRESS

CAMBRIDGE
UNIVERSITY PRESS

University Printing House, Cambridge CB2 8BS, United Kingdom

Cambridge University Press is part of the University of Cambridge.

It furthers the University's mission by disseminating knowledge in the pursuit of education, learning and research at the highest international levels of excellence.

www.cambridge.org
Information on this title: www.cambridge.org/9781107136304

First published 2016

Printed in the United States of America by Sheridan Books, Inc.

A catalogue record for this publication is available from the British Library

Library of Congress Cataloguing in Publication data
Connell, Sophia M., author.
Aristotle on female animals : a study of the generation of animals / Sophia M. Connell.
New York : Cambridge University Press, 2016. | Series: Cambridge classical studies |
Includes bibliographical references.
LCCN 2015033488 | ISBN 9781107136304 (hardback)
LCSH: Females. | Reproduction. | Aristotle. | Aristotle. On the generation of animals.
LCC QP259 .C66 2016 | DDC 591.56–dc23
LC record available at http://lccn.loc.gov/2015033488

ISBN 978-1-107-13630-4 Hardback

For my mother and my daughters

CONTENTS

LIST OF FIGURES AND TABLES

Figure

Tables

ACKNOWLEDGEMENTS

This book is a development of my work for an MPhil and a PhD in the Faculty of Classics, University of Cambridge. It has gone through many changes since that time. I am most especially grateful to my PhD supervisor, Geoffrey Lloyd, who has always believed that this work would eventually be complete. I would like to thank him for his profound insights and his generous support. I would also like to acknowledge the assistance of Myles Burnyeat, Robert Wardy, Nicholas Denyer, David Sedley and Malcolm Schofield, who advised me during my early research, and David Charles, my external PhD examiner.

I have presented earlier versions of the chapters of this book in various settings and have particularly benefitted from the criticisms and suggestions of Allan Gotthelf, James Lennox, Robert Bolton, Marguerite Deslauriers, Devin Henry, Mariska Leunissen, Joseph Karbowski, Philip van der Eijk, Andrew Coles, Sabine Föllinger, Chad Weiner and Richard King. Most recently, I presented part of Chapter 6 at the Sorbonne, University of Paris. I would like to thank David Lefebvre, Greg Salmieri, James Lennox and Andrea Falcon for their questions on that occasion. I am also very grateful to Liba Taub and Rae Langton, my generous colleagues at Newnham College, for reading over and commenting on drafts of chapters.

Academic institutions that have supported me over the years include the Social Sciences and Humanities Research Council of Canada, Darwin College, Cambridge, St John's College, Cambridge, the Faculty of Philosophy, University of Cambridge and Newnham College, Cambridge. I am very grateful to those involved in the editing and production of this book. From Cambridge University Press I thank especially Michael Sharp, Elizabeth Hanlon, David Mackenzie and two anonymous readers for their extremely helpful comments and criticisms. Thanks also to Ramya Ranganathan for project management, Kevin Hughes for copy editing, Ellisif Wasmuth for help with the Index locorum and Chiara Blanco for proofreading the Greek text.

I am grateful to my friends and family for their many kindnesses, particularly my mother and father, Rosanna Omitowoju, Susanna Avery-Quash, Jan-Melissa Schramm, Christopher Schramm, John Short, William Elliott, Daria Chrin-Elliott, Anthea Connell, Derek Matravers, Alix Cohen, Florian Steinberger, Jane Heal, Fraser MacBride, Simon Harrison, Lubomira Radoilska, Helen Dolling, Katherine Woodard, Melinda Rose, Katie Tomkies, Erika Hutchinson and Daryll Hughes. They will all know how much it means to me to publish this book. Finally, I owe the greatest debt of gratitude to my husband, Phil Connell. If there is any virtue in having completed this project, I must share it with him, a true friend. 'For what our friends achieve is, in a way, achieved through our own agency' (*EN* 1112b27).

ABBREVIATIONS

Aristotle's works

A. Po.	*Analytica Posteriora (Posterior Analytics)*
A. Pr.	*Analytica Priora (Prior Analytics)*
Cael.	*De Caelo (On the Heavens)*
Cat.	*Categoriae (Categories)*
de An.	*de Anima (On the Soul)*
Div. Somn.	*De Divinatione per Somnum (On Divination by Dreams)*
EE	*Ethica Eudemia (Eudemian Ethics)*
EN	*Ethica Nicomachea (Nicomachean Ethics)*
GA	*De Generatione Animalium (Generation of Animals)*
GC	*De Generatione et Corruptione (Generation and Corruption)*
HA	*Historia Animalium (History of Animals)*
IA	*De Incessu Animalium (Progression of Animals)*
Insomn.	*De Insomniis (On Dreams)*
Juv.	*De Juventute et Senectute [De Vita et Morte] (On Youth and Old Age [On Life and Death])*
Long.	*De Longitudine et Brevitate Vitae (On Length and Shortness of Life)*
MM	*Magna Moralia*
MA	*De Motu Animalium (Movement of Animals)*
Mem.	*De Memoria et Reminiscentia (On Memory and Recollection)*
Metaph.	*Metaphysica (Metaphysics)*
Mete.	*Meteorologica (Meteorology)*
PA	*De Partibus Animalium (Parts of Animals)*
Ph.	*Physica (Physics)*
Poet.	*Poetica (Poetics)*
Pol.	*Politica (Politics)*

Resp.	*De Respiratione (On Respiration)*
Rh.	*Rhetorica (Rhetoric)*
Sens.	*De Sensu et Sensibilibus (On Sense and Sense Objects)*
Somn.	*De Somno et Vigilia (On Sleep and Waking)*
Top.	*Topica (Topics)*

Works by other Greek authors

Hippocrates

Aër.	*Airs, Waters, Places*
Aph.	*Aphorisms*
Carn.	*On Fleshes*
Cord.	*On the Heart*
Epid.	*Epidemics*
Genit.	*On Generation*
Loc. Hom.	*Places in Man*
Morb.	*On Diseases*
Morb. Sacr.	*On the Sacred Disease*
Mul.	*On Diseases of Women*
Nat. Hom.	*On the Nature of Man*
Nat. Puer.	*On the Nature of the Child*
Steril.	*On Sterility*
Vict.	*On Regimen*

Galen [Claudius]

Sem.	*De Semine (On Semen)*
UP	*De Usu Partium* (*On the Usefulness of the Parts of the Body*)
Ut. Diss.	*De Uteri Dissectione* (*On the Anatomy of the Uterus*)

Hesiod

Theog.	*Theogony*

Plato

Phlb. *Philebus*
Rep. *Republic*
Ti. *Timaeus*

Pseudo-Aristotle

Mu. *De Mundo (On the Universe)*
Pr. *Problemata (Problems)*
Spir. *De Spiritu (On Breath)*

Theophrastus

CP *De Causis Plantarum* (*On the Causes of Plants*)
HP *Historia Plantarum (Enquiry into Plants)*
Metaph. *Metaphysics*

Other abbreviations

DK H. Diels and W. Kranz (1903) *Die Fragmente der Vorsokratiker* (Berlin, Weidmann).

KJV The Holy Bible (1987) *Authorized King James Version: Containing the Old and New Testaments* (Cambridge: Cambridge University Press).

L Hippocrates (1839–61) *Oeuvres Completes d'Hippocrate*, 10 vols., ed., trans. and comm. E. Littré (Paris: Librarie de l'Academie Royale de Medicine).

LSJ H. G. Liddell, R. Scott and H. S. Jones (1996) *A Greek-English Lexicon*, 5th edn. with supple (Oxford University Press).

References to Aristotle's texts follow Bekker pagination and Peck's emandations, unless otherwise specified

References to Plato's texts follow Stephanus pagination.

Translations of the *Generation of Animals* are adapted from Peck in Aristotle (1942). Translators of other works are cited in footnotes.

INTRODUCTION

All animals grow old and die; the only guarantee of life's continuance lies in reproduction. Aristotle was so struck by this fact that both his biological and his metaphysical work continually allude to the issue. Furthermore, his *On the Generation of Animals* (*GA*) focuses exclusively on animal reproduction and is one of the most comprehensive works on the topic surviving from the ancient world. Two sets of scholars have taken a special interest in the theory of reproduction found in the *GA*. The first are those who are interested in Aristotle's views on gender and his 'sexism' – various feminist schools of theory and criticism. The second set are classicists and historians of philosophy who hope to discover more about Aristotle's metaphysics and philosophy by looking at this mature theoretical treatise. The approaches of these two sets of commentators widely diverge, and so do their conclusions concerning Aristotle's intentions and philosophical achievements. For the former, there is much to disparage in Aristotle's attitude towards women; for the latter, the theoretical content is problematic, perhaps representing contradictory or incoherent positions to the reader. Both present any modern commentator with unique challenges. Given this history of interpretation, the question remains: why might this text continue to interest us? One of the most important reasons to read Aristotle's *GA* is that it provides us with unique access to his views on female biology, which lead to a greater understanding of his philosophy. Thus, those interested in Aristotle's philosophy will find study of his *GA* to be profitable and so will those interested in gender.

As is obvious from even the most basic of empirical observations, the continuation of life through reproduction is only possible due to the *female* body. Indeed, to any casual observer, the female body's ability to produce young, and to provide nourishment for their survival, may seem quite wondrous. In comparison, the ability of the male to produce semen (the function of which was not understood until relatively recent times) does not seem so crucial. Indeed, in some cultures semen is not deemed to be of any importance at all.[1] Certain accounts of human pre-history postulate a transition in emphasis from women to men as primary generators of offspring. This changeover is then taken to fit with transitions occurring in social, cultural and religious practices; when fertility goddesses were superseded by male divinities, real men took over power from real women.[2] This narrative can be easily questioned, starting with the assumption that respect for female fertility necessarily entails respect for actual women.[3] Its endurance in the popular imagination, however, can help to explain how many frame Aristotle's theory of reproduction. Aristotle (as with others in early written history) comes to represent an endorsement of the male as the primary sex through 'discounted female importance in the one area where the primitive and uneducated mind suspects female superiority' (Horowitz, 1976, 185–6). Seemingly despite clear evidence to the contrary, Aristotle thinks that the male provides life and soul to an incapable female system. Thus his theories can be made to mark the 'scientific' underpinning of wider reaching cultural phenomena – the assertion of a 'father right' that undermined the freedoms of real women.[4] This is, in part, why his views have been of interest to many feminists.

Feminist critiques of Aristotle that draw on his role in establishing the male as the primary sex often look to his biology. Linking his biological works with other texts, many try to show that Aristotle was in the business of justifying patriarchy through his depiction of female bodily and reproductive incapacity. An initial survey of Aristotle's views on women and females from

[1] For example, Trobiand Islanders. Malinowski (1929).

[2] Along with this transition, women may have lost power over their own fertility – proscriptions were brought into place to control women, for instance with regard to premarital sex, marriage, wearing of veils, adultery, contraception and abortion. Bachofen (1861), Frazer (1922), Briffault (1927), Neumann (1956), Gimbutas (1956) and Rich (1977) ch. IV. Ruether (2005), however, argues that theories of early matriarchy were constructs of nineteenth century European scholarship (Introduction and ch.1).

[3] Some second wave feminists have been accused of fabricating a matriarchal golden age. See particularly the criticisms of Davis (1971) in Hackett and Pomeroy (1972).

[4] See, for example, Keuls (1985).

various works is helpful. In his *Politics* he says that they are naturally ruled by men and that their virtue consists in obeying their husbands (*Pol.* 1254b12–15, 1259b4–9, 1260a23). He states that women lack an authoritative deliberating faculty – they cannot think adequately on their own and so must be told what to do by someone better than themselves (*Pol.* 1260a20–30). As for their behaviour, he considers them to be weak, cowardly and conniving in temperament.

All females are less spirited than the males, except the bear and the leopard . . . [they] are softer, more vicious, less simple, more impetuous, more attentive to the feeding of the young, while the males on the contrary are more spirited, wilder, simpler, less cunning. . . . Hence a wife is more compassionate than a husband and more given to tears, but also more jealous and complaining and more apt to scold and fight. The female is also more dispirited and despondent than the male, more shameless and lying, is readier to deceive and has a longer memory; furthermore she is more wakeful, more afraid of action, and in general is less inclined to move than the male, and takes less nourishment. The male on the other hand, as we have said, is a readier ally and is braver than the female. Even in the case of cephalopods, when the cuttlefish is struck with the trident the male stands by to help the female; but when the male is struck the female runs away (*HA* 608a32–608b19).

Aristotle also held that men contribute form and women matter in reproduction. Comparing the process to carpentry, he argues that there must be something equivalent to the timber that constitutes chairs and beds. The 'timber' in generation is the menstrual blood of the female, which will be moulded into the offspring by the male who acts as the carpenter.

Thus grasping the widest view of each, [the male principle] as maker and mover, and [the female principle] as that which is acted on and moved, the thing that comes to be is not made one from these, except as a bed is from the carpenter and the timber, or as the sphere is from the bronze and the form (*GA* 729b14–19).

For Aristotle, the female is distinguished by a lack of power or ability; it is weak and cold (*GA* 726b31–35) and, thus, unable (*adunamia*) to concoct pure semen (*GA* 728a18–20, 765b9–19, 766a31–5). For this reason, at one point he even characterizes it as 'like a deformed male' (*GA* 737a28). Together these areas, the political, the socio-biological and the reproductive nature of women, are thought to represent a misogynist or 'sexist' programme. Aristotelianism, especially in its medieval guise, has given us systematic sexism, linking together the various fields noted above and denying women an equal role in society on the basis of reproductive incapacity. However, on close analysis of the key texts it becomes clear that Aristotle himself did not have such a system in place; crucially, he did not form any justificatory arguments as later thinkers did.

When it comes to his biological works on their own, and particularly the *GA*, the usual feminist critique argues that Aristotle regards the female as contributing nothing positive to generation, no movement or ability, being like passive or empty matter. This characterization of the female role is then combined with the idea that the female is not a 'real' or true parent of offspring and that the human female is not fully human. I will argue against these intertwined ideas by first exposing the persistent and entrenched habit in many of these commentaries of getting crucial details wrong about Aristotle's theory of reproduction. In particular, in the *GA* Aristotle offers positive views of the female role. She contributes very specific materials to generation: materials that contain movements ensuring resemblance to herself and her ancestors (of both sexes). When analysing Aristotle's biology the fact that his theory of the sexes is 'sexist' ought not to obscure sound scholarship which hopes to capture an accurate account of his theory of the female role.

Meanwhile, scholars of ancient philosophy and classical texts, who study the *GA* in order to learn more about Aristotelian philosophy more generally, find it curious that Aristotle seems to change his mind about the female role, first noting its passivity and then explaining its positive contribution to heredity. Although this has been an important focus for the study of Aristotle's theory of reproduction it turns out that his position need not be seen as problematic or contradictory with regard to the female role. Through in-depth analysis of his position, it is possible to understand how he moves from a general overview of issues to a more detailed analysis. There is no doubt that for Aristotle the male is superior and that the sexes stand in a hierarchical relation. However, he also espouses a reciprocal or complementary model of the sexes, meaning that in some respects their contributions are comparable and not exclusive of the properties of the other (not like pure form and pure matter, which would be impossible in his ontology in any case). It is a misunderstanding to imagine that these two positions need contradict each other – however, a subtle exposition of his metaphysics and philosophy is required in order to explain their complementarity.

Close readings and contextualization will be the main techniques used to interpret Aristotle's writing in this book. Through this method, it will be seen that the role he assigns to the female is complex and significant. Furthermore, through careful consideration of the female role in the *GA*, we are able better to understand the sophisticated and intricate association of soul and body in Aristotle's philosophy. His view of the living world as striving for perfection, action and completion is a theme that runs throughout the work, but there is also an awareness on Aristotle's part of the effects of external, non-purposive forces on the life, health and reproductive capacities of living beings.

In Part I, I intend to set out two general approaches to the study of Aristotle on females, namely those emerging out of feminist concerns and those that attempt to defend Aristotle and set his philosophy apart from sexism. I will end Chapter 1 by noting the different ways in which Aristotle's sexism has been characterized, setting out which of these best captures his attitude. I then go on in Chapter 2 to discuss where the *GA* is traditionally placed in Aristotelian scholarship. I argue that the methods that Aristotle employs in order to discuss and explain various aspects of generation, while generally falling in line with his methodological recommendations in other more familiar works, also deserve our special attention as a source of his philosophy.

Part II begins a more specific study of Aristotle's description of how male and female contribute towards the generation of a new animal. Interpretations which obscure the important work of the female in generation often focus on Aristotle as a 'one-seed' theorist, who believed that only the male contributes towards generation. In Chapter 3 I argue that this is a misinterpretation of Aristotle's theory, which ought, in fact, to be counted as a 'two seed' theory. Chapter 4 looks closely at the texts which describe how the female contributes 'matter' to generation and argues that this cannot be empty or passive, in that 'matter' must be understood in terms of its relation to the soul of the adult animal and the specific potentialities it contains. Careful analysis of the text and related ideas in Aristotelian biology reveal that the female supplies an extremely specialized material. Moreover, the material role ought not to be viewed in a static manner; matter and form in the context of the generation of animal must be taken dynamically.

Part III considers the ways in which the male role has been characterized by commentators and asks which is the most plausible interpretation, given the text of the *GA*. I argue that, although the male contributes no material, his role is not transcendently divine, as it came to be understood in the medieval period. By conveying form, the male facilitates the continuation of type from one generation to the next, but Aristotle is not directly connecting the male role to godliness. Instead its association with theology must be understood with reference to Aristotle's particular ideas about what divinity consists in. In this section I will also reject reductions of the male role in Aristotle to mechanistic models, which misleadingly suggest that his science is closer to ours than is actually the case. These models have often resulted in a tendency to undermine the female role in his philosophy. The female role becomes that of inert and empty post-Newtonian 'matter', which is not what Aristotle intended, and thus these models must be put aside.

The final part of the book, Part IV, is a study of generation in lower animals and particular instances, Chapter 7 focusing on the former. The *GA* includes detailed analyses of non-human reproduction, including that of bees, fish, birds and those animals that are incapable of generation and must rely on it happening spontaneously. These discussions display how carefully Aristotle considers the empirical evidence, without forcing it to fit with preconceived ideas, particularly of gender roles and relations. It also reveals more about the content and capacities of the female contribution through a detailed study of eggs. This fits well with the reciprocal model of the sexes so prominent in *GA* IV–V. Chapters 8 and 9 discuss *GA* IV.1–3 on sexual differentiation and hereditary resemblance. The two are related in so far as both require that the male and female contributions bring their own unique potentialities to the mixture which results in the new animal. The male and female contributions must be in the correct proportional relationship (*summetria*) with each other in order for generation to occur (*GA* I). By *GA* IV Aristotle has refined his analysis of the interaction of male and female in order to account for particular instances of generation. The process of development in generation results not only in an animal of the type required but also in a unique sexed individual, possessing characteristics of both parents, and sometimes of ancestors on both male and female sides. In his explanations of sexual differentiation, a robust drive towards femaleness is apparent. The complementary model of the sexes becomes prominent in this context and allows him to expand on the idea of the female contribution as a material possessing specialized capacities. In heredity, many of these capacities stand in direct opposition to similar capacities in the male contribution.

In the final Chapter (10), I consider other discussions within the *GA* which reinforce my interpretation of the female role as a natural product of the body, connected to the soul and working together with the male to ensure the development of the new animal. The section is framed by questions concerning teleology, since deformity and variations in generation occur when both male and female souls have been unable to subdue and control environmental forces. Once again it becomes clear through looking closely at the text that the female does not contribute recalcitrant materials; instead, the ultimate goal of generating to type is disrupted by factors outside of the realm of the specific purposes of any particular species.

* * *

A series of Medieval and Renaissance illustrations depict Aristotle being dominated by a woman (Figure 1.1). The woman, called Phyllis, is sitting on his back, as if he were a mule. According to myth, Phyllis sought revenge on Aristotle after he told her husband, Alexander the Great, not to neglect public affairs on her account. She made sure that Aristotle fell in love with her, and was then able to humiliate him. The original message of the illustration seems to have been both anti-Aristotelian and anti-woman: Aristotle is shown as someone who cannot abide by the

Figure 1.1 'Weibermacht' (The power of women), copper engraving, Southern Germany or Switzerland, sixteenth–eighteenth century © INTERFOTO / Alamy.

principles he espoused, while Phyllis is cunning and insincere. If originally misogynistic, the idea of Aristotle being ruled by a woman has also been attractive to feminist critics, as a symbol of their power over this philosophical great.[5] The image could also

[5] See Horowitz (1976) 189–91.

capture another important truth which is that, despite the many ways in which he characterized men as superior to women, Aristotle was also noticeably struck by female influence in the natural world. As a philosopher of nature, literally a lover of the wisdom gained in studying the natural world, he did not so much endure female nature as embrace it with his characteristic wonder and curiosity. Female nature (like Phyllis in the myth) can be seen to powerfully influence Aristotle's theories. However, in considering its wonders, Aristotle did not thereby rescind his male-centred and male-bias point of view. He never completely fell in love. Thus, he need not be viewed as someone who cannot abide by his principles. The feminist cannot dominate him on these grounds. However, she can illuminate his intricate theory of the female in the belief that this is an important part of the history of philosophy and one which helps us to better understand ancient attitudes towards female nature.

Preliminaries concerning Aristotle's study of nature

Aristotle's philosophy of nature comprised a series of detailed and technical studies. It will be necessary to assume some knowledge of the general structure of his ideas in this area. This section gives a brief *précis* of five topics: (1) explanation of change, (2) matter and form, (3) the four causes, (4) potentiality and actuality and (5) the generation of animals.

(1) Explanation of change. The first challenge a natural philosopher faced in the time of Aristotle was to give an account of how change occurs (*Ph.* I). For Aristotle something has to remain unchanged throughout a transitional process in order for an explanation of it to be stable and knowable. He posits that there are three factors in any instance of change, the two poles between which the transition occurs and the stable substratum. Thus, if I sit in the sun, a human will remain as substratum while cold and hot exchange for each other. This is an example of qualitative change; the opposites exist in the category of quality. Aristotle thought that change could take place in four categories: quality, quantity, place and substance. Quantitative change is either in the direction of growth or diminution in size. Change in place is best known by the Latin 'locomotion'. Change in substance is quite a different type of change and

understanding how it works is of fundamental importance to Aristotle's philosophy. A substance in Aristotle's metaphysics is generally believed to be 'what something is': its form or essence. But when Aristotle refers to substance in his early account of change it probably means some definite object, rather than a class. So, for instance, the table I am leaning on is a substance. Substantial change occurs when a substance comes into being where it did not exist beforehand, or a substance ceases to exist where it once did. To explain this process adequately, Aristotle needed to introduce matter (*hulê*), form (*eidos*) and privation (*sterêsis*). A transition occurs between the poles of form and privation (of form); the stable substratum of this transition is the matter. In the case of this table coming to be, an unordered state of the materials (privation of form) exchanges for an ordered state (form); the timber (matter) underlies the change and continues to exist throughout.[6]

Aristotle often uses craft analogies to illustrate his theory. However, items such as tables were not his favoured objects of study and probably do not represent true substances (*Metaph.* VII.10–11). Aristotle's *Physics* is the study of nature (*phusis*) and natural objects are fundamentally different from artificial ones. The difference lies in the fact that items with a nature have an internal source of change and rest. (If they are animals or plants, this source of change and rest is the soul.) Artificial objects rely on an external agent to create, operate and maintain them. Natural objects do not.[7]

(2) Matter and form. Aristotle's distinction between matter and form can be contrasted with that of Plato. Platonic Forms are transcendent and the relation they bear to the world as we know it is obscure, sometimes resulting in the 'two worlds' problem, that is, it sometimes seems that for Plato there are two separate and incommensurable worlds: the world of Forms and that of sense objects or 'particulars'.[8] Plato falls into this difficulty in part because the Forms are objects of knowledge and as such must be immutable. Objects in this world, in contrast, undergo (almost constant) change and transition. Aristotle shares Plato's epistemological concerns to some extent: forms must be immutable in order to be properly investigated and known. However, he avoids any two-world problem by making his forms immanent, that is, they exist as part of the normal objects we experience. The form of a table is within the table itself and not some separable, transcendent

[6] See *Ph.* I. [7] *Ph.* I.1–2. [8] Annas (1981) 193–4.

reality. The combination of form and matter, such as this particular table here, is what Aristotle refers to as a 'composite substance'; it is the main type of object we inquire into and attempt to gain knowledge of.[9] Thus, forms are immutable in type or kind for Aristotle but do not have to exist separately from their particular instantiations (*Metaph.* VII). The most important substances are living things, since they display a certain fundamental unity.[10] In living substances, the form is the soul and the matter is the body. Aristotle defines soul as the actualization of a body potentially possessing life (*de An.* 412a19–21). Without the organic natural body, there would seem to be no soul or form.

(3) The four causes. When Aristotle explains that all natural objects can be analysed in terms of the four causes, he uses the craft analogy to illustrate this. The material cause is like the timber out of which tables are made. The efficient cause is the tools and the impetus that puts these tools into action in order to effect change; it is like the carpenter. The formal cause is what is to be the end product, the form 'table'. The final cause is 'that for the sake of which' the change occurs, the good end (*telos*) that is brought about by the process. The *telos* of a table would be its usefulness to humans. Aristotle's teleological metaphysics is made explicit by the four causes, which he thinks are present in most natural products and processes. In general, it is safe to assume, he thinks that almost all natural objects have a good end, something towards which they strive.[11]

(4) Potentiality and Actuality. Another way to illuminate the striving attitude of the natural world is by the use of the concepts potentiality and actuality. Every instance of change can be seen as a potential situation transitioning into an actual one. Natural objects are in a constant state of striving to complete themselves, to actualize their potencies (*Metaph.* VIII). Aristotle refined the use of this distinction in several places, most notably in the *de Anima* (his treatise on the soul) to include two levels of potentiality. First potentiality is characterized as the ability to acquire natural capacities and second potentiality as the ability to use these (*de An.* II 5). On one interpretation of Aristotle's metaphysics, potentiality and actuality are fundamental to our understanding of matter and form in his ontology. After all, each natural substance is composed of matter and form and the whole is unified through the fact that matter is the potentiality and form the actuality of substance (*Metaph.* 1045b18–21). Thus, Aristotle's distinction of matter and form within a substance is always dynamic.[12]

9 Substances are primary being (*Metaph.* 1028a31–33).
10 Barnes (1995b) 101.
11 Although this idea will be much refined in the *Generation of Animals*.
12 Kosman (1984).

(5) The generation of animals. The generation of an animal is an instance of substantial change. A new substance comes to exist where once it did not. You might imagine that the matter which underlies this transition is the female menstrual blood, the form is that of the type of animal involved and the privation of that form is the state of the materials before they have been formed. But Aristotle does not use the cruder analysis in this instance, opting instead for an account in terms of the four causes. The material cause is the maternal contribution, the efficient cause is the action of the individual father, the formal cause is the species form (perhaps taken universally) and the final cause is the good of the individual or the entire kind.

Aristotle also describes the process of animal generation in terms of potentiality and actuality. Both male and female contributions are potentially alive; and even once they combine, the embryo is in a state of potentiality. Only when it becomes a functioning adult does the new animal have a fully actualized form. However, there are significant stages along the way that can count as actual in a sense. When the embryo can live on its own, it is actually alive. When it is born and separate from its parents, it is actual in a fuller sense and when it gains all its proper capacities and functions, it is closer still. It is, however, helpful at times to view every immature creature as merely potential. Once fully developed, animals still actualize potentials, only on different levels. They use their faculties, such as those of sight and the other senses, in order to live.[13]

[13] See Connell (2001) 319–23.

PART I

METHODOLOGY IN THE STUDY OF ARISTOTLE ON FEMALE ANIMALS

Introduction

Part I aims to explain and then overcome two potential methodological difficulties in the study of Aristotle on female animals and our main source for this topic, his *Generation of Animals*. The first is the problem of its sexism and how any modern thinker, in light of feminism, ought to approach the text.[1] The second is how to interpret the work as a whole given its perceived contradictory content with respect to the female role. How the second issue is dealt with affects the first one; for the views of many modern feminist critics rest on traditional readings of the text, which decide to resolve the seeming contradiction by ignoring or subordinating Aristotle's positive characterizations of the female role in generation.[2]

Issues of consistency and coherence exercise scholars of many of Aristotle's works – most of whom feel it is best to resolve seeming conflicts, whether they exist (1) across texts or (2) within a given text.[3] An example of the second sort is found in

[1] 'By "feminism" I mean the belief that women should not be disadvantaged by their sex, that they should be recognised as having human dignity equal to that of men, and that they should have the opportunity to live as fulfilling and as freely chosen lives as men can', Möller Okin (1993) 10. We must keep in mind how difficult it is to fit this idea to the ancient context, particularly since the concepts of freedom and human dignity were not apparent or at least different from the ones with which we are now more familiar.

[2] See especially Preus (1975), Morsink (1982), Dean-Jones (1994), Coles (1995) and Balme in Aristotle (1972). Balme uses various techniques to try to avoid assigning conflicting views to Aristotle, including positing that Aristotle changed or modified his view in the later parts of the *GA*. The positive influences of the female are most often discussed only with reference to hereditary resemblance. Coles (1995), Tress (1992), Dean-Jones (1994), Balme (1987b), Cooper (1988) and Matthews (1986), to give but a selection of examples. More recent work in particular by Henry (2006a, 2006b, 2007b) and Gelber (2010) attempts to solve the problem of reconciling conflicting views on the female role by employing a more thorough approach to the text as a whole.

[3] An example of the former would be the soul as Aristotle describes it in *de Anima* and soul as he describes it in the *Parva Naturalia*. See Menn (2002) and Bos (2003).

much scholarship on the *Nicomachean Ethics* where it is hard to avoid the issue of how Aristotle could advocate a life of political virtue within a community on the one hand, and a cloistered contemplative one on the other.[4] Meanwhile scholarship on the *Metaphysics* Book Zeta (VII) works hard to discern what he means by primary substance, when the requirements he specifies for ontological and epistemic primacy seem to conflict.[5] A focus on perceived inconsistency can provide a handy impetus for scholarship but it can also be a problematic because the creation of consistency within a given text tends to aim at appeasing modern philosophical disquiet rather than at better understanding Aristotle's point of view.[6] A modern philosophical approach tends to view seemingly conflicting points of view as incommensurate. And if two points of view or doctrines are assumed to be so, the modern scholar is strongly inclined to privilege one view and diminish the importance of the other.[7] For instance, in her book on Aristotle's *Physics*, Helen Lang employs what she terms a 'method of subordination' in order to find that 'the coherence of Aristotle's arguments is their primary characteristic'.[8] The seemingly conflicting account of the female role in the *GA* has also inspired this sort of response. If one takes the two descriptions of the female role as incommensurate and wishes to make sure that Aristotle is not contradicting himself, one can be tempted to subordinate his more positive characterization.[9]

[4] Ackrill (1981) 135–41, Lear (1988) 152–64 and Richardson Lear (2004).

[5] Code (1984), Witt (1989), Woods (1968) and Connell (2001).

[6] For example, in interpreting Aristotle's *Metaphysics* Mary Louise Gill explains what she hopes to achieve. 'I believe it is possible to extract a coherent metaphysical theory from these Books, a theory that succeeds in offering satisfactory answers to questions that are as central and legitimate metaphysical concerns nowadays as they were when Aristotle raised them' (Gill 1989a, 11). Here, it seems, there is a deliberate attempt to 'extract' what might appeal to modern philosophical tastes.

[7] This can also be done chronologically, although the developmental picture of Aristotle has become deeply unfashionable, but see Wians (1996).

[8] Lang (1998) 18.

[9] This sort of interpretation is most apparent in older commentaries such as those of Peck in Aristotle (1942) and Platt (Aristotle 1912) and is also evident in the work of Cooper (1988) and Boys-Stones (2007). The details of these interpretations will be discussed in due course. The importance of mentioning the traditional reaction to differing descriptions of the female role lies in the fact that many non-specialist scholars, including feminists, take these commentaries as their source for Aristotelian doctrine and are thus less well informed about the more positive features of the female role in the *GA*, which

Feminist scholars often imbibe second-hand scholarship coming from this interpretative tradition. The idea that Aristotle represents the very foundation of Western misogyny depends in part on a disregard for his more positive characterizations of the female role in generation in the *GA*. Those who attempt to eliminate any seeming contradictions by disregarding or subordinating female influence also tend to disparage the feminist point of view on the text. Underlying this scholarly tension is a concern with proper procedures in the study of the history of philosophy. A 'present-centred' approach focuses on issues of concern to the author currently; an historical text is probed in order to discover what the author's views might have been with regard to such issues. A 'past-centred' approached, in contrast, focuses on the context in which the original ideas were formulated and tries to understand the historical philosopher, as much as possible, on his or her own terms.[10]

While it may be the case that anyone who reads Aristotle will have to have both past– and present-centred concerns in mind, many scholars try to emphasize one more than the other. Thus, most feminist historians of philosophy, while most attentive to modern issues when investigating historical texts, possess a looser grip on the historical context. Meanwhile, those in the scholarly field of classical philosophy can become so immersed in the historical context that they are less conscious of the need to be aware of what makes the history of philosophy useful and interesting to us today – the modern ideas that drive forward our investigations. I will urge that a balance must be achieved. Scholarship which attempts to understand Aristotle on his own terms must also endeavour to reassess what his writings can tell us about issues of interest to us today.

This chapter aims to disentangle Aristotle's theory of the female role in generation from various feminist critiques, many of which are based on misunderstandings of the content of the *GA*. Chapter 2 will suggest that the *GA* can be studied on its own merit without continual reference to better known or popular

can lead to exaggerations and inaccuracies. I do not mean to suggest that modern scholars of Aristotelian biology make this type of error.

[10] The distinction can be found in Rorty (1984). See also Lloyd (2000) 246–7.

works of Aristotle; this serves to undercut a tendency to dismiss the later Books and also allows a clearer view of the actual methods and procedures Aristotle employs in his study of generation. After undertaking an overview of these methods, I will end the chapter by briefly sketching out how this helps to explain his differing characterizations of the female role in generation from the beginning to the end of the treatise.

1
FEMINISM, SEXISM AND ARISTOTLE

1.1 Feminist histories of philosophy

In principle a present-centred component is an important part of any analysis of Aristotle, However, it is also important to assess whether this approach retains coherent connections with the extant texts, that is, whether the necessary past-centred component is present. Some feminist readings of Aristotle, in attempting to address pressing modern issues, involve inaccuracies with respect to the historical texts. Added to this, some of these mistakes have become thoroughly ingrained in certain scholarly traditions. These mistakes, which I will argue can even serve to undermine certain feminist goals, come from several different lines of feminist critique, and from several strands of feminism. It will be necessary, therefore, to set out some of the different ways in which feminism has treated Aristotle, grouping feminist histories of philosophy into second wave, third wave/post-modern and analytic.

The second wave of feminism, which began in the 1970s, has perhaps the most coherent and forceful interpretation to offer of the history of Western philosophy. In attempting to raise consciousness about the sexism inherent in many long established institutions, such as marriage and the family, this movement took in aspects of the history of philosophical and political thought. In most cases, the goal of these investigations was to uncover a systematic and comprehensive denigration of women in the 'Western tradition of thought' or the 'Western mind-set'. By uncovering a network of misogyny, it was perhaps hoped that radical changes would be brought about, reforming thousands of years of sexist thinking. A favoured focus of attention fell on the apparent tendency in Western philosophy to dichotomize existence into hierarchized dualisms. These dualisms were often considered to consist of a great variety of politically and conceptually influential terms: nature/culture, private/public, emotion/reason, body/mind, passive/active and, of course, female/male.[1] Using this type of analysis these feminist historians of philosophy attempted to locate and criticize the symbolic systems, gendered at heart, that supposedly underlie Western metaphysics.

Third wave, or post-modern, feminism took up the idea that philosophers are always influenced by societal precepts[2] and, on these grounds, came to challenge objectivity itself.[3] In general, post-modern feminism eschews the idea that knowledge is unitary, focusing instead on the context and the language of philosophical discourse.[4] Some then combine psychoanalytic techniques with the idea that gender informs Western metaphysics.[5] One of the main proponents of this type of approach, Luce Irigaray, sets out to uncover the feminine within Western systems of thought, which are dependent on what she terms the 'unacknowledged mother'.[6] By revealing the hidden power of the female in opposition to the male, she strives to construct a new feminine philosophy.[7] Since the 1980s much feminist historical analysis has sought to reveal the very deep roots of sexism not only within specific philosophies of the past but also at the very core of abstract reasoning or thought.[8]

One final technique which is rooted in the present-centred approach of analytic philosophy has begun to find favour. This is the attempt to extract and use tools from past philosophy in developing modern feminist theories.[9] Feminists whose

1 See, for example. Merchant (1980), Ortner (1974) and Whitbeck (1973–4). It is frequently assumed that Aristotle was partly responsible for coming up with this dualistic metaphysics. See especially Berman (1989) 230–3 and Whitbeck (1989). This line of thought is still influential in some areas of contemporary feminist research in philosophy. See particularly Meynell (2009) 8–10.

2 Schott (1997) 3–5, Tuana (1994b) Preface, Fricker and Hornsby (2000) Introduction and James (1998).

3 On the 'third wave' see James (1998) 581–2. On the variety and diversity of feminist methodologies see Jacobsen (2000) 1.

4 On post-modern feminism see Nicholson (1990) 1–18 and Holland (1997) 1–22.

5 On the psychoanalytic feminism see Irigaray (1985a), Tuana and Tong (1995) and Fricker and Hornsby (2000) 3.

6 Whitford (1991a) 7 and (1991b).

7 For example, Bordo (1987) on Descartes. See also Garry and Pearsall (1989) 2.

8 Harding and Hintikka (1983) ix. See also Antony and Witt (1993) xiii and Bordo (1987).

9 Aristotle, Kant, Foucault, Hegel, Derrida, Nietzsche and Merleau-Ponty have all been seen as useful to feminism. Freeland (1998a), Hekman (1996) 6–7, 11, Holland (1998) 2, Mills (1996) 21, Oliver and Pearsall (1998) 2 and Schott (1997) 3. For the use of Merleau-Ponty see Mackenzie (1992) and Maclaren (2009).

sympathies lie with the continental tradition, such as the postmodernists just described, often treat the work of past philosophers in a somewhat impressionistic manner, using the ideas to frame and add rhetorical strength to their own theories, without thereby attempting to interpret past philosophy on its own terms. Accuracy is not the point.[10] Analytically inclined feminists, however, are interested in the possibility of extracting arguments and theories in order to advance their own work and so take a more piecemeal approach. Consonant with this is the idea that the work of past philosophers is only partially sexist. Certainly this brand of feminism would not reject the logical basis of Western philosophy. The portion of past philosophy it proposes to cleanly extract can then be used to aid current feminist philosophy.[11]

These three techniques were variously applied to the study of Aristotle after his ideas on women and the female in nature came to the particular attention of scholars. While teaching Aristotle many modern university lecturers felt that they could not in good conscience ignore this disparaging attitude towards women.[12] In order to rightly connect a student's (or teacher's) own experiences with what they were encountering in their reading, it seemed quite natural that modern feminist issues should be addressed while looking at Aristotle's philosophy. The drive to further study Aristotle's views on women looked to be an interesting and promising sort of inquiry – but instead of developing this field, feminist scholars initially set out to reject the Aristotelian *oeuvre* in its entirety.[13] They reasoned that even if it were possible to separate out useful bits of theory, it would not be appropriate to use the work of such an arch-misogynist. Aristotle must be exonerated

[10] In her commentary of Irigaray's interpretation and use of Aristotle, Freeland (1998b) points out that one learns more about Irigaray and her views than about Aristotle from reading her work (59–92).

[11] Some argue that feminism should not attempt to fall in line with the theories of prominent males of the past in this manner. See for instance Poster (1998) 343.

[12] Freeland (1994) notes 'recently I have also come to feel more uneasy in teaching Aristotle. I've had a harder time trying to decide what to say to students about his notorious claims about women' (145). Many other female Aristotle scholars have felt compelled to write about, comment on or at the very least teach courses about Aristotle's view of women. See, for example, Deslauriers (1998), Nussbaum (1998) and Nielsen (2008).

[13] Horowitz (1976), Whitbeck (1973–4), Pomeroy (1975), Cantarella (1987) and Lange (1983).

from these charges – or, at the very least, his ideas about women must be found to be largely innocuous to us now – before modern feminist sympathizers can be comfortable with examining his views on female animals.

For certain feminists who search for the origins of female oppression, Aristotle's philosophy has become 'the very source of patriarchal thought' and 'fundamental to the Western world-view'; and Aristotle himself is seen as the 'fundamental architect of the Western world[view]'.[14] The simplicity and power of this idea is undeniably attractive and has thus influenced many scholars with feminist leanings to adopt it.[15] However, before succumbing to its charms, it would be a good idea to consider whether it is really correct to characterize Aristotle or his philosophy in this manner.

This characterization originates in large part in the second-wave project of uncovering the latent sexism in Western thought, which often treated Aristotle as a convenient figure to start from. Different sets of dualistic oppositions are commonly attributed to him by these feminists, but they all appear to agree that the system is based in the fundamental opposition of male and female.

> The basic dualism poses women and men not only as opposites and oppositional categories but as subordinate and dominant relationships, which also underlie and correspond to other dualisms that have been central to Western philosophies, science, ideologies, and institutions: from man over woman follow culture over nature, mind over body, reason over emotion, objectivity over subjectivity, creativity over procreativity, the public and political life over the private life (Bleier 1984, 197).

The idea that Aristotle originated and forcefully perpetuated a set of sexist oppositions rests on a number of key assumptions.

Systems of oppositions in metaphysics supposedly:

D1. represent mutually exclusive categories.
D2. are hierarchical.
D3. comprise a rigid system.
D4. comprise a system which serves to positively bolster or justify the political position of men.
D5. comprise a system which has been maintained throughout history.

[14] Bleier (1984) 197, Tuana (1993), xi and Green (1992) 71.
[15] Allen (1985), Keuls (1985), Tuana (1989), Green (1992) and Zeitlin (1996).

Aristotle's philosophy is seen as falling under the general tendency to set up systems of dualistic opposition (D1–D3) which he is taken not only to have originated but also to have purposefully perpetuated (D4–D5).

Clearly Aristotle can be found to oppose the sexes and associate them with other opposites. One example comes from the *Generation of Animals* in the context of his attempt early on in the work to explain male and female principles.

If we consider the matter on general grounds, we see that when some one thing comes to be from the conjunction of that which acts and that which is acted upon, then the actor is not situated in the emerging entity. More generally, it comes from that which changes and that which is changed. Now the female, as female, is acted on and the male, as male, is that which acts and that from which the change originates. Taking, then, the widest view of each, as actor and changer, and as that which is acted on and changed, the thing that comes to be is not made one from these, except as a bed is from the carpenter and the timber, or as the sphere is from the wax and the form (*GA* 729b9–19).

Although male and female are opposed here, this does not show that the Aristotelian opposition of male and female entails D1–5 above. In the next section I will show how close readings of the texts make it difficult for us to accept that for Aristotle male and female can be characterized in terms of D1–3 (1.2). Problems also arise when attempting to connect D4 and D5 with D1–3 (1.3). Ultimately, we shall see that although legitimate feminist questions remain in our study of Aristotle, he becomes an inappropriate feminist target as the originator of a misogynist mind set. Therefore, rather than reject his philosophy wholesale, the feminist historian of philosophy is entitled to and ought to look carefully at his views on females.

1.2 Unravelling erroneous associations Part 1: the metaphysics of oppositions

For the mainstream of feminist critics through the past half-century, Aristotle is regarded as thinking that the sexes: (D1) represent mutually exclusive categories, which (D2) are hierarchical and (D3) comprise a rigid system. However, an analysis of Aristotle's view on the ways in which male and female can be opposed to one another reveals a much less comprehensive picture.

Four different types of opposition are detailed in the *Categories* (11b15–20).

O1. correlatives (*ta pros ti*).
O2. contraries (*ta enantia*).
O3. privative and positive (*sterêsis kai hexis*).
O4. affirmation and denial (*kataphasis kai apophasis*).

Correlatives can only be understood in relation to one another, such as double, which must be understood as double of what is half of it (*Cat.* 11b25–8). Contraries are opposites which either admit intermediaries or not. For instance, white and black have the intermediate state of greyness, whereas odd and even do not have anything in-between. In the latter case, one or other opposite must adhere in a given subject. 'For example, both health and disease may be said to be naturally present in the bodies of all living things, and in consequence one or the other must be present in animal bodies' (*Cat.* 12a4–6).[16] The term 'privative' (*sterêsis*) indicates the absence of something possessed (*hexis*). In the case of 'blindness', for instance, it is the privation of sight, which is a positive state. Finally, affirmation and negation are a matter mainly of language, since only statements such as 'he is sitting' and 'he is not sitting' can be opposed in this manner.

On some occasions it seems that female and male represent privative and positive for Aristotle: the possession of a faculty (*hexis*) as opposed to its absence (*sterêsis*). *GA* I and II tend to speak only of the male principle positively as ability and of the female principle as a privation of this – as inability (*adunamia*). At times the rich description of the male principle is enhanced by repeated comparisons with the lack of maleness the female exhibits.[17] Aristotle also links male and female with other oppositional categories which appear to represent positive and privative.

[16] Health and disease are contraries in the *Cat.* but elsewhere they are privatives.

[17] Many feminists focus only on this strand of thought. 'Aristotle repeatedly expressed the view that what is essential to being female is a lack of certain attributes . . . Aristotle was one of the first to build a detailed biological theory upon this idea [of female lack]', Whitbeck (1973–4) 56. 'Female characteristics are described in two ways; by analogy with male and by comparison of inferiority with the male body', Sissa (1992) 65.

So for instance at *GA* 729b15 (quoted earlier), it seems that by showing that the male is agent and mover, the corresponding fact is also demonstrated: that the female is patient and moved. Similarly, when Aristotle decides that the male is active and form, he requires that the female be passive and matter. Male and female have generation in common, but it seems at this point to be as different as negative is to positive.

There are many instances, however, when it clearly does not make sense to view Aristotle's female as the privation of male. The female is complementary to the male in that both have abilities which complement each other in generation. At the very beginning of the *GA*, when Aristotle talks in terms of female animals rather than the female principle in the abstract, he focuses on their function and ability, since entire animals on Aristotle's view cannot be lacking or incomplete.[18] Male and female have different plans (*logoi*) and functions (*ergoi*) rather than one lacking what the other has (*GA* 716a18–20). Aristotle goes on to explain that every function requires tools or organs, which are parts of the body – both male and female have these organs, the female has the uterus and the male the penis and testes (*GA* 716a23–35). Later on in the text he compares the female organs, along with the male organs, to functioning eyes.

> There must be an organ both for the male and for the female; hence the male has his genitals and the female has the uterus. Nature gives each one its organ simultaneously with its ability, since it is better done thus. Hence each of these regions of the body gets formed simultaneously with the corresponding secretions and ability, just as the ability to see does not get perfected without eyes, nor the eye without the ability to see (*GA* 766a3–9).

On rare occasions, Aristotle mentions the fact that males simply cannot do what females do (e.g. gestate in *utero, GA* 738b36–739a2, 764a16–20). Although the female condition is sometimes described in terms of its inability, Aristotle never calls it a 'privation' (*sterêsis*) of humanness or even maleness. Thus, the sexes are not only or even primarily opposed for him as positive and privative.

[18] Although, metaphorically speaking, some animals can be viewed as 'incomplete' according to Aristotle. See Lloyd (1983) 26–43 and Chapters 7 and 10.

One might be tempted to view the opposition of the sexes as contraries that do not admit intermediaries. Perhaps a natural body can only be either male or female, just as it can only be either healthy or diseased. This is reinforced somewhat by Aristotle's description of human hermaphrodites as only possessing one functional organ (*GA* 772b27–9). As he explains in his *Metaphysics*, 'all things which cannot be present at the same time in that which admits of them both are called opposites' (*Metaph.* 1018a22–3). This type of opposition is mutually exclusive – the opposing entities are unable to admit one another. However this cannot be the only or even the primary way in which Aristotle opposes the sexes. First of all, when discussing those animals which are naturally hermaphroditic (unlike human beings which are only unnaturally so), such as the bee, he notes that they do contain functional male and female systems within the same natural body.[19] Furthermore, male and female are contrasted at many points in terms of temperature (the male is hotter, the female colder), a type of contrariety that admits of degrees.[20]

Finally, let us consider the sexes as correlatives, which are co-dependent and defined in terms of one another, in Aristotle's thought. Correlative opposites can privilege one member over the other or they can be complementary. Clearly, male and female are co-dependent yet it often seems that Aristotle emphasizes and values the male and defines female in terms of a lack of maleness rather than the other way around. Thus it looks as if the sexes could be non-complementary correlatives.[21] Yet Aristotle also describes the sexes in a complementary manner, especially when he emphasizes the proportional blend (*summetria*) necessary in generation.[22]

This brief survey shows that Aristotle has a number of different ways of contrasting the sexes as 'opposites'. Fundamentally the

[19] King (sic.) bees 'contain within themselves the male as well as the female' (*GA* 759b30).

[20] For example, see *GA* 728a18–22, 738a13–14, 766a31–4, b18–22. Thus Aristotle allows for the possibility that there can be more manly (or hotter) women and more effeminate (or colder) men.

[21] The technique of describing the more important partner by continually underlining its antithesis to the less important one is used elsewhere by Aristotle. See for instance *Pol.* 1252a31–5 on ruler and ruled and his *Metaph.* VII.7 and XII.7 on potentiality and actuality.

[22] *GA* 767a16–29, 723a29–32, 729a16–19.

female is slightly colder than the male, which is a difference in degree, admitting of more or less, and so they are seldom depicted as 'mutually exclusive' (D1). Furthermore, the way that Aristotle opposes the sexes is not always hierarchical (D2), for example when they are described as symmetrical. Aristotle's intricate discussion of the sexes shows he does not draw a 'sharp distinction' which would mark his methodology as 'rigid dichotomous thinking' (D3).[23] Thus, it is inaccurate to regard the opposition of the sexes as (D1) mutually exclusive, (D2) hierarchical and (D3) entirely rigid for Aristotle.

1.3 Unravelling erroneous associations Part 2: biology and politics

Accepting that, in a loose manner, Aristotle holds to a set of dichotomous categories that include 'male' and 'female', we must next consider whether the presence of such a set of divisions is enough to prove any misogynistic thinking. It is for this very reason, I take it, that the feminist critic endeavours to discover a connection between sex-biased metaphysics and an intention to oppress women. Recall that Aristotle's dichotomies according to second wave feminist critics have to:

D4. comprise a system which serves to positively bolster or justify the political position of men.
D5. comprise a system which has been maintained throughout history.

Let's first consider D4: did Aristotle attempt to use his biological views on female nature to justify the political position of men? The majority of feminist readings of Aristotle attempt to locate such thinking in his works; in order to argue against and undermine such a point of view, one has to first find it. In this analysis, I will first consider several interpretations of Aristotle's biology that make a speculative link between what he says about the 'female' and the effect these ideas have had on real women. One idea, derived from his views on sex difference, is that for Aristotle

[23] Aristotle is taken to possess a 'uniform', 'consistent', 'coherent' system of sexism. Green (1992) 70–2. In contrast to this view see Freeland (1994).

women represent something antithetical to reason or intellect, be that body, nature or emotion (Clack 1999, 31). One influential feminist, Genevieve Lloyd, believes that the association of female or woman in Western metaphysical thought with the opposite of reason had the social effect of excluding them from rational activities. Further, she locates the origins of these associations within the reproductive metaphysics of Plato and Aristotle (Lloyd 1996, 42–3).

Interpretations of this type appear to get their inspiration, to some extent, from Irigaray, who speculates that for Aristotle men are divine and godlike in their superior ability to produce form. Meanwhile, women represent the unintelligible realm of matter, from which men try to separate themselves (Irigaray 1985, 162). Many feminist commentators accept and reinforce this picture of the masculinity of reason in Aristotle.

> Aristotle's world is characterized by hierarchical dualisms – that is, polar opposites within which one side rules *over* the other; for him Soul rules *over* body, Reason *over* emotion, male *over* female, and so on. For him, Pure Mind ('Nous' only possible for males) is connected with 'divine' Soul, which is supreme of all earthly things. The male Mind is therefore *higher* and *holier* than all matter, even higher than the believed Apollonian (read, male) body; certainly the male Mind and Reason rule over and are 'more divine' than the female body because she (being ruled by emotions and body functions) is not capable of Mind or Reason, and so on (Wilshire 1990, 93).[24]

The general idea is that by assigning males the formal role in generation, Aristotle also associated men with knowledge and rationality; likewise by assigning females the material role he condemned women to an emotional and irrational state (Haraway 1988, 592).

Another view, attributed to Aristotle with reference to female nature, is that women have no autonomy.[25] A person possessing autonomy, valued as a rational self-determining agent, can be contrasted with someone who is to be treated in a purely instrumental manner, in order to serve others' ends. Irigaray's early

[24] Cf. *Ibid.* 43–4, Grosz (1990) 155 and (1994) 5.

[25] 'Autonomy' was not a recognized concept until at least the time of Kant, but the general idea is similar to the more Aristotelian one of 'rational self development' (or developed practical reasoning capacity [*phronêsis*]).

work saw her characterizing women, according to Aristotle's biology, as the latter: she exists 'for the sake' of someone other than herself; there is no good reason for her in and of herself (1985, 165–7). In 1983 Lynda Lange offered a similar idea as a literal interpretation of Aristotle's theory of generation. Biologically, females only serve the purposes of males, who alone are properly human.

> The individuals who seek to partake in the eternal and the divine by reproducing themselves are by implication only the males. *Her* partaking is quite different. She is instrumental to species eternity, and potentially rather than actually human. The actuality of her human potential is the incubation of a male child. . . . the female has been defined as virtually an organ of the male body (Lange 1983, 11–12).

Lange's characterization of Aristotle's biology is echoed in many other feminist scholars.[26] The idea is that women lack any control over their destiny, or that they lack a viable independent human existence for him.

A feminist post-modernism perspective underlies Lange's view. However, there is some failure to temper it in light of how speculative and suggestive this line of inquiry is. Irigaray, for example, does not pretend to offer concrete interpretations of Aristotle's texts or seek to set out how precisely his sexism manifests itself into any system. Others who attempt to do so are many times led into errors. One particularly striking mistake is to assume that because he associates the female principle with passivity and matter, he must think that women are themselves passive and material.

> Though indispensable, the female contribution is one of matter, *with which woman is identified*. And the contribution of the woman-matter is passive by nature, while the male contribution, man being form and spirit, is active and creative (Cantarella 1987, 60; emphasis mine).[27]

Aristotle himself would think it nonsensical to regard an animal as 'matter'. There is simply no way in which a woman (as a female animal) could be merely material for him. Within the text of the *Generation of Animals*, he is careful to abstract reproductive

[26] Grosz (1990) 155, Sissa (1992) 60–5 and Cantarella (1987) 60–1.

[27] See also Horowitz (1976) 194 and the origins of such a view in Irigaray (1985a) 133, 141, 162–4 (which is a more imaginative interpretation).

functions from the individual animals in which they reside, stating that the female *principle* is matter or that the female animal *contributes* matter, never that the female herself is material.[28] Furthermore, the idea that the female principle is matter must be very carefully interpreted within the context of that discussion. In all of his works, form and matter are relative to one another (*Metaph.* 1044a18, *Ph.* 194b9). There can be no such thing as matter on its own (as some characterize Aristotle's female role to be). Thus, Aristotle does not identify woman, or even the female contribution to generation, with matter in any straightforward manner.[29]

When one looks closely at Aristotle's writings, it is clear that he never explicitly puts forward the view that women lack autonomy. Many passages suggest that for Aristotle both female and male animals represent the species and both have the same form.[30]

> In all animals that move around, male and female are separate. While one animal is female and another male, they are identical in species. So, for instance, both [men and women] are human beings and both [mare and stallion] are horses (*GA* 730b33–731a1).[31]

Since both male and female are required for the generation of their kind, it would not make sense to view one as merely serving to reproduce the other. Male and female come together in order to generate 'because this is something that concerns them both' (*GA* 732a10–11). All this tells against the idea that Aristotle thinks that the female is not a real member of its kind, and exists merely for the sake of another.[32]

Within the Aristotelian biological account of female nature women are not depicted as less rational or more emotional than men. Furthermore, form is not associated with intellect in the way

[28] The female animal, like all others, has a body and a soul; it has both form and matter. Sparshott (1985) 182.

[29] The matter contributed by the female must be matter of a particular type (*hê toiautê*) (*GA* 733b32).

[30] Male and female do not differ essentially but only *per accidens*. See *Metaph.* 1058a29–b25 and Deslauriers (1998). See also Plato *Rep.* 454c-e for a similar point.

[31] Ἐν μὲν οὖν τοῖς ζῴοις πᾶσι τοῖς πορευτικοῖς κεχώρισται τὸ θῆλυ τοῦ ἄρρενος, καὶ ἔστιν ἕτερον ζῷον θῆλυ καὶ ἕτερον ἄρρεν, τῷ δὲ εἴδει ταὐτόν, οἷον ἄνθρωπος ἢ ἵππος ἀμφότερα·

[32] See Henry (2007b) and Chapter 8 of this book for more thorough defences of this position.

it has been since the later medieval period. Even the idea that matter is unintelligible must be understood within its context.[33] In any case, it is hasty to conclude that Aristotle's habit of 'associating' women with indeterminate matter requires that they be incapable of reason or knowledge. The Pythagoreans placed 'female' in the same category as the 'unlimited' and the 'bad', but we know that they did not regard women as intellectually inferior to men. They famously included women in their philosophical cult, teaching and guiding them as they did men.[34] Overall, then, it has been shown that the claim that by associating women with matter Aristotle thereby must have undermined their autonomy or disallowed them intellect is spurious.

Other feminist critics attempt to find firmer connections between Aristotle's texts on biology and his writings on the place of women in society. This requires finding links between what Aristotle says about females in his biology and what he says about them in his *Politics*. Key passages from the *Politics* must be included in this analysis. Two of these are as follows.

(P1) It is in a living creature, as we say, that it is first possible to discern the rule both of master and statesman: the soul rules the body with the sway of a master, the intelligence the appetites with constitutional or royal rule; and in these examples it is manifest that it is natural and expedient for the body to be governed by the soul[35] and for the passive part to be governed by the intellect, the part possessing reason, whereas for the two parties to be on an equal footing is harmful in all cases. Again, the same holds good between humans and the other animals: tame animals are superior in their nature to wild animals, yet for all the former it is advantageous to be ruled by humans, since this gives them security. Also, as between the sexes, the male is by nature superior and the female inferior, the male ruler and the female subject[36] (*Pol.* 1254b3–15).

(P2) For he rules over wife and children, over both as free persons, but not with the same style of rule: over a wife he rules in the manner of a statesman, over

[33] Matter is unknowable only in so far as it cannot be defined, since it does not represent a unified substantial entity on its own (*Metaph.* 1040b5–9). This does not mean that nothing can be known about it; in Aristotle's biological ontology a great deal can be said about the powers and content of matter.

[34] Some of the earliest sources for women philosophers are the writings of Pythagorean women. Waithe (1987). See also Irby-Massie (1993).

[35] κατὰ φύσιν καὶ συμφέρον τὸ ἄρχεσθαι τῷ σώματι ὑπὸ τῆς ψυχῆς.

[36] ἔτι δὲ τὸ ἄρρεν πρὸς τὸ θῆλυ φύσει τὸ μὲν κρεῖττον τὸ δὲ χεῖρον, καὶ τὸ μὲν ἄρχον τὸ δ' ἀρχόμενον.

children in that of a king; for by nature the male is more fitted to be in command than the female, unless conditions in some respect contravene nature;[37] and the elder and fully grown is more fitted than the younger and incomplete. Now in most cases of rule by statesmen there is an interchange of the role of ruler and ruled, since they tend to be equal by nature and not to differ at all . . . But that is the permanent relationship of male to female (*Pol.* 1259a39–b10).

According to these passages, woman cannot live a public life like a man and she has to obey her male relatives. Aristotle appears to suggest that this is because actual men are superior to actual women and naturally rule over them. It is unclear, however, why he thought this since he does not provide any justification for placing women in this inferior social and political position except to say that it is 'natural' – a term employed in a variety of different ways in his works. The idea of naturalness suggests to the modern mind a direct connection with biology, which is why feminist critics turn to Aristotle's biological works where they seek (and often claim to find) arguments for female inferiority.

In order to create such connections feminists have sometimes taken what is expressed in passage (P2), the idea that women are fitted by nature to be ruled over, and linked this to Aristotle's description of the female role in generation. A further passage from the *Politics* is similarly linked to Aristotle's reproductive biology – wherein is contained the idea that women are somehow intellectually impaired.

(P3) We have an immediate guide to our position in the case of the soul, where we find natural ruler and natural subject, whose virtues we say are different—that is, one belongs to the rational element, the other to the non-rational. Well then, it is clear that the same applies in the other cases too, so that most instances of ruling and being ruled are natural.[38] For rule of free over slave, male over female, man over child, is exercised in different ways, because, while the parts of the soul are present in all of them, they are present in them in different ways. The slave is completely without the deliberative element; the female has it, but it has no authority; the child has it, but it is incomplete[39] (*Pol.* 1260a4–14).

37 τό τε γὰρ ἄρρεν φύσει τοῦ θήλεος ἡγεμονικώτερον, εἰ μή που συνέστηκε παρὰ φύσιν.

38 ὥστε φύσει τὰ πλείω ἄρχοντα καὶ ἀρχόμενα.

39 ὁ μὲν γὰρ δοῦλος ὅλως οὐκ ἔχει τὸ βουλευτικόν, τὸ δὲ θῆλυ ἔχει μέν, ἀλλ᾽ ἄκυρον, ὁ δὲ παῖς ἔχει μέν, ἀλλ᾽ ἀτελές.

For Aristotle, the soul can be divided (for practical purposes) into a rational and an irrational part. The rational part can be further divided and described as a combination of practical and theoretical reasoning; practical reasoning is required for self-determining action and rests mainly in the ability to properly deliberate, while aiming at the correct or good ends of action (*NE* VI). The fully virtuous agent must possess practical wisdom (*phronêsis*). The claim here seems to be that women are unable to deliberate and so cannot become practically wise and fully virtuous.[40] Significantly, it is also implied that they would not be able to make their own choices– thus would not be self-determining agents (or autonomous in modern parlance). Many feminist critics find some support from this passage, together with (P1) and (P2), for a connection between bodily and mental incapacity in Aristotle's woman. The incapacity of the deliberative faculty is understood to be crucially linked to their physical inability (*adunamia*) to concoct male semen (Green 1992, 85; Clack 1999, 2–3).

Feminists who want to make sense of Aristotle's views in his *Politics* often turn to his biological account of female nature.

> Passivity in reproduction is one of the factors that Aristotle uses to *justify* the social and legal inferiority of women (Cantarella 1987, 60; my emphasis).

The claim here, and elsewhere in this sort of literature, is that Aristotle uses his stance on the biological nature of women in order to *justify* his comments about their position in his *Politics* (D4).[41] Because Aristotle does not refer to his biological theory and claim it as support for the views expressed in his *Politics*, those who allege such justification must take it to be implicit.[42]

[40] However, whether this interpretation stands depends on how one understands the term 'unauthoritative' (*akuros*) in the passage. I will discuss this in the next section (1.4).

[41] 'Since the time of Aristotle in the West, men have thought it important to *justify* their social dominance by appealing to ostensibly natural differences between males and females' (my emphasis) (Longino and Doell 1983, 207).

[42] They also assume that 'science' is used to justify female inequality. Keuls (1985) 145 remarks that Aristotle 'sought to prove scientifically . . . that the male is superior to the female'. Meanwhile Clack (1999) 30 notes that, '[i]n *GA*, Aristotle grounds his misogyny in his "scientific reflections"'. And finally, Tuana (1994a) argues that '[t]he image of woman as incubator . . . is given the scientific credibility of systematization in Aristotle's *HA* and *GA*' (192). This idea is anachronistic, since there was nothing like a scientific establishment with epistemic authority in antiquity.

The lack of any explicit mention of biological sex difference in the *Politics* leads some feminist critics to 'add missing clues from the biological writings'. And when considering the biological works, which do not discuss the possible social consequences of sex difference, we are encouraged to fill in these 'gaps' using the *Politics* (Green 1992, 79–80).[43]

The supposed connection between sexist metaphysical dichotomies and political policies affecting actual women is unsatisfactorily vague; but when, recognizing this, such theories take the further step of attempting to find specific connections in Aristotle's theories they encounter further difficulties. Forcing such connections results in incautious use of his actual writings and mistakes in interpreting these texts. One example comes from the well-regarded feminist academic Nancy Tuana, who tries to connect the two by claiming that Aristotle argued for female intellectual inferiority on the basis of the smallness of the female brain. She cites *GA* 784a3–5, which says that 'men have the largest brains of all animals', but this fact is not used to claim their intellectual superiority. In any case, what the quote actually says is that human beings (*hoi anthropoi*), not just males (*tes andres*), have the largest brains. She also cites *HA* 608b9–19 and *GA* 775a14–21, neither of which passage mentions brain size or even brain development (1989, 148, 1992, 68). Still, Tuana insists that, somewhere, Aristotle *argues* that women's role in reproduction causes her intellectual inferiority (1994a, 189). Her idea also rests on a fundamental ignorance of Aristotelian physiology.[44] The brain has nothing to do with mental or even perceptual capacity for Aristotle (*PA* 652b2–7).[45] The function of the brain is to cool the body, counteracting the heat of the heart (*PA* 652b17–23). Men may well have larger brains than women, since they are warmer and so have hotter hearts. Thus it is plausible to suppose that brain size is a sign of male superiority (if Aristotle ever commented on it, which we have no evidence of). However, it is inaccurate to claim that Aristotle anywhere *argued* from the smaller brain size of women *to* their intellectual inferiority.

43 See also Elshtain (1981) 44–5.

44 Mayhew (2004) also criticizes Tuana on this point.

45 The senses are located at the heart; *nous*, the intellect, does not have any specific bodily location (*GA* 736b23–9).

Another example of an attempt to find that female intellectual inferiority has a biological basis in Aristotle is Judith Green's article in which she decides that the bodily temperature of an animal creates intellect (*nous*) through movement (1992, 84–5). Thus, the hotter man is able to develop *nous*, whereas the cooler woman is not. This theory does not appear in any form in Aristotle and seems to be, in fact, Green's own creation. The mistakes made by Tuana and Green are rather more blatant than most. But a general lack of attention to detail is apparent in many feminist critiques and leaves them open to the legitimate criticisms of specialists. The idea that a woman's biological nature has made her incapable of reasoning, or has caused a defect in her soul, is not available in the *Politics* or the biology and the connections made by Tuana and Green can be easily dismissed as specious.

It does not follow from these misinterpretations, however, that it is illegitimate to probe Aristotle for answers about why he considered women to be inferior to men. The difficulty is not that this type of speculation takes place, for many other anachronistic questions are asked of Aristotle. The sticking point is that this analysis often tends to be conducted without acknowledging that there is nothing within the *GA* or other biological texts on reproduction that serves to bolster men's supposed intellectual or political superiority. Aristotle himself does not posit or argue for a connection between women's biological or physical role in generation and their social role.

Although we find no explicit justifications for the oppression of actual women in Aristotle's biology (D4) some feminist historians of ideas argue that this is not important. These writers try to show how Aristotle influenced various harmful views of womankind which have affected their social position throughout the ages and depend in part on the idea that this supposed system of misogyny has been uniformly at play in Western cultures (D5 above).[46] This is a matter I do not wish to investigate, mostly because it would require a detailed and nuanced study of a broad expanse of history. But it is worth noting that their view leaves

[46] Green (1992) 70 for the view that Aristotle's ideas about women were 'historically harmful'. Horowitz (1976) 192 speaks of the women 'whose lives have been adversely affected by' Aristotle.

Aristotle's own culpability out of the picture.[47] One might even propose to exonerate Aristotle for his sexist views on the grounds that he was not able to extract himself from the cultural context in which his outlook developed.[48] But it seems that one need not go this far. At the very least, if we accept that he is not responsible for the way in which his ideas were later used, this will allow the right-minded feminist to study his works without anxiety.

1.4 Combining the concerns of feminist and traditional scholars of Aristotle

Having disentangled Aristotle the biologist from the idea of him as misogynist metaphysician, one might wonder if it is at all profitable to read his work from a feminist point of view. The feminist commentator on Aristotle so far presented appears at best to simplify his ideas to fit a modern agenda, and at worst to distort his philosophy. Why not abandon any hint of feminist concern and return to viewing Aristotle's works without considering the emancipation of women from centuries of social and political oppression? Initial responses of classical scholars to feminist critiques of Aristotle saw this as the obvious way to proceed.[49] In general, classical scholars have tended to be rather unsympathetic to feminist approaches to Aristotle and when they address these writings they often combine three types of response.

(i) An attempt is made to form a more accurate picture of Aristotle's views on women from his texts, by both looking closely at what is said there and also by placing his views within the ancient Greek socio-political context.

47 It rests rather on the way in which other thinkers took up his ideas. On the issue of blame, see Cook (1996).

48 '[W]e give less than full responsibility to persons who, though acting badly, act in ways that are strongly encouraged by their societies . . . many male chauvinists of our fathers' generation, for example. Their false beliefs in the moral permissibility of their actions and the false values from which these beliefs derived may have been inevitable, given the social circumstances in which they developed. If we think that the agents could not help but be mistaken about their values, we do not blame them for the actions those values inspired', Wolf (1987) 56–7. This view is not without its detractors. See, for instance, Unger (1996) ch.1.

49 See especially Morsink (1979) in response to Horowitz (1976).

(ii) It is posited that one ought not to investigate Aristotle's works with any type of feminist agenda in mind as this would create distortion.
(iii) An attempt is made to establish that Aristotle was not sexist and so ought not to be of interest to feminism.

(i) counteracts the tendency among some feminists to inaccurately describe Aristotle's views on women. (ii) suggests that feminist questions are inappropriate and so asserts interpretative authority for the opponents of feminist scholars, who being supposedly undistracted by modern issues, are able to be more accurate. Finally, (iii) tries to dismiss feminist critiques wholesale on the grounds that, even if feminist questions were relevant there is nothing of interest to the feminist thinker in Aristotle's writings, since there is no sexism.

Many traditional scholars incorporate all three of the above strategies when addressing feminist critiques. I will describe some of these with reference to interpretations of Aristotle's *Politics* and thereby show that (ii) and (iii) are not viable defences and so ought to be abandoned, leaving (i) as a legitimate response. In establishing that (i) is the best way to respond to and actually supplement feminist concerns, one is also able to acknowledge that an awareness of gender issues is appropriate when studying Aristotle. Ultimately this is because, contrary to (iii), it is easy to accept that Aristotle was indeed sexist and this alone gives feminists (and any modern thinker) good reason to remain critically interested in his ideas from a feminist point of view. Employing method (i) leaves open the possibility, even probability, that Aristotle employed sexist presuppositions gleaned from his culture in many of his discussions of women and the female. But it does not make it necessary to ascribe any ideological intentions to him, nor does it presuppose that he had a sexist system or a systematic justification for any sexism. I will argue that feminists who wish to interpret ancient philosophy ought to adopt technique (i) and more accurately describe Aristotle's views on women and female animals. This gives them enough interpretative authority to fend off the invalid moves of (ii) and (iii), which push feminism out of Aristotelian scholarship. Thus feminist ancient philosophers can counterbalance more exaggerated or careless readings of Aristotle's philosophy which have tended in the past to undermine the legitimacy of feminist approaches to classics and philosophy.[50]

Let us turn now to interpretations of what Aristotle says about women in his *Politics*, which was outlined in several passages in the previous section (P1–P3 above). The focus here will be on (P3). When it comes to feminist complaints about (P3), traditional scholars begin to defend Aristotle using the first technique listed above (i); typically they will point out that the term *akuros* (used to describe a woman's deliberative faculty) never means incapacity or even inferiority in Greek.[51] The term is the opposite of *kurios*, a word used of the male head of a household who, by Greek law, ratified all the decisions of his dependants.[52] The term, as used of women, indicates that decisions do not have conventional authority, because they are not enforced or considered valid. Thus, a woman is *akuros* because she has a different role to play from the man in the household and in the city. This is taken to be a mere convention, which does not imply any fundamental psychological deficiency.[53] To reinforce the possibility that Aristotle viewed women's deliberative faculty as psychologically sound or whole, references are made to other passages within his ethical and political works, in which he holds that women are capable of virtue and happiness and thus also practical wisdom (*phronêsis*), the perfection of the rational soul.[54] It is also noted that Aristotle considers women to be free persons, who can determine their own lives to some extent.[55] Defenders of Aristotle (against the charge

[50] See, for example, Freeland (1994) and Tress (1992) 341 for the technique of close reading as used by feminist scholars of ancient philosophy.

[51] For example, Sparshott (1985) 180–9 and Mulgan (1994) 197–200.

[52] Women within a household were required to seek approval from the male head of the household for their decisions. Schaps (1979) 48, Mulgan (1994) 200 and Saxonhouse (1985) 49.

[53] It is of course possible that Aristotle also thought that women have a mental deficiency in comparison to men, which he could hold along with the empirical claim that they do not in fact have decision making authority. However, the one need not imply the other. Part of the traditional defence is to assume that Aristotle is observing and making note of the sexist structure of the city rather than aiming to justify it. Sparshott (1985) 183–4 and Mulgan (1994) 199.

[54] For example, *EN* 1162a15–29, *Rh.* 1361a1–10, Saxonhouse (1985) 51 and Mulgan (1994) 198.

[55] *Pol.* 1260b18; Mulgan (1994) 187. This may be because the male head of the household turns over authority within the house to his wife, who then makes decisions about its internal management. *EN* 1160b32–4; Sparshott (1985) 186–8.

of blatant sexism in the *Politics*) sometimes also point out that he never, in the *Politics*, uses the idea that women are unintelligent or mentally incapable, even when this would have helped his arguments.[56]

It is important to discern what Aristotle means when he says that the female's deliberative faculty is *akuros*. And because of their expertise, classical scholars can give a more nuanced account of the term. But critics of feminist scholars do not stop there; most are keen to claim in addition (ii) that to approach an ancient text (such as Aristotle's *Politics*) with feminist questions in mind is inappropriate. The idea is that it is the anachronistic feminist agenda itself which causes feminist critics to 'mis-state essential points' while only scholars who do not focus on feminist concerns can 'determine what Aristotle actually thought about women'.[57]

Part of the argument seems to be that feminist questions are problematically anachronistic. How precisely modern political and philosophical issues such as feminism form a part of modern interpretations and analyses of ancient texts is a debated topic. Following a certain tradition in analytic philosophy, in the *Cambridge Companion to Aristotle* Jonathan Barnes encourages readers to study Aristotle from the point of view of modern problems in philosophy:

> [T]he *Companion* is primarily concerned with the philosophical parts and aspects of Aristotle's thought—that is to say, with those parts and aspects which we now should be inclined to judge philosophical ... The *Companion* thus regards Aristotle's works from our point of view and not from the point of view of their author and of his contemporaries ... (Barnes 1995a, ix).[58]

Barnes believes, however, that acceptance of modern concerns into the study of past philosophy creates a dilemma. He thinks that we can study Aristotle with philosophical interest only in the context of modern philosophical problems.[59] However, it turns out that Aristotle's views on all these issues are outdated and inadequate, such that any 'philosophical reason for the study of

56 Sparshott (1985) 181 and Saxonhouse (1985) 45–6.
57 Mulgan (1994) 182. See also Sparshott (1985) 177, 192.
58 This also leads Barnes to be reluctant to investigate Aristotle 'as a man of science'.
59 The idea is that we ought to approach the work of an ancient philosopher as if it had appeared in last month's issue of Mind. See Grice (1986) and Williams (1994) 20.

Aristotle is not to be urged' (Barnes 1995a, xviii). Thus, a philosopher who is really interested in the modern issues must never take any real *philosophical* interest in past philosophy for the philosophy found there is not as good as what has been produced more recently. And so it is that Barnes decides that one should study Aristotle purely out of historical interest.

It will be clear from any brief look at modern scholarship on Aristotle's ethics, metaphysics or philosophy of mind that this attitude is not a popular way to approach his works.[60] If modern philosophers attempt only to reinforce or explain their own views with reference to the authority of Aristotle there is of course a danger that anything authentically Aristotelian will get left behind. However, such crude analyses are rare: most scholars acknowledge both the importance of carefully considering Aristotle's own views before any attempt to incorporate them into modern analyses of similar issues, and that the fit is not only imperfect but also, thereby, illuminating. In his review of *The Cambridge Companion to Aristotle*, David Charles argues, in response to Barnes, that the distinction between past– and present-centred approaches is inadequate, especially if this means the interpreter must choose between philosophical and historical interest.

> There is no reason why [the historian of philosophy] should take what is currently fashionable as the sole criterion of what is philosophically important or worthwhile. Indeed, if she is tempted to do this, she is working not too much but too little as a philosopher. Part of her task is to devise the appropriate conceptual apparatus with which to understand the issues which both Aristotle and her contemporaries are addressing with varying degrees of success. This requires distance from, rather than the uncritical acceptance of, the assumptions of one's own day (Charles 1997a, 234).

For Charles, part of the job of the historian of philosophy is to undertake a re-conceptualization of modern concerns in light of the broader perspectives gained by engaging with past philosophy.[61] One must first be aware of past ideas within their

[60] See MacIntrye (1981), Sandel (2007), Nussbaum (2007), Hursthouse (1999) on Aristotle's ethics, Loux (1991) on his metaphysics and Nussbaum and Putnam (1992) and Caston (2007) on his theory of mind.

[61] This attitude is similar to the one found in Williams (1994). There he argues that Rylian 'triumphant anachronism' is very harmful to philosophy as it makes philosophers less critical of their own views. The study of thinkers of the past ought to be taken seriously

own context, and really try to understand Aristotle and his contemporaries' points of view on matters of importance to them. One must also acknowledge an interest in modern issues. Furthermore, even when approaching an historical text without current intellectual trends directly in mind, such issues cannot be avoided. In the present context, it is common to find that when Aristotle discusses or touches on gender difference, the modern reader cannot help but begin to compare and contrast his ideas with more recent views. This naturally leads the reader, whatever task he or she is busy undertaking, to wonder where the sexism in the text comes from and how it works, perhaps even with a view to somehow reformulating the questions and issues that inform the modern field of gender research.

Let's return then to a certain type of response to feminist readings of Aristotle, which not only criticizes interpretative carelessness (i), but moves beyond this to suggest that feminist concerns ought not to be a part of Aristotelian scholarship (ii). The claim that feminist critics anachronistically assume that Aristotle must have had a reason for saying that men have perpetual rule over women cannot establish (ii).[62] It does not damage modern interpretations to admit that we can hardly avoid asking the sorts of questions that most interest us when attempting to understand any past thinker even when these are the kind of questions that they could never have asked themselves.[63] In assuming that their own interpretation is truer to what Aristotle actually thought, critics of feminist scholars imply that theirs is a purely past-centred methodology – finding out what Aristotle really thought about women. And yet the types of question that the modern reader asks of the text of the *Politics* are the same whether she be a

as philosophy, which then has the positive effect on the discipline of making 'the familiar seem strange' (20–1).

62 Aristotle 'did not think the issue of how and why women were subordinate to men to be a very important one' Mulgan (1994) 200–1. Here Mulgan also suggests that Aristotle was more interested in justifications for slavery because Greek free men were themselves in danger of becoming slaves (if they were defeated in war). They were, however, never in danger of becoming women and suffering under the arbitrary rule of others in *that* manner.

63 Why ought we to think about whether Aristotle's political thought is relevant to modern liberal individualism, as Mulgan does (1999), while disregarding issues of importance to feminist politics?

feminist scholar, mainly interested in the role of women in society, or a classical scholar, brought up in a tradition of gender equality. Either party is likely to end up wondering why Aristotle places women in an inferior position within the family and the city, comparing and contrasting these views with more recent ideas. And any modern reader might feel the need to find some kind of answer to that question which makes sense from a modern perspective. Because the texts of Aristotle do not give us any direct answers, any interpretation has to resort to speculation. This means that traditional scholars cannot be more accurate about Aristotle's views on this matter. They differ from feminist scholars because rather than focus on possible sexism in Aristotle, they instead consciously attempt to provide a non-sexist rationale for his assigning an inferior role to women. One speculation of this sort is that husbands rule over wives in the household because the wives are so much younger than their more experienced partners (Sparshott 1985, 188). Another commentator suggests that their deliberation is not authoritative because, just like labourers, they do not have the leisure to develop intellectual skills.[64]

Certain feminist commentaries, in contrast, do not attempt to deny or explain away Aristotle's sexism. However, in their efforts to find a solid theory of oppression they can be led to make erroneous connections between Aristotle's politics and his biology (see previous section). The need to do so is very much linked to both the attempt to satisfactorily answer the question of how Aristotle developed and justified his sexism and the feeling that the goal of ending female oppression would thus be aided. We might all agree that Aristotle had no reason to say that men naturally rule over women, simply assuming this to be the case (Mulgan 1994, 200–1). It does not follow from this that there is no possible question to pose and no possible explanation as to why Aristotle was sexist. Furthermore, the attempt to find such connections between Aristotle's views on female biology and his views on their political position does not constitute a misunderstanding of Aristotle's philosophy. For Aristotle, souls are

[64] Saxonhouse (1985) 49. Scholars can feel very motivated to exonerate Aristotle from the charge of sexism because they think this is the best or the only way to continue to respect his work. See Chapter 1.5 for further analysis of this phenomenon.

the actualization of particular bodies (*de An.* 414a19–28); so that if a body suffers incapacity, the soul would appear to also do so. Since the female body is lesser or lacking compared to the male one, it would seem to follow (even though he nowhere says so) that the female soul must be similarly impaired. We can see, then, why certain feminist critics look so intently to Aristotle's biology of the female for a coherent rationale lying behind what he says about women in his *Politics*.[65] These questions themselves are legitimate ones and do not on their own distort readings of the text. Any distortion comes not from the questions asked but from a careless treatment of the material, when the answers come too easily.

When classical scholars defend Aristotle in his *Politics*, they tend to insist that he never explicitly writes about the incapacity of the female mind or soul and so to make such connections is deemed by them to be 'inaccurate'. But accuracy is not at issue – neither making a connection between biology and politics in Aristotle nor denying that one can be made is accurate – neither point of view is definitively settled by looking at the texts. The key difference between the sort of feminist critique which imputes to Aristotle a view about female physical and related mental infirmity and the traditional defence which denies this, is that the former are keen to explain Aristotle's apparent sexism while the latter are at pains to deny any sexism exists (iii). Inaccuracy only appears if either view is imposed on the texts that we have. Thus either could be 'inaccurate' if they include the claim to have discovered what Aristotle actually thought on the issue of female mental capacity/infirmity; on this, the scholar can merely speculate. Furthermore, any disagreement between feminist critics and scholars of ancient philosophy on what Aristotle thought about women need only be one of degree: both have to include both past- and present-centred elements in their analysis. A proper balance of the two would lead

[65] We can contrast this with Plato whose psychology could not provide any rationale for placing women in an inferior political position. For Plato minds are separable from bodies and so women's minds can be perfectly sound despite any supposed physical infirmity. The *Republic's* injunction that women with philosophical souls can rule the city (Plato, *Rep.* V.4–6) seems coherent with this variety of dualism.

to a healthier and clearer debate.[66] A focus on gender issues ought not to pull us away from seeking to get as close as possible to what Aristotle actually thought, which requires that we not impose any theories onto him which he does not clearly put forward himself. Feminist critics ought to be encouraged to take care to learn the details of Aristotle's physiology (i), for example. Meanwhile more traditional scholars should consider that, as with the mind/body problem or the principle of non-contradiction,[67] the question of the origins of sex-bias is a legitimate one to put forward in interpreting Aristotle, rejecting (ii). In light of these questions, it is not necessary to always try to find non-sexist reasoning to explain away any apparent sexism, thus also abandoning (iii), which can be just as speculative as forcing connections where none are available.

1.5 Aristotle's sexism

It is difficult when reading Aristotle to push aside the question of the extent and nature of his sexism.[68] Despite this there continues to be some inclination on the part of classical scholars and philosophers, some of whom are detailed in the previous section, to respond to feminist critiques by denying that Aristotle was sexist.[69] Any brief reading of certain portions of the *GA* cures the misplaced desire to entirely exonerate him from this charge. Take, for instance, the stark and undeniable sexism which enters his explanation of the division of the sexes in *GA* II. Ignorant of evolution,

[66] 'Any good reading of the historical material involves both an awareness of the context in which the writer worked and an awareness of one's own concerns and goals,' so that 'the past text and the present issues are mutually illuminated' Lloyd (2000) 253.

[67] On non-contradiction see Łukasiewicz (1971).

[68] It is interesting to contrast earlier commentaries on Aristotle's *GA* with more recent work. Morsink (1982), Balme in Aristotle (1972), and Peck in Aristotle (1942) all appear to find the issue of Aristotle's sexism irrelevant while Coles (1995), Tress (1999), Mayhew (2004) and Henry (2007b) feel they must address it.

[69] Many hold the view that Aristotle had good 'philosophical' or 'scientific' reasons for his theory of the sexes, for example, Morsink (1979) 84: 'Aristotle accepted this apparently sexist theory not for sexist reasons but for good scientific ones.' Tress (1992) 310: 'what imbalance there is in his treatment [of male and female] arises not from misogyny or sexism, but from particular philosophical problems Aristotle is addressing in the treatise'. Mayhew (2004) 116: 'Aristotle's views about women and other females, however mistaken, were largely the result of empirical science – of reasoning based on observation – not of misogyny and ideological rationalization . . . Basically, he was an honest and objective scientist.'

Aristotle cannot see the need to explain why males came to exist but instead seeks good reasons why males have always existed separately from females (*GA* 731b23–4). The goals of generation cannot explain it: hermaphroditic animals successfully continue their formal existence from one generation to the next. His explanation is that the male is superior and so ought to be separate from the inferior female:

Whenever possible and so far as possible the male is separate from the female, since it is something better and more divine in that it is the principle of change for generated things, while the female serves as their matter (*GA* 732a7–10).

The idea that the male 'is something better and more divine' than the female, in this passage, is surely sexist.[70] And there are many more such remarks in Aristotle's works.

Once one admits that Aristotle and his works count as sexist, one can try to discern the extent and nature of this sexism. Investigations into this issue often lean heavily upon the question of whether Aristotle was ideologically motivated or whether he adhered to or promoted a sexist ideology. There is, however, great ambiguity in the term 'ideology' in this context. In its modern philosophical use the term originates in the political philosophy of Karl Marx and is first employed to describe aspects of Aristotle's works by Lloyd in his 1983 book *Science, Folklore and Ideology*. The term is also employed with reference to Aristotle's views on slavery in his *Politics*, where a helpful analysis is provided by Schofield (1990). Finally, Mayhew has made it the focus of a monograph entitled *The Female in Aristotle's Biology* (2004). Their uses diverge widely. Lloyd understands by the term something reasonably broad, akin to the propensity in a thinker to accept and incorporate contemporary cultural and social norms. Mayhew, in contrast, takes it to be a fully conscious effort to enforce unfair power relations through scientific theory. He uses the term in this strong sense in part because he hopes to present Aristotle as 'an honest scientist and philosopher,' as someone who does not employ 'ideological rationalisation'.[71] If

[70] For a longer and more comprehensive discussion of the above passage see Henry (2007b) 13–17.

[71] Mayhew (2004) 5: '[I]deological bias always involves evasion and/or dishonesty – psychological or cognitive states that are under one's control. To accuse a thinker of

one of these divergent uses is employed, the term is not as helpful as it might at first seem in descrying Aristotle's sexism, being either too narrow or too broad to capture it. Mayhew's understanding of the term makes it almost impossible to find any ideology in Aristotle's works: we have no access to his intentions and so we cannot in good faith assign bad ones to him – but may always allow the conclusion that his sexism is based on mistakes rather than misogyny (which, on the whole, is what Mayhew tries to do). Meanwhile, Lloyd's definition makes it impossible not to find ideology ubiquitously throughout Aristotle's works (and also the works of almost all other scientists and philosophers).[72] There is, however, the option of Schofield's more subtle and detailed understanding. He characterizes the 'broadly Marxist conception of ideology' in the following manner:

(i) a widely held body of ideas systematically biased towards the real or imagined interests of a particular sex or social group or class within society;
(ii) believed by adherents not because of rational considerations . . . but rather because of either (a) social causation or (b) a desire to promote the interests indicated . . . (1990, 1).

On this understanding of ideology, it is possible to regard it as unconscious or unintended, stemming from upbringing or other forces outside the individual's direct control (ii, a). Although this fits best with Aristotle's position, several qualifications still need to be made. First of all, it is perhaps not possible to capture what is wrong with sexism if no blame can be apportioned for it. Second, along with another popular interpretation of Aristotle's sexism, this view posits that Aristotelian ideological bias and his philosophy can be entirely separated. This is also assumed by Schofield in the contrast he sets out between beliefs (1) that are socially

ideological rationalisation is to imply that he could have done otherwise, that he could have come to other conclusions if he had not evaded, been dishonest . . . '

[72] This is especially so if ideology means promoting what one thinks to be the case as Lloyd indicates at one point: 'All ancient science is, no doubt, ideological in the sense that the different groups of those engaged in various types of inquiry were more or less actively engaged in legitimating their own positions.' He goes on to explain: 'But not all of ancient science was seen as in the service of a morality or directly linked to a notion of the good life, let alone geared to underpinning the moral and political attitudes implied in the dominant ideology', Lloyd (1983) 215.

conditioned and involuntary as opposed to those (2) that occur due to philosophical arguments which are taken to be rational and pure. The following discussion finds that Aristotle's position is very close to ideology in Schofield's [(ii) (a)] sense while maintaining that both the above assumptions require modification.

Denying that Aristotle was sexist is not a viable option but neither does it seem profitable to return to the early feminist attempts to characterize his sexism as comprehensive and ideologically motivated in the strong sense. A midway position between these two is desirable: a position which acknowledges Aristotle's sexism and the influences of his culture on his ideas (ideology in the weaker sense) but which does not thereby assume that he consciously aimed to further denigrate women. One might, then, deny that Aristotle is to be held to account for his sexism by positing that any undesirable effects of his sexism were due to the entirely unforeseeable consequences of his views.[73] Although it is plausible to suppose that his ideas had negative effects on real women, there are two key problems with characterizing his sexism in this manner. First of all, it takes us away from any interpretation of the texts themselves and towards complex socio-historical studies.[74] Second, it does not seem fair to blame Aristotle for what was done with his ideas, if he did not intend these consequences and so fails to capture something important about sexism, which is that it is essentially unfair, wrong-headed and to some extent blameworthy – a sentiment felt by almost all modern thinkers.

There is another way in which to characterize Aristotle's attitude towards gender that does not deny his sexism, but does attempt to bypass ideological motivation and indeed seems, furthermore, to deny that Aristotle was culpable for his sexism. This stance, which has been popular among commentators on Aristotle's biological works, affirms that his rational philosophy and science can be separated from his sexism. The main idea is that

[73] Morsink (1979) 84: 'Aristotle's biology, though not now sexist in conception and origin, was still sexist because of the consequences it had.'

[74] Even a skilled historian is likely to find the task of recovering such proof almost impossible. However, that fact cannot entirely exonerate Aristotle from the charge of sexism, thus I do not concur with Mayhew's claim that '[t]he critic of Aristotle must show how Aristotle's claim would lead to some kind of injustice or maltreatment' (Mayhew 2004, 8).

although Aristotle was undeniably sexist, this is only ever to a limited, and innocuous, extent. Two moves are apparent in interpretations of this sort. (1) It is taken that Aristotle's philosophical and scientific commitments lead to a value-neutral general theory. Only when he must choose to assign roles to male and female in the context of reproduction does sexism enter the account.[75] (2) Next, one might thereby consider that Aristotle's philosophical and scientific commitments, reasoned and, in this sense, justified, conflict with his unreasoned or 'unreflective' belief in female inferiority.[76] In this vein, Deslauriers contends that in his *Metaphysics* Aristotle assigns essence to both male and female of a species while statements within the biology suggest that the male is the true essence and the female fails to achieve this ideal (1998). It appears, then, that Aristotle had a more reasonable view of the human essence in portions of his *Metaphysics*. A possible parallel case is the way in which Aristotle sometimes reasons that, in terms of teleology, the female is functional. But when in a more sexist mode of thinking, the female becomes an accident of nature: a flawed male.[77] There are two problems with this analysis, however. First of all, it cannot be extended to cover all instances of Aristotelian sexism, such as the *GA* II passage above. Second, it is based on the assumption that metaphysics is objective and separable from value judgments. The difficulty of insulating scientific practice from the beliefs and attitudes of the scientists engaged in its construction or maintenance, is

[75] Furth (1988) 133, 139–40, Preus (1975), Cook (1996) 52, Matthews (1986) 22–3 and Mayhew (2004) 52–3.

[76] 'Aristotle does have a positive, reasoned, philosophical commitment to the essential sameness of male and female in any species, which works against any misogynist prejudice' Deslauriers (1998) 139. Cf. Lennox (1985a), Nussbaum (1998), Mayhew (1999) 14–17, Wardy (1993) 23 and Mayhew (2004). Schofield's interpretation of Aristotle's theory of natural slavery follows a similar pattern: 'This [the belief that slaves are barbarians] is a nasty case of false consciousness. But it does not infect Aristotle's theory of slavery itself. In working out his theory he does not make the supposition that he is describing contemporary slavery or even that what is is saying is applicable to it. The false consciousness goes to work when Aristotle stops thinking' (1990) 22.

[77] These normative distinctions are recurrent in Aristotle's explanatory framework. See Freeland (1994).

acknowledged by many branches of modern philosophy.[78] In the case of Aristotle's views on female nature, the idea that his philosophy can be separated from his sexism in a neat manner is questionable. A male-biased attitude pervades many parts of Aristotle's thought. So, for instance, when he relaxes his intent theoretical focus and conducts empirical investigations, he is not always overcome by the need to privilege males and characterize females negatively.[79] Furthermore, Aristotle's sexism does not always appear to be irrational; some of his 'good' science, when his thinking is most acute and 'rational', falls in line with his sexism.[80] Aristotle's sexism is variegated – occurring in different contexts, sometimes conflicting with the facts and sometimes in line with his science.

The other difficulty with the view that rational science and sexism are separable in Aristotle's works is that we are left feeling that he gets off too easily, like a naughty child whose irrational moments can be ignored and pardoned. As I have shown, the idea that Aristotle had a thorough-going sexist metaphysics does not sit easily with his biological writings. In any case, the focus of feminist critics was never solely the possession of such a system but the explicit attempt to use his gendered metaphysics or his biased biological theories of sex difference in order to create, or at the very least reinforce, gender inequality in society and culture.

[Aristotle] sought to prove scientifically not only that the male is superior to the female, but also that the female, despite her nurturing of the fetus during

78 Scheman (1993) 180 and Papineau (1996) 1–20. Attempts to present only the bits of Aristotle's work that would be acceptable to the 'rational' inquirer can distort what we find in his writings. Aristotle did not engage in metaphysical inquiries by separating them from other forms of inquiry. Even in the largely abstract treatise on primary philosophy, the *Metaphysics*, references not just to biology but also to animal generation frequently occur. *Metaph.* 983b26, 988a5–8, 1013a7–8, 31–2, b23–4, 1032a24, 31–2, 1034a35–b1, 1049a2, 14–15, b18–22, 1050a3–6, 1070a7, 28, b34, 1071a15, 22, b30, 1092a16.

79 Aristotle allows for the possibility of female only wind-eggs (*GA* 741a17–19) and an exclusively female fish species (*GA* 741a33–741b2) despite this clashing with a sexist attitude towards the primacy of the male generative role in nature.

80 Consider that in many empirical investigations the female does appear weaker and less muscled (in large mammals at least) than the male of the kind (*HA* 538a22–538b24). There is nothing irrational about this observation or some of the conclusions that Aristotle comes to draw from it, for instance about what parts of the body are articulated by heat and what receptacle will be required for nutritive residues in each sex (*GA* 727a16–17, 764b32–5). Yet all of these assumptions rest on sex bias.

pregnancy, has no genetic input into procreation, thus making the father the only real parent. The desire of the male to arrogate himself the procreative function and to reduce the female to mere incubator is the most fundamental cause of repression and sex antagonism (Keuls 1985, 145).

It has now been sufficiently shown that it is impossible to find that Aristotle himself intended to 'cause ... repression and sex antagonism'.[81] In terms of ideological bias in the strong sense, then, there is no evidence of this in Aristotle. But we do have another way to find that his sexism was not wholly pardonable. An unconscious (Schofield's [ii] [a]) or involuntary (Schofield's [ii] [b]) desire to promote one group's interests over another's is not pardonable.

Three possible options are available in terms of understanding the import of Aristotle's sexism. One could say (a) that Aristotle's biology was influenced by the sexism apparent in his society and culture, which he could not have avoided (the naughty child interpretation); or (b) that his biology was influenced by that sexism but he could have avoided it; or (c) that his biology was misogynistic, that is, strongly intended to reinforce or strengthen the sexism found in his culture (ideology in Mayhew's strong sense and the interpretation favoured by many feminist critiques). We find plenty of (b) but very little evidence for (a) or (c). So although (c) remains a possibility, it rests on speculation and so is quite easily dismissed by those who defend Aristotle from feminist critiques. But those who try to defend Aristotle against the charge of engaging in (c) are not thereby able to dismiss (b) or simply

81 Regrettably, simplified versions of Aristotle's views on women continue to dominate mainstream thought. Consider the following passage from a recent article on modern theories of sex difference. 'Aristotle affirmed the essential difference between the sexes: men's brains were bigger, women were more inconstant, emotional and compassionate, at least in part because they do not produce semen – whence men's and women's different behaviour and place in the social order. Symbolically, at least, biology's long, continuing and often lamentable history of using its authority to define women's nature, in order to justify attributing her with inferior status, begins here', Rose and Rose (2011) 17–18. In this, the first paragraph of their article, Hilary and Steven Rose write with assured authority, making the following by now familiar errors. Like Green (1992), they wrongly assume that brain size plays a role in sex differentiation in Aristotle. Next, they posit explicit connections between his biology and politics while also imputing to him the same kind of general 'symbolic' associations as did second wave feminists. And finally, they see him as the first to justify the oppression of women using biology. All of these views are misguided.

convert it to (a).[82] Aristotle ought not to be fully exonerated from any blame and treated as just an 'honest philosopher and scientist' even if he did not fully intend to be sexist. The reason is this: there is some fault we can recognize in imbibing a problematic *status quo* without questioning it. At the very least, this counts as a failure to act as one ought. And so we can blame Aristotle for such a failure, especially in light of the fact that his own teacher did challenge the prescribed role of women in society.[83]

The idea that Aristotle was in some sense innocuously sexist – that his sexism was forgivable, unavoidable or innocent – ought to be abandoned. The weaker Lloydian sense of ideology can be applied to Aristotle if it is modified to indicate the acceptance of cultural norms which one could, and perhaps should, have resisted. Thus one can attach blame to a past thinker, not because they actively pursued gender inequality but rather because of their passive failure to initiate any reform which could have aided women.[84] Although when we label the assumptions or points of view in his biology 'sexist' we do not attribute to Aristotle any ideology in the strongest sense (c), we do not thereby imply that Aristotle could not help thinking what he did about women (a). He could have thought otherwise, and indeed much of his closer

82 This is the *non sequitur* I take Mayhew (2004) to be attempting most of the time. Occasionally he seems aware that this will not work. For instance, when admitting Aristotle's failings, he acknowledges that 'not all of Aristotle's mistakes were unavoidable'. There was, he admits 'occasionally . . . a (sometimes faint) whiff of ideological influence or motivation' (116). Use of the term 'ideology' seems to confuse his point. Surely what he means is not something as strong as this but something like 'culpable sexism'.

83 Plato *Rep*. V. 454c-456e.

84 One could conceptualize this as failing to meet a positive duty to aid rather than failing to respect a negative duty not to interfere. The former is less morally significant than the latter. For positive and negative rights and duties see Foot (1978). What Schofield (1990) terms Aristotle's 'false consciousness' concerning slavery is not, for him, as blameworthy since it is a 'sin of omission' rather than a sin of 'commission'. This sort of false consciousness is distinct from the Marxist one, says Schofield, which may lead us to some confusion yet again about the applicability of ideological bias in this context. False consciousness ought to indicate some sort of purposeful subterfuge or 'insincerity'. And the fact that we cannot definitely find anything of the sort in Aristotle's views on women in his biology leads Mayhew, for example, to conclude that no ideology is present. However, if ideological bias (as displayed in false consciousness) does not require this, and can be merely a sin of omission, then we cannot say that it is not here. Furthermore, unlike Schofield, it does not seem to be something that can be neatly separated from the philosophical work in the text.

empirical research ought to have indicated to him that the female in nature is the equal in value to the male.

One might still worry that it is distasteful and repugnant for a feminist to read a work containing the varieties and examples of sexist thinking that Aristotle's undoubtedly does. Indeed, this seems to be the motivation of those who argue (as in [1] above) that only when he was being irrational and unscientific was he being sexist. For feminists who choose to view Aristotle's sexism in this way, it validates their use of the supposedly non-sexist and rational parts of his thought which one separates from the irrational and illogical sexist portions.[85] Seeing how questionable it is to make such a conceptual separation, it is good that none is necessary for a person with feminist sympathies to continue to study Aristotle. It need not be worrying for him or her to read sexist and blameworthy Aristotelian works because the content of these works can no longer hurt women. Many thinkers in the past have defended their own views with reference to the authority of some great philosopher. Those who used Aristotle's biology for misogynist purposes (e.g. Albertus Magnus, Aquinas) when the science was still acceptable tended to grossly exaggerate the sexism, and link it to systems of argument they themselves put in place to serve political ends.[86] Obviously no one could now use the texts in this manner to actively subordinate women. Therefore, Aristotle's sexism in his biology has become politically inconsequential, which allows the feminist to feel unthreatened when analysing it. But can we go further than this and suggest that a feminist not simply refrain from reading Aristotle's biology, but actively and contentedly engage with the texts?

The answer to this question is complicated. The key original goal of feminism is the emancipation of woman. One cannot hope to reach this goal directly through the study of Aristotle. For in order to do so, one would have to find arguments for female

[85] '[T]he failure [of Aristotle] with respect to women is a failure to argue well, to apply his methods of scrutiny well. And this means, I think, that while we should not forget what Aristotle said here, we may proceed to appropriate other elements of his thought without fear that they are logically interdependent with his political and biological misogyny', Nussbaum (1998) 250.

[86] See Chapter 6.2.

inferiority within his texts and break these down, and there are no such arguments. Creating or reconstructing connections between biology and politics tends to give the impression that Aristotle denigrated the female at all levels of existence. These feminist strategies must now be abandoned not only because they come up with inaccurate portrayals of Aristotle's views but also because they do not help to achieve what they set out to. It is not possible, in the way that they thought, to show how wrong Aristotle is (about being sexist) through his own works.

All is not, however, lost. Once one is able to properly describe the extent and nature of Aristotle's sexism, it may be that this can serve to aid feminist research not into feminism itself but into Aristotle. Feminism benefits through the fact that a serious discussion of gender in Aristotle's works opens up, and highlights the importance of, areas of his philosophy that have been consistently marginalized in the past. This marginalization of key texts on women and the female in the Aristotelian corpus was actually reinforced by second wave feminist critiques. Because the goal was to find him to be wholly sexist in the most detrimental way, which is not consonant with careful reading of the relevant material, only very simplified versions of his theory were accepted.[87] A focus on traditional readings, which paste over positive characterizations of the female in Aristotle, strengthen the scholarly trend among ancient philosophy experts to consider gender in philosophy to be of no real significance to the study of Aristotle. Another way of reading Aristotle from a feminist perspective would run counter to both of these techniques in that it would urge close and careful analysis of all of his statements and theories concerning female nature (in terms of their role in the generation of animals, for example) even when these are not as negative as one would expect from an 'arch misogynist'. This type of focus then attempts to bring such discussions into the mainstream of Aristotelian scholarship.

[87] This strand of feminism made use of earlier articles on Aristotle without any real engagement with the texts in question. Shildrick (1997), who presents Aristotle as someone who completely erased female influence on reproduction, cites MacLean (1980), Laqueur (1990) and Tuana (1989) as sources for this fact. Many more use Horowitz (1976) and Lange (1983) as their only sources for Aristotle.

Despite the optimistic view that it is 'no longer acceptable to read Aristotle's works while ignoring issues of gender', most of the time gender is still treated as irrelevant in scholarship on Aristotle.[88] A feminist reading of Aristotle's biology seeks to counteract this academic trend by close analysis of many neglected texts coupled with acknowledging male bias and sexism within them (and thus resisting the urge to deny this, which so many Aristotelians find difficult to do). By conducting such academic work the scholar with feminist sympathies not only helps to broaden our knowledge of Aristotelian philosophy but also challenges the common practice among scholars of Aristotle of extracting what is most abstract and general and disregarding his more detailed analysis.[89] By taking seriously the entirety of Aristotle's biological works on female nature, and realizing that many of these explanations are not negative, he or she thereby promotes the study of women and the female and provides evidence of the importance of such study to the field of ancient philosophy. No longer will it be so easy, or in fact in any way desirable, for the historian of philosophy to dismiss discussions of gender in philosophy as either inaccurate or unimportant to the study of Aristotle.[90] Aristotle's account of women and female animals ought to be of interest to us.

[88] Freeland (1998a) 15. Some feminist interpretative work argues that studying gender in philosophy can help us to better understand that philosophy. See Oliver and Pearsall (1998), ch.1 on Nietzsche and Fricker and Hornsby (2000) 'Introduction' on analytic philosophy.

[89] See, for example, Needham (1959), which encourages the reader to disregard the 'messy' later Books of the treatise. Just a few examples of his more detailed account where he characterizes the female in a more positive fashion include his discussion of the female generating on her own (*GA* 741a33–741b2), providing nutritive soul on her own (i.e. generating 'up to a point'; Aristotle, *GA* 741a19–25), and the female contributing *kinêseis* (sources of change), which ensure differentiation of sex and resemblance to members of female blood lines (*GA* 768a10–21).

[90] Recently, several other scholars have acknowledged the importance of studying the entirety of the text of the *GA* while keeping an eye on gender issues. See, for example, Henry (2007b), Sier (2009) and Gelber (2010).

2

CONSISTENCY IN THE *GENERATION OF ANIMALS*

2.1 Methodology in scholarship on Aristotle and the *Generation of Animals*

We have now established that focusing on Aristotle's views on females is acceptable and productive. This will require taking seriously the later Books of the *Generation of Animals* where the account of female animals gets its fullest treatment. However, scholars often consider these Books to be problematic, particularly when attempting to give a coherent picture of Aristotle's theory of generation. This is why, for many, the starting point of interpreting the work as a whole involves attempting to reconcile divergent discussions of the female role in generation within the text.[1] An overview of the topics and content of the *GA* highlights why this issue dominates. *GA* I begins with a review of the four causes (material, efficient, formal and final) and the question of how animal generation is brought about (*GA* 715a4–7). Male and female are the principles (*archai*) of generation, and so the starting points of this investigation (*GA* 716a5–6). The two sexes must be present in all cases of animate generation, whether they exist separately in different animals or not, and generation is caused in some way by semen (*sperma*) coming from the parents. After considering the views of various opponents Aristotle decides that semen is nutriment in its final form, which is ready to be distributed, or to become each of the parts of the body (*GA* 726a26–8). The female does not emit semen like the male: her

[1] The positive influences of the female are most often discussed only with reference to hereditary resemblance. Coles (1995), Tress (1992), Dean-Jones (1994), Balme (1987b), Cooper (1988) and Matthews (1986), to give but a selection of examples. Katayama (1999) 98 who sees himself as 'defending Aristotle from a charge of inconsistency in the *GA*', notes 'I have presented his embryology as a consistent and coherent theory'. This assurance comes too easily given that Katayama does not discuss the female role or *GA* III–IV at all.

contribution is the less concocted menstrual blood (*GA* 726b31–3). Furthermore, the female principle is the material cause and the male the efficient and formal cause of generation (*GA* 729a10–12). The male semen works on the female menstrual blood, like a carpenter works on wood to produce a bed; the female principle is passive, the male active (*GA* 729b9–19).

At the beginning of *GA* II Aristotle says that male and female often exist separately in different animals because it is better for the superior formal factor (the male) to be separated from the inferior material one (the female) (*GA* 732a5–10). In what follows he also characterizes the sexes in terms of the contribution of body and soul respectively (*GA* 738b26–7), emphasizing the male role as the formal cause of generation. Towards the end of this Book Aristotle begins a more specific study of particular types of animals and their methods of generation, noting the exceptional case of hybrids. *GA* III continues the specific investigations covering birds, insects and testacea. Next, *GA* IV details sexual differentiation and heredity, describing reciprocity between the sexes and how one prevails over the other to determine sex and inheritance. After this, a discussion of monsters and deformities centres on a complementary model of the sexes' contributions. Finally, *GA* V discusses variations in the features of animals, such as eye and hair colour, and explains how these are due to similar causes.

A general contrast between sections of the *GA* seems apparent. The early Books emphasize more a hierarchical model of the sexes: the male animal is superior, active, hotter and the conveyer of form. The female animal is inferior, passive, colder and the conveyer of matter. The later Books give a complementary or reciprocal model of the interaction of the male and female contributions in which they oppose one another's powers. In this context the male and female contributions share the following characteristics: both possess a degree of heat and both contain specific potentialities to form parts of the body. The question then becomes: can Aristotle consistently maintain these two positions?

As previously stated, issues of consistency exercise many scholars of Aristotle'.[2] Resolution is attempted both when such

[2] See Part I, Introduction.

supposed conflicts exist (1) across texts and (2) within a given text. If the two points of view or doctrines are assumed to be incommensurate, the modern scholar is strongly inclined to privilege one view and diminish the other.

The next three sections of this chapter will argue that it is best to resist these scholarly tendencies which have in the past served to undermine Aristotle's account of the female. Across texts (1), the *GA* is often used to assist other reconciling projects – in particular, the desire to find a coherent theory of scientific method in Aristotle's analytical works. This is not only distracting but pulls the reader away from more complicated and detailed portions of the *GA*. Within the text of the *GA* (2), the attempt to reconcile the supposed crisis concerning the female role can also result in the subordination or suppression of important aspects of the later Books of the work. I will argue that, although we ought to endeavour to establish that Aristotle's works are not wholly incoherent or grossly contradictory, we have to be careful not to dismiss or obscure important discussions in order to achieve this aim. There is a fine line, then, between understanding subtle moves in Aristotle's complex methodology and breaking away from his original intention in order to secure rigid consistency.[3]

Often the biological texts are consulted in order to help resolve problems that appear in the *Metaphysics* and *Analytics* and are therefore not treated as informative of Aristotle's philosophy in their own right.[4] When these better known and popular texts are privileged, this can at times inhibit one's understanding of the *GA*. For instance, a commentator might have a set idea of what matter and form must represent for Aristotle before reading the *GA*, and therefore not choose to acknowledge the nuances found there, thus subordinating these to more straightforward and

[3] I have sympathy with Gill's stated approach in her work on Aristotle's *Metaphysics*. She writes, 'I assume that Aristotle had a broadly coherent and interesting account to offer, one well worth trying to reconstruct, and that conflicting claims within the account *more probably reflect inadequate understanding on our part than confusion on Aristotle's*' (1989a, 11, my emphasis).

[4] Katayama (1999) 8: 'I rely on passages from the De Anima and especially the Generation of Animals to solve problems in the Metaphysics'. For other examples see Connell (2001) 298–9. However, some recent work is beginning to counteract these tendencies, see especially Lennox (2001), Charles (2000) and Henry (2005).

familiar characterizations of the distinction (gleaned from better known texts). When treating the seeming inconsistency, then, within the *GA* passages, the account that appears to fit best with more generalizing treatises, like the *Metaphysics*, are often privileged and those that qualify and set out a more intricate analysis of such concepts, especially when these views are unfamiliar, may be unduly subordinated. Within the *GA*, many modern scholarly readers, while taking the two descriptions of the female role to be completely incommensurate, feel that they must save Aristotle from the charge of contradicting himself. Therefore, they take the account found earlier on in the treatise to be primary, since it is set out in a more formal or familiar manner. Once the abstract arguments of *GA* I and II have been analysed, the rest of the work is subordinated to that early account. The more prominent role of the female in the later Books is then explained away or suppressed.[5]

With reference to interpretations across texts (1), the *GA* ought to be allowed to stand on its own as a resource of Aristotelian doctrine, sometimes as a refinement or elaboration of the views found in other more popular works. And when interpreting the *GA* on its own (2), if we do not assume that the portions which fit neatly with familiar texts are more philosophically interesting, then there seems no good reason to privilege these over other parts of the work. Thus, I will urge that the *GA* ought to be read as a whole, with equal emphasis placed on each part rather than using the technique of subordination, and in doing so problems with coherence and consistency are greatly diminished.

In the next section, I will give a brief overview of the connections between the method in the *GA* and the general epistemological guidelines Aristotle puts forward in his works on method. I will then place the *GA* within the context of Aristotle's study of nature. Concentrating on the *GA*, I will argue that Aristotle moves from more general to more specific discussions, which helps to explain how he can incorporate seemingly conflicting

[5] This methodology is most apparent in Cooper's interpretation of *GA* IV.3 and his outright rejection of a positive role for the female in Aristotle's theory (1988). See Chapter 9.

views into one account. I will run through the text section by section, mapping out this method. The final section explains how understanding the treatise in this manner mitigates against the charge of inconsistency with respect to the female role in generation.

2.2 The *Generation of Animals* and the *Analytics*

At the beginning of the *Posterior Analytics*, Aristotle explains the acquisition of knowledge (*A. Po.* 71a1–9). *Epistêmê*, which is often translated as 'scientific knowledge', can only be gained through demonstration (*apodeixis*). A demonstration is a particular type of logical form with two premises and a conclusion. With previously existing knowledge of the premises, one can gain the knowledge of the conclusion. Thus, Aristotle argues that knowledge must be constructed from incontrovertible logical demonstrations. He presents these logical forms or syllogisms in his *Prior Analytics*. The acquisition of scientific knowledge requires a particular type of syllogism whose terms are universals (Table 2.1).

Knowledge of the premises of a demonstration can be gained either primarily or as the conclusion of an earlier syllogism. But it is particularly important to discover those with which to begin and these must be 'true, primary, immediate, better known than, prior to and causative of the conclusion' (*A. Po.* 71b21–2). Primary premises are definitions and axioms. Crucially, they must be explanatory of any conclusion (*A. Po.* 71b16–33).

In the *Analytics* Aristotle is insistent that to know something is to know that it will always be the case and cannot ever be otherwise. To put it another way, once you know something through demonstration you will never change your mind about it (*A. Po.* 72b4).

Table 2.1 *The syllogistic form of scientific knowledge*

	All As are Bs
	All Bs are Cs
therefore	All As are Cs

> If any fact is the object of unqualified knowledge, that fact cannot be otherwise than it is.[6] Now whether there is also another way of knowing we shall say later; but we say now that we do know through demonstration (*A. Po.* 71b15–17).[7]

We might imagine that the *Analytics* explores abstract or *a priori* knowledge as opposed to empirical understanding gained through a looser method such as induction. However, many are attracted to the idea that the *Analytics* contains Aristotle's overall assessment of all types of scientific understanding and so it ought to be the case that what he says there fits with his physical, biological and zoological treatises.[8] As in other areas of scholarship on Aristotle, commentary on his theory of knowledge tends to be concerned to overcome any perceived tension or conflict. Most generally, the worry is that although Aristotle advocates strict logical procedures for inquiry in his *Prior* and *Posterior Analytics*, this is seldom if ever carried out in his scientific works. In the past, the debate concentrated on the discrepancy between the *Posterior Analytics* and the *Physics*. In the latter, reasoning is based on dialectic, or assessing arguments, rather than on syllogistic logic.[9]

The central focus of the debate later shifted from the *Physics* and more abstract treatises, to the biological works and their empirically based content. Partly in order to vindicate these as philosophically significant and partly in order to reconcile or accommodate different portions of Aristotle's *oeuvre*, one focus of scholarship on the biological works concentrates on finding a place for the *Analytics* model of inquiry.[10] Sometimes the biological works are scoured for any indication of *Analytics*-style

[6] ὥστε οὗ ἁπλῶς ἔστιν ἐπιστήμη, τοῦτ' ἀδύνατον ἄλλως ἔχειν.

[7] See also *EN* 1139b19–25: 'Now what knowledge is, if we are to speak exactly and not follow mere similarities, is plain from what follows. We all suppose that what we know is not capable of being otherwise; of things capable of being otherwise we do not know, when they have passed outside our observations, whether they exist or not. Therefore the object of knowledge is of necessity. Therefore it is eternal; for things that are of necessity in the unqualified sense are all eternal (πάντες γὰρ ὑπολαμβάνομεν, ὃ ἐπιστάμεθα, μηδ' ἐνδέχεσθαι ἄλλως ἔχειν· τὰ δ' ἐνδεχόμενα ἄλλως, ὅταν ἔξω τοῦ θεωρεῖν γένηται, λανθάνει εἰ ἔστιν ἢ μή. ἐξ ἀνάγκης ἄρα ἐστὶ τὸ ἐπιστητόν. ἀίδιον ἄρα· τὰ γὰρ ἐξ ἀνάγκης ὄντα ἁπλῶς πάντα ἀίδια, τὰ δ' ἀίδια ἀγένητα καὶ ἄφθαρτα).'

[8] Gotthelf (1997a, 1997c, 2012c) and Lennox (2001).

[9] Owen (1961). Cf. Taylor (1990) 125.

[10] See Gotthelf and Lennox (1987) Part II, Lloyd (1990) and (1996) ch.1, Gotthelf (2012a) Pt II and Lennox (2001) chs. 1–3.

analysis.[11] However, most acknowledge that fitting biology to logic in Aristotle is hardly ever straightforward. The first problem is this: the metaphysical transience of the objects of biological study seems incongruous with the *Analytics*-based goal of establishing incontrovertible truths. Since nature acts as a cause *always or for the most part*, any premise concerning natural phenomena will be contingent rather than necessary. For instance, if Aristotle tried to explain why rain falls in winter in a demonstration, he would have to use the premise: all rain falls in winter. But this is not the case. There may be a drought, or some rain might fall in summer (as in *Ph* II.8). Rain does not fall always in the same way because objects below the moon do not act in a completely regular manner, unlike the perfect motions of the physical heavens (*Cael.* 304b25, 305a12, 269b13–17, *GC* 338a2–b5, *Mete.* 339a11, 25).[12] In the *GA* context, Aristotle gives us an account of monsters and other anomalies which may occur in the course of embryonic development (*GA* IV). Sometimes, under normal conditions, parents generate deformed offspring (*terata*) and although these are mistakes they are also a part of nature. It would seem difficult to fit this within demonstrative science since he says there, 'you cannot understand through a demonstration what happens by chance' (*A. Po.* 87b19). Furthermore, because natural objects undergo generation and destruction, there can be no eternal demonstrations for them. So, although in the *A. Po.* Aristotle states 'we must proceed in the study of animals by establishing what attributes belong to every animal' (98a4–5), any attributes he found would never be as necessary as the mathematical examples he is so fond of there.[13] This is why those who study most closely the way in which the biological works use the recommendations of the *Analytics* insist that Aristotle allows that it is possible to provide demonstrations of things that are not eternal. Although there cannot be eternal

[11] Detel (1997) and Gotthelf (1997a). But see Charles (2000) ch. 11 for a more moderate approach.

[12] The primary sense of 'necessary' is what cannot be otherwise (*Metaph.* 1015a34-6, 1015b10-15).

[13] Lloyd (1990). I do not believe, however, that most commentators who explain the deeper connections between Aristotle's biology and logic are guilty of ignoring this point.

demonstrations, there can be demonstrations of a more relaxed variety that fit the subject matter at hand.[14]

Other difficulties with finding bald connections between logic and biology can be briefly sketched.[15] In order to apply his theory of demonstration to biology, Aristotle must first locate some primary premises from which to build up syllogisms. In his *A. Po.* and *Metaphysics*, he gives us three axioms which are absolutely certain principles: the principle of non-contradiction, the law of excluded middle and the mathematical premise that if equals are taken from equals, equals remain (*Metaph.* IV.3, 996b26–30, *A. Po.* I.11). These axioms are not specific to a particular subject area but apply to any type of study. Biological axioms are also very general, such as the idea that 'nature does nothing in vain' (e.g. *de An.* 434a30, *PA* 695b19–20, *IA* 704b16, 708a9, *GA* 741b4–5). Although they can be found to be of some use in biology, they are not specific enough to get the student of nature very far.[16]

Definitions would seem to be more useful as a starting point for demonstration in biology. Aristotle's attention to animal *differentiae* points to an interest in obtaining something like definitions of the essences of certain animals. In *A. Po.* II definitions are to be gained through a process of dichotomous division. The definition consists of the nearest classificatory group (or *genus*) and the *differentia. Differentiae* are supposed to distinguish a species as different from the rest in the larger group. So, for instance, if the genus is 'bird', the *differentia* 'large beak' picks out the species 'toucan'. This makes the definition of toucan, 'large-beaked bird'. When he states in the *HA* that he will 'take the differentiae and attributes of all animals' this recalls the language of logical divisions detailed in the *Analytics* and *Metaphysics*.[17] Furthermore, the facts collected in the *Historia Animalium* in many ways correspond to the type of starting points

[14] See Lennox (1991) on 'partial demonstrations' (275–83).

[15] For a fuller exposition see Lloyd (1990).

[16] Lennox (1997), reprinted in *idem* (2001) ch. 9, argues that 'nature does nothing in vain' is an explicit premise used by Aristotle in scientific explanations. For the use of axioms in Aristotle's biology more generally see also Gotthelf (2012b).

[17] See *Metaph.* VII.12 and *HA* 491a9.

of investigation recommended in *A. Po*. II.[18] However, in certain respects, finding definitions in biology will be tricky. In the context of the *Analytics*, Aristotle very often uses the definition of human being as an example, which he gives as: 'two-footed animal' (*A. Po*. II.14, II.22). This definition does not, however, provide us with a *differentia* unique to humans, so as to pick them alone out of the larger group. Birds also have two feet. My own example is equally flawed, since many birds, other than the toucan, have large beaks. In biology, it is a challenge to provide the necessary *differentia*. Even finding out what general group or genus some sort of animal belongs to can be difficult in practice. According to Aristotle, some animals have characteristics of more than one *genus* and so 'dualise'. The bat, for example, can be classed as both flier and land animal.[19] Although such definitions might be hard to pin down, these sorts of moves can be found by careful study of Aristotelian biological texts, while keeping in mind a general view of what demonstrations Aristotle might be seeking.[20]

The project of fitting biology to logic requires a detailed and nuanced knowledge of both sets of works and a willingness to conscience the idea that Aristotle modified his views.[21] The *PA*, for instance, appears to follow on from the recommendations of the *A. Po*. and to take these further – refining them. Thus we find in the *PA* that real kinds are marked off by several *differentiae*, and not ascertained through dichotomous division

[18] See Kullman (1974), Lennox (1987, 1991), repr. in *idem* (2001) chs. 1 and 2, Gotthelf (1997c), Charles (2000) ch.11 and Balme in Aristotle (1991) 13–26.

[19] '[S]eals and bats are in an intermediate position. Seals are between land-animals and water animals, bats between land animals and fliers: thus they belong to both classes or to neither' (*PA* 697b1-4). Human beings may be in a similar position. See Lennox (1999). We might also need to consider the overlap of modes of locomotion with modes of generation (*GA* II.1). Weiner (2010).

[20] See Lennox (1991) and Gotthelf (1997c) who argue that Aristotle sought numerous *differentiae* with the aim of providing explanations in zoology. Gotthelf (2012c) only finds 'partial definitions' in the *HA*, but argues that the *A. Po*. requires no more than this (178).

[21] Lennox (2001) urges that Aristotle's philosophy of science, found in the *Analytics*, provides guidelines which are inevitably restructured in order to fit with new discoveries in biology (*Introduction*, 5–6). See also Lennox in Aristotle (2001): '[in many ways] the *Analytics* conception of science must be enriched and supplemented in order to be applicable to a biological context . . . it is the purpose of *PA* I to provide this enrichment and supplementation' (xiv).

(*PA* 643b12–13). Within the *Analytics* itself a certain tentativeness and openness to revision and innovation is apparent. Procedures must be adapted to suit the subject matter.[22] The *Analytics*, in so far as that text encourages its readers to discover the essential attributes of kinds, which are properly 'necessary' to their subject (*A. Po.* 74b5–10), ought to allow for multiple *differentiae* for defining essences.[23]

Despite such positive progress, there is a danger that by focusing on matching theory and practice, commentators may disregard or discredit certain portions of the biological works. If they cannot find a way for parts of these texts to fit, they are inclined to downgrade the importance of the content, even to the extent of subordinating it to the supposed more acceptable doctrine they find in the *Analytics*. So, for instance, Charles (2000) makes a good case for why the biological works are important for elucidating Aristotle's essentialism. However, by focusing primarily on logic, he becomes increasingly uneasy about what he finds in the biology, stating that much in these works represent 'a crisis in Aristotle's essentialism'. Meanwhile, after spending a considerable time attempting to fit biological texts into axiomatic structures, Detel (1997) concludes: 'From all this it follows that Aristotle . . . is relying on a pretty flexible notion of essentiality that is barely compatible with the simple view that the essence of a species or a genus is captured by one immediate definition-formula pointing to a single unified cause of all the other properties that are supposed to be scientifically proper explananda' (83).

[22] Ward (2008) 190. 'Aristotle's revised conception of the method of class definition in An. Po. Beta thus begins with what are termed "major kinds", each of which contains several differences together that define the kind, instead of having one difference that defines the class (649b9ff, 644b1ff.)' Ward (2008) 193.

[23] As 'odd' necessarily adheres to the number three (*A. Po.* 84a20-1). Charles (1997b) 41 suggests that Aristotle 'modified' his *Analytics*-based assumptions and so did not strictly adhere to them, but also did not 'radically reject' these. See also Charles (2000). For the idea that demonstration may be present in a more relaxed form see Gotthelf (1997a) and Detel (1997), and against this view see G.E.R. Lloyd (1996) ch.1. Such studies mitigate greatly any crude claims of Aristotelian procedural rigidity because often those who criticize Aristotle's procedures on feminism grounds (e.g. as those of an uncompromising phallocentric logician) simplify and misrepresent his methodology. For an account of this see especially Freeland (1994) and Hass (1998).

The *GA*, as with other biological work, contains much that cannot be easily connected to the recommendations of the *A. Po.* Explanations of what occurs in generation depend on many different facts, some of which are not closely related to first principles, for instance the idea that a blend or balance must occur in generation (*GA* 767a16). One general connection between recommended procedure and practice does suggest itself with reference to the *GA*. In the treatise Aristotle regards an account of generation in terms of material interactions on their own to be inadequate and is concerned to sketch out teleological accounts. This fits with a very general point made in the *Posterior Analytics*: the middle term of scientific demonstrations must be explanatory of the conclusion (and not merely factual); science prioritizes 'why' over 'how' questions.[24] The most important lesson we can take from the *Analytics* in relation to our study of the *GA* is that an ideal procedure focuses on finding principles. These principles are primarily explanatory and most closely linked to essence. Such principles, and the method of reaching them, are a very important part of the procedure in the *GA*. We must remain aware, however, that those principles typically considered to be primarily explanatory are not the only ones operative in the work.

As many have pointed out, the role of the *Analytics* in the actual practice of science is unclear.[25] Is the student/reader supposed to think that it is a procedure to be learned thoroughly before any attempt is made to gain scientific knowledge or is it intended as a way to organize information after such investigations are complete? Alternatively, it could represent some mid-point between these two – providing suggestions for how

[24] *A. Po.* 71b9-11, 94a20-36, 98a30-b24. See also *Ph.* 194b17-23, *Metaph.* 983a25-31. Within the text of *A. Po.* the way in which the 'why' questions might be adequately addressed is not specified clearly, which is one reason why scholars probe the biological works – wishing to discover more detailed accounts or examples of scientific demonstration.

[25] Commentaries on the *Posterior Analytics* often include some explanation of the purpose of the work in light of the fact that its recommendations do not appear to have been carried out in the rest of the corpus. For some, the *Analytics* represent an ideal, which can never be practically achieved. For others, it indicates an incomplete research programme, a way of writing out results or a pedagogic method. Barnes in Aristotle (1994), Lennox (2001) 5 and Lloyd (1990) 29–34.

to begin investigating while not requiring that all the steps be filled in until later. In any case, Aristotle need not think that the content of the *Analytics* must always be at the forefront of the biologist's mind. Given this there is no imperative to connect stated methodology to actual practice. Although the importance of perceptible and clear connections between Aristotle's works is undeniable, at times it is also crucial to concentrate on his goals and methods within a given work and thereby avoid the tendency to dismiss certain discussions as irrelevant. When reading the *GA*, then, I will not engage directly in the difficult challenge of fitting the recommendations in the *Analytics* to Aristotle's practice in biology, thereby leaving more parts of the text open for analysis.[26]

2.3 Situating the *GA* in Aristotle's study of nature and living things

In order to see how to characterize the method found in the *GA*, apart from in the *Analytics* style, one must first locate it in Aristotle's study programme. Clearly, the treatise is part of his study of nature (*phusis*). The work entitled 'On Nature', the *Physics*, begins with some very general guidelines concerning this area of research.

> In all disciplines in which there is scientific understanding of things with principles, causes or elements, it arises from a grasp of those:[27] we think we have knowledge of a thing when we have found its primary causes and principles, and follow it back to its elements. Clearly, then, understanding of nature must start with an attempt to settle questions about principles (*Ph.* 184a10–16).

Gaining knowledge of the natural world requires focusing on 'principles' (*archai*) and 'causes' (*aitiai*).[28] The *Physics* goes on to explain that we must begin inductively, gathering information which is easier to understand, and working towards first principles, which although truer and 'more knowable by nature'

[26] It is possible that by elucidating certain arguments found in the *GA*, this book can support endeavours to find ways to connect it to the *Analytics*.

[27] Ἐπειδὴ τὸ εἰδέναι καὶ τὸ ἐπίστασθαι συμβαίνει περὶ πάσας τὰς μεθόδους ὧν εἰσιν ἀρχαὶ ἢ αἴτια ἢ στοιχεῖα, ἐκ τοῦ ταῦτα γνωρίζειν.

[28] Aristotle uses the term 'principle' (*hê archê*) in several different ways (*Metaph.* V.1).

are further away from our immediate cognition (*Ph.* 184a16–27). These first principles will be quite general – insofar as they will be able to explain and account for the messier facts around us, which will not at first seem clearly connected into any pattern. However, principles come at different levels of specificity and different disciplines will require different explanatory starting points. Principles which cover all existent things or beings (τῶν ὄντων, *Metaph.* 1025b3–17) will not be the same as those that will be discovered and used in other more specific subject areas.

In the case of the *GA*, the subject matter is much more specific than all beings. The first step towards that more specific level of analysis is Aristotle's discussion of existent things which have a nature (*phusis*) in his *Physics*; this is the study of all things that exist by nature (τὰ φύσει, *Ph.* 185a13) – that is the four elements, plants, animals and the heavenly bodies (*Ph.* 192b9–11). In this work Aristotle is concerned to discover key broad principles by which to set out the activities and operations of all or most natural objects; these include the four causes, the related illustration of craft production, and an underlying teleological outlook (*Ph.* II.3; cf. *Metaph.* 1044a32–1044b20, *PA* 639b11–21). Importantly nature is a source of change and rest contained within the natural object; it is directed towards the ends of the individual acting 'for the sake of which' (the final cause). Aristotle narrows down the field of natural objects into separate studies. So, for example, the *Meteorologica* IV is a study of the elements. Living beings, which include animals, plants, and the heavenly bodies, are the more specific focus of the treatise *de Anima*, on the soul. *De Anima* explains generally what it is that makes something alive and the various capacities of soul. Aristotle's zoological works concern only one type of living being, animals.[29] The *GA* appears to be the culmination of one branch of zoological study, that which undertakes to describe the affections that animals display, and so forms the final part of the group which includes *The Movement of Animals* (*MA*), *The Progression of Animals* (*IA*) and the *Parts of Animals* (*PA*). The latter is most strongly tied to the *GA*, since, as he

[29] Aristotle leaves aside plants and heavenly bodies. Plants were taken over by his student, Theophratus, in his *HP*, while the heavens are the focus of the treatise entitled *De Caelo*.

explains there, it concerns the being as opposed to the coming into being of living things (*PA* 640a3–6, a13–19).[30] The description of the generative parts and the account of the coming into being of animals (*GA* 715a11–14) is the province of the *GA*.

Within each treatise on natural philosophy, Aristotle often begins by setting out answers to the most general questions concerning the subject matter and finding more specific principles. So, for example, after having given a survey of the views of his predecessors in *de An.* I, he puts forward a preliminary definition and account of soul at the beginning of *de An* II. He initially takes in a broad array of specific examples and endeavours to bring them under one general account. In the *PA* he endeavours to explain how each of the parts of an animal serves the functions of the whole. Both the *IA* and *MA* also come up with general principles to account for their more specific subjects. The *GA* likewise establishes a general account of generation in the first two Books. The beginning of the work actually attempts to connect the phenomenon of generation to a general metaphysical framework. This is why animal generation is set out in its most general and comprehensive manner possible with reference to already familiar terminology. However, Aristotle also wishes to elucidate generation as it occurs in many more detailed aspects of generation, such as heredity and deformity and in varieties of animal which do not all fit the standard model. As he proceeds through the *GA*, he can be found to develop his account of generation in these two ways.

After establishing a general account in terms of principles more specific to the subject matter of generation, he refines and elaborates this as he elucidates more aspects of the phenomena (procedure I). As in other zoological works, another procedure is present which overlaps with the above. Initially focusing on the best or paradigm type or types of animal, Aristotle moves on to describe the phenomena in question in less perfect kinds. He often ends with a single quite exceptional type, which requires a very particular explanation (procedure II). I will set out how the *GA*

[30] The *PA* ends with the promise to next 'deal with the question of [animal] generation' (*PA* 697b30).

incorporates both of these and will end by sketching how procedure (I) helps to explain the different characterizations of the female role in generation found in the text. First of all, the text will be broken down into sections for ease of exposition.[31]

Sections of the Generation of Animals *broken down by topic*

A. Book I 715a1–721a26: Analysis of non-uniform generative parts: uterus and penis.
B. Books I and II 721a31–739a6: Analysis of uniform generative parts: semen and menstrual blood.
C. Book II 739a7–746a29: Conception and embryology in the live-bearing kinds.
D. Book II 746a29–749a7: Exceptional live-bearers: hybrids.
E. Book III 749a10–763b16: Egg-layers (birds, selachians, fish), larvae-producers, cephalopods, crustacea and testacea.
F. Book IV 763b20–767a35: Account of sexual differentiation.
G. Book IV 767a36–769b10: Account of inherited resemblances.
H. Book IV 769b3–776a14: Account of deformity and birth defects, superfoetation, imperfect offspring, twins and *mola uteri*.
I. Book IV 776a15–778a12: Account of milk production and gestational periods.
J. Book V 778a16–789b20: Inessential features (*pathêmata*) in parts, particularly eyes, hair, skin, voice and teeth.

Aristotle begins the work with some of the general principles he had already established in other works, for instance the four causes (*GA* 715a3–11). He also reviews the distinction between different parts of animals (*GA* 715a8 11), thus returning to the *PA* and its more detailed study of this topic. This is why Sections (A) and (B) commence with an account of generative parts. Section (B) contains two key elements. Aristotle first attempts to connect the project in the *GA* with broader metaphysical principles, and then moves towards first principles in the specific study of animal generation. The first element is most apparent when Aristotle engages in debates with those who imagine that parts located in generative fluids are sufficiently explanatory of reproductive phenomena (*GA* 721a31–724a13). This debate concerns quite overarching metaphysical issues – such as what ought to count

[31] These do not always align with the chapter heading we have inherited.

as properly existent; which is why a more general point of view is presented, in line with his more general works on the natural world and ontology (e.g. *Posterior Analytics, Metaphysics, Physics* or *de Anima*). The shift to abstract ontology is signalled at several points by the use of 'generally'.[32] Aristotle also attempts to move towards first principles that are general to all instances and types of animal generation, but specific to animal generation as opposed to other aspects and attributes of animals, in section (B). These principles include the specification or definition of male, female and semen.

The generalizing tenor of section (B) is best captured by looking at a series of passages. In the first, the inquiry into what semen is, as a general principle in the specific subject area of animal generation, employs an analysis of language use rather than zoology.

There are, however, numerous senses in which one thing is formed or comes into being 'from' another: (1) as we say 'from day comes night' and 'from boy comes man', meaning that the one come after the other; (2) as a statue is formed from bronze, or a bedstead from wood and all those cases where we describe things as being formed from something material; here the finished whole has been fashioned into a certain shape from something which was there to begin with; (3) as a person may become uncultured from being cultured or ailing from healthy, i.e., all cases of a contrary coming from its contrary; (4) as in a 'cumulative' passage in Epicharmus: e.g. from slander comes abuse, from abuse a fight; in all these cases 'from so-and-so' means that so-and-so is the source of change (*GA* 724a19–31).

Aristotle has not yet established how it is that the semen is a cause, and so only employs the very broad idea of it as 'that from which' a new animal or plant comes to be. He notes that 'it is generally held that all things are formed and come to be out of the semen, and semen comes from the parents' (*GA* 721b6–7; Cf. 716a10–12). This definition is narrowed down considerably when at *GA* 725a24–6 it is said to be the surplus residue of nourishment, distributed to the whole body.

Aristotle connects these more specific principles to those that hold his bigger world picture together. At *GA* 729a9–34 he links his quest for a definition of male and female principles to more general ideas in his philosophy, including matter and form and

32 *GA* 729a24, b9: καθόλου; 732a20: ἐπὶ τὸ πᾶν.

active and passive principles. These connections are most apparent in *GA* 729b8–19.

The second suggestion [i.e. that the male plays no material part in generation] is clearly the right one, as is shown both by reasoning and by observed fact. If we consider the matter on general grounds, we see that when some one thing is formed from the conjunction of an active partner with a passive partner, the active partner is not situated within the thing which is being formed; and we may generalize this still further by substituting 'changing' and 'changed' for 'active' and 'passive'. Now of course the female, *qua* female, is passive, and the male *qua* male, is active – it is that whence the principle of change comes. Taking, then, the widest formulation of each of these two opposites, viz., regarding the male *qua* active and causing change, and the female *qua* passive and being changed, we see that the one thing which is formed is formed from them only in the sense in which a bedstead is formed from the carpenter and the wood, or a ball from the wax and the form.[33]

It is clear that *GA* I in particular finds Aristotle keen to link the phenomena of animal generation to his analysis of all natural objects (e.g. from the *Physics*). Let us now look at sections other than (B) where we find a different pattern of analysis. Rather than moving to and setting out general principles in embryological theory, and linking these to first principles in the broader fields of metaphysics and natural philosophy, Aristotle seeks to shift from paradigm cases to explaining more and more exceptional ones.

2.4 Generation in different animal varieties

Particularly at the very beginning of the treatise (A), the methodology of the *GA* is strikingly similar to that found in the *PA*. Indeed, this section is constructed using the method Aristotle made explicit there (*PA* 639a19–26, 30–b5).[34] Thus it is helpful to turn to the *PA* discussion of zoological methodology.

[33] κατά τε δὴ τὸν λόγον οὕτω φαίνεται καὶ ἐπὶ τῶν ἔργων. καθόλου τε γὰρ ἐπισκοποῦσιν οὐ φαίνεται γιγνόμενον ἓν ἐκ τοῦ παθητικοῦ καὶ τοῦ ποιοῦντος ὡς ἐνυπάρχοντος ἐν τῷ γιγνομένῳ τοῦ ποιοῦντος, οὐδ᾽ ὅλως δὴ ἐκ τοῦ κινουμένου καὶ κινοῦντος. ἀλλὰ μὴν τό γε θῆλυ, ᾗ θῆλυ, παθητικόν, τὸ δ᾽ ἄρρεν, ᾗ ἄρρεν, ποιητικὸν καὶ ὅθεν ἡ ἀρχὴ τῆς κινήσεως. ὥστε ἂν ληφθῇ τὰ ἄκρα ἑκατέρων, ᾗ τὸ μὲν ποιητικὸν καὶ κινοῦν, τὸ δὲ παθητικὸν καὶ κινούμενον, οὐκ ἔστιν ἐκ τούτων τὸ γιγνόμενον ἕν, ἀλλ᾽ ἢ οὕτως ὡς ἐκ τοῦ τέκτονος καὶ ξύλου ἡ κλίνη, ἢ ὡς ἐκ τοῦ κηροῦ καὶ τοῦ εἴδους ἡ σφαῖρα.

[34] Having discussed most of the parts of the bodies of animals in the *PA*, Aristotle now notes a need to elaborate on the generative parts: 'of the parts, it remains to describe those which subserve animals for the purpose of generation, about which I have so far

> The following question about how one ought to carry out an examination would not be overlooked – I mean the question of whether one should study things in common according to kind first, and then later their distinctive characteristics, or whether one should study them one by one straight away (*PA* 639b3–6).

Aristotle decides on the former proposal in the *PA*, starting with a paradigm kind of animal and moving on to those that differ by the more and less. Analysis of each part begins with a description of that part, which will explain how it operates and how it is structured in most animals that possess it. It then proceeds to note variations and differences (in size for instance) from that standard model and eventually ends with exceptions. At the very end of the treatise (*PA* IV.11–13), we find descriptions of peculiar classes such as the snake, whose neck, unlike that of any other animal can turn backwards while the rest of his body remains still (692a3–4). Also, among birds, only the 'crook-taloned' cannot perch because of the shape of its claws (694a20). In such cases, Aristotle often explains that the peculiar difference in the part is due to necessity (694a23–30) or material causes, such as fluidity and heat.[35]

A similar methodology appears also in the *GA* (Sections A and C–I). Aristotle starts with the paradigm or best animal type then moves on to describe differences from this case in attributes, parts or affections. Since they differ by 'the more and the less' (*PA* 644a16–18) the first variations discussed are those that differ in the smallest degree. *GA* I starts with general points about generative parts, covering all or most animals. For longer discussions of non-standard cases, the reader is directed elsewhere (e.g. *GA* 719a8–10 to *HA* 510b5–511a34, *GA* 731b14 to *GA* III.11, *GA* 728b14 to *HA*

said nothing definite' (*GA* 715a11-13). Aristotle has placed discussion of the generative parts 'at the end of our account of the parts' and has put 'the beginning of the account of generation immediately after them' (*GA* 715a16-17). The end of the *MA* also points towards the *GA*: 'we have now dealt with the reasons for the parts of each animal, the soul, and also sense perception, sleep, memory, and general movement. It remains to deal with the generation of animals' (*MA* 704a3-704b3).

35 Aristotle ends the treatise with an exceptional class, the Libyan ostrich (697b14-30), which displays the method of dealing with the general cases first and the most extreme exceptions last. As Lennox notes, 'Aristotle is dealing with legitimately unusual animals here and is doing so by making explicit use of his philosophical systematics' (Aristotle 2001, 343). On this sort of methodology see Gotthelf (1997a) 91–3; reprinted in Gotthelf (2012a) 193–4.

572b29–573a26). Aristotle then moves onto other classes of animal. The larger groupings ought to be familiar to the reader from other biological works and include the blooded and the bloodless (*GA* 715a21–2, 30), number and/or type of feet (717b5–13), mode of locomotion (i.e. swimming, flying or walking) and the presence of a lung (716b21). In the context of generation, the paradigm animal is the hottest and the most productive of semen (732a20, b33). It is the human kind (737b26–7).[36] A progression then takes place towards the coldest and earthiest animals, displaying a gradual decline in efficacy of methods of generation: from live births of offspring like parents, to those unlike, then onto the oviparous, both the perfect egg-layers and the imperfect egg-layers, and finally reaching larvae-producers (offspring completely unlike parents and less developed than eggs). Section (A) contains many instances of the methodology by which the most general class is described before more specific ones, for instance with respect to the location of the growth of the embryo or egg in the upper or lower body (718b15–32). This pattern is also prominent in section (C), for instance when he starts by explaining that the upper portion of the body is formed first but also explains exceptions such as what occurs in insects and cephalopods (741b25–34). Thus, Aristotle fulfils his injunction that, '[a]s far as animals are concerned, we must describe their generation just as we find the theme requires for each several kind as we go along, linking our account on to what has already been said' (*GA* 716a2–4).

As he moves through the different types of animal under discussion, Aristotle will also give exceptions within that kind before proceeding to the next set of animals. So, for example, in Section (C), his account of conception and early embryonic development in the live-bearers, exceptional live-bearing hybrids occur at the very end. And at this juncture an even more exceptional sub-class of live-bearers is discussed, the infertile mule. Hybrid animals (Section D) do not participate in proper species-forms; form cannot explain how they come to be – and certainly not the formal cause as the father's role. Thus, more specific facts and different principles than teleological ones must feature in the account of

[36] For its size, humans produce the most semen and menstrual blood (*GA* 776b26-8, *HA* 582b29-583a14).

their generation.[37] *GA* III (Sections E-I) shows Aristotle moving on to a different set of animals, eventually covering generation in all lesser animals, starting with egg-layers (birds, selachians and fish; 749a10–757b30) and then bloodless cephalopods, crustaceans, insects and testacea (757b31–763b16). Aristotle explains his procedure in the text.

> The phenomena of generation here are on the one hand similar to those which obtain in the animals that walk, so that the same statement will serve for all of them; on the other hand, these animals exhibit certain differences not only as between themselves, but also when compared with the animals that walk (*GA* 749a13–15).

At the very beginning of *GA* IV, that is, the very end of this section (E), Aristotle notes 'now the formation of animals has been dealt with' (763b20–26). This statement indicates that a quite general account of generation in all types of animals has now been achieved. At this point, then, he must explain other phenomena associated with generation, by focusing on what occurs in particular instances.

2.5 More detailed accounts of the generative process

Many have hoped to find that the *GA* is a unified and coherent work and some have suggested that this project can be aided by bringing to mind the distinction between more general and more specific discussions or levels of analysis.[38] So, for example, in the last few pages of his 1972 commentary on *GA* II, David Balme notes a few apparent inconsistencies in the text – one concerning whether or not intellect is bodily (160–1) and the other whether or not the female contributes nutritive soul (165). In both cases,

37 *GA* 738b23-36 will be discussed in the next chapter in relation to the female contribution of matter to generation.

38 See especially Balme in Aristotle (1972) 161–5 and van der Eijk (1997) 235–7 and (1999) 494–5, 499. By 'general' and 'specific' I do not mean to indicate the much debated Aristotelian distinction in zoological investigations between genera (*genos*) and species (*eidos*). I mean to refer, rather, to studies that take in a broader spectrum of objects and phenomena as opposed to those that have a narrower scope and take account of a smaller set of objects or phenomena. I also use the terms comparatively to indicate the many different levels of investigation that Aristotle undertakes, including some which overlap.

Balme argues that Aristotle's conflicting pronouncements on these matters do not constitute any true discrepancy. Rather they represent some of the many instances in the corpus in which Aristotle describes the same or similar phenomena both at a general level and at a more precise one. This type of distinction is not novel. A contrast between general starting points and more detailed investigations has long been employed to help explain seeming inconsistencies, for instance in the contrast between the description of elemental qualities in *On Generation and Corruption* and that in the *Meteorology* IV. In *GC* the material elements (earth, air, fire and water) are entirely passive and potential. Form, that is, an animal or plant soul, is active and brings about change by transforming the passive elements into something ('being' or substance). However, a contrasting approach to the elements is detailed apart from form and soul in the *Mete.* IV: here Aristotle explains that each element possesses active powers (i.e. of fluidity, dryness, cold or heat). Thus, in *GC* the elements are entirely passive and in *Mete.* 4 they have active powers. For some who face this conflict, one or other of the descriptions must predominate.[39] In order to accept the validity of both texts, however, others argue that this is only a seeming discrepancy since the texts constitute two different types of analysis. Thus, Alexander of Aphrodisias explains that the account given in the *GC* is 'more general and universal' while that available in the *Mete.* 'is more particular and physical'.[40] Also, Düring held that while the *GC* is a 'purely dialectical discussion', the *Mete.* IV concentrates on the

[39] Eric Lewis privileges the theory that appears in *Mete.*, which he regards as Aristotle's 'real theory': thus, he thinks that Aristotle changed his theory of the physical elements. See Alexander of Aphrodisias (1996) 2 and 15.

[40] Alexander of Aphrodisias writes in full as follows: 'Generation has been said [in *Mete.* IV] to be a change of the underlying matter [brought about] by the active powers in their relation to the passive powers. This occurs when the active powers are in such proportion to the passive powers that they master them. Yet in the *GC* Aristotle characterized generation as a change from being in potentiality to being in actuality. There he gave a more general and universal account of it; here the account is more particular and physical (κοινότερον ἐκεῖ καὶ καθολικώτερον ἀποδοὺς αὐτῆς τὸν λόγον, ἐνταῦθα δὲ προσεχέστερον ἤδη καὶ φυσικώτερον)' (Hayduck 1903, 185/13–19; trans. Lewis in Alexander of Aphrodisias 1996, 72). By 'more physical' (*phusikôteron*) Alexander means that it has more to do with 'nature' (*phusis*) than with logic. At *Cael.* 283b18 Aristotle talks of less general and more 'physical' arguments and inquiries (Καὶ φυσικῶς δὲ καὶ μὴ καθόλου σκοποῦσιν).

concrete operations of hot and cold.[41] A similar distinction between 'more precise and his general statements' has also been made in Aristotle's psychology.[42]

It is not at first obvious how the contrast between general and more detailed discussions helps us to better understand the content of the *GA*. It is certainly the case that Aristotle qualifies and refines more general pronouncements in the later Books of the work. So, for example, he proclaims that 'conception is impossible without the male emitting semen' (*GA* 739a27) yet also undertakes a lengthy analysis of how male insects *do not* need to emit semen (730b10, 25–26, 731a15). In another case Aristotle's remark that 'no fish at all of any sort comes into being apart from the semen of the male' can be contrasted with the interest he shows in the probable existence of female-only fish species (*GA* 741a33–741b2).[43] The fact that he begins with statements that appear to exclude exceptions but later offers ways to include them shows the beginning of the refinement of his account. Aristotle does not just note the exceptions, though. He attempts to give an account of them employing a more complicated explanatory framework often based on more specific facts and principles.[44]

The contrast between general or abstract discussions and more detailed or intricate ones, which rely on different kinds of explanation, resembles the difference between the way in which one would explain atomic structure when introducing the basics of chemistry as opposed to the same topic as part of a more advanced course of study. So, in the first instance, the student learns that electrons are negatively charged atomic particles. Only later will she need to conceptualize them as less particle-like and as

41 Düring in Aristotle (1944) 66.

42 In discussing the controversy surrounding Aristotle's theory of soul, Bronstein remarks, 'it is often the case that in Aristotle's treatises certain views undergo conceptual refinement, not because Aristotle changed his mind, but because different treatises operate at different levels of explanation' (2006) 426. See also van der Eijk (1997).

43 Aristotle also considers the possibility of parthenogenesis in insects at *HA* 637b18, although the authenticity of this book has been doubted. See Chapter 3.3.

44 '[H]aving to consider the implications of generation ... he finds these more precise distinctions necessary ... These differences of precision do not lead to contradictions. The distinctions necessitated by one subject matter could be misleading in another,' Balme in Aristotle (1972) 164.

occupying clouds or orbitals, differing in shape, in which an electron may exist at any given time. The latter more detailed description is necessary in order to explain some of the particular chemical reactions the student will encounter in her more advanced studies. Similarly in Aristotelian biology, explanations can be geared to the difficulty or intricacy of the issue at hand.

In section (B) Aristotle reaches a general account of how generation comes about. His account refers to teleological principles – the male is the efficient cause and the female the material cause. It is also based on how the process takes place in the best animals, which are live-bearing and internally gestate. These animals are hot enough for the male to produce semen and the female to have menses. This initial movement towards general first principles[45] in a specific area of study is then supplemented by a further attempt to elucidate other aspects of the phenomena, some of which could be exceptional, occurring in very few instances and many of which will require different principles of explanation. For instance, in section (C), Aristotle explains that not all the parts begin small and then get larger as one might expect, but in all instances of development one part does not follow that pattern: the eyes. They begin larger and shrink, which is due to 'necessity' as he explains in some detail (743b33–744a19). Section (C) also gives the reader a series of slightly anomalous seeming empirical observations such as males that do not emit semen at all (i.e. insects) and females that do not experience pleasure during sexual intercourse (739a30). Later on he brings up the phenomenon of wind-eggs, which show that the female contribution contains soul (741a18). This then leads him to wonder if there are any female-only species, something that he, perhaps surprisingly, does not attempt to resolve one way or another (741a33–b2). Although more detailed explanations for these variations are not yet provided, one comes to understand that all such points will require further, more subtle, analysis and some reference to necessity.[46]

45 'General' can of course refer to many different items: universals, definitions, explanations and models or paradigms. I will be mostly concerned with general models and explanations in the *GA*, which are meant to cover all or most instances, without explaining any particular details or differences.

46 As Aristotle notes here, natural objects ought to be investigated with reference to many principles of explanation (*GA* 742b34-35).

Aristotle's discussion of the mule in Section (D), which forms part of his plan to cover generation in all live-bearers first, shows him reaching for different, more detailed, explanations when faced with tricky cases. The fault with rival accounts of why mules are infertile is that they are too broad (*GA* 747a23–30). Democritus, for instance, says that in mules the genital passages are destroyed during gestation because their parents differ in type. The fact that all other hybrids *do* generate means that Democritus' explanation is incorrectly focused. In order to be properly informative, this explanation would have to apply to *all* hybrids and it cannot (*GA* 747a28–34). Aristotle elaborates on this idea when he argues against another explanation, which he calls too abstract (*logikê*). This theory states that since the definition of 'mule' is 'that which is produced by the synthesis of horse and ass parents', then two mules could not be understood to produce a mule (*GA* 747b28–748a8).

> [The argument] is too general; there is nothing in it, because there is nothing in any argument which does not start from the first principles belonging to the particular subject. Such arguments may appear to be relevant, but in fact they are not. For a geometrical argument, you must start from geometrical principles, and the same applies elsewhere; that which is empty, which has nothing in it, may appear to be somewhat but in fact is nothing at all. But also, this argument is false, because many of the animals that are produced from parents of differing species are fertile, as was said earlier. No; this method of inquiry is as wrong in natural science as it is elsewhere. We shall be more likely to discover the reason we are looking for if we consider the actual facts with regard to the two species, horse and ass (*GA* 748a7–17)

There are several ways in which the accusation of being too abstract or general might operate here. Aristotle says that a geometrical argument must start from geometrical principles. A passage from the *de Caelo* explains that in such instances, thinkers use the wrong principles and are too stubborn to recognize this.

> These philosophers find themselves, in a discussion about phenomena, making statements with which the phenomena conflict. This is because they have a wrong conception of the primary principles, and try to bring everything into line with hard-and-fast theories. For surely the principles of sensible things are sensible, of eternal things eternal, and of perishable things, perishable: put generally, a principle is of the same genus as what falls under it. Yet out of affection for their fixed ideas these men behave like speakers defending a thesis in debate: they stand on the truth of their premises against all the facts, not admitting that there

are premises which ought to be criticized in the light of their consequences, and in particular of the final result of all (*Cael.* 306a6–18).

One might have a legitimate complaint if principles applying more particularly to animals (such as male and female principles) were being applied to the heavenly bodies (i.e. stars and planets), since more specific sub-principles in one subject area will not apply to another. However, this is not what Aristotle complains of in his rivals' explanation here in the *GA*. It is not that the subject matter does not fall under the principles employed but rather that these are too general to cope with the specific problem under analysis. First of all, this means that the opponent moves (or is forced to move) towards a generalizing conclusion that all hybrids are infertile on the basis of one particular instance.[47] Thus, the conclusion is incorrect because it conflicts with the facts – the evidence of fertile hybrids. The explanation also lacks adequate specificity. As sketched earlier, different areas of study falling under a more general area require more particular principles. For example, astronomy, which falls under physics (i.e. the study of nature), requires the use of more specific principles, such as *aithêr* (*Cael.* I.2).

Sometimes when one applies principles of explanation that are general to all beings or all living beings in more specific contexts one loses sight of what needs to be explained. Here, Aristotle thinks that other theorists have used the wrong principles because they fail to recognize that exceptional phenomena will often require reference to different principles within the given science which cannot be fixed by setting one's sights on more general principles. For Aristotle, the natural philosopher must be sensitive to the situation at hand and be prepared to consider what facts and principles of explanation are appropriate in that setting. Those that cover all instances of a kind of animal or type of phenomena are

[47] Aristotle often complains in the *GA* that other theorists do not take sufficient heed of a number of different instances which fall under a general heading (e.g. 756a4-6, 765a26-29, 765b4-7, 788b10-12). One may be able to fit Aristotle's point here with a similar complaint against Bryson's proof which squares the circle: 'such arguments prove in virtue of a common feature which will also hold of something else: and so the arguments also attach to other items which are not of a kind with them. Hence you do not understand the item as such but only incidentally – otherwise the demonstration would not attach to another kind as well' (*A. Po.* 75b39-76a4).

not always the ones that are required. In the case of mule sterility, facts about horse and ass kinds plus the principles surrounding material explanation (or necessity) have to be the focus of any adequate explanation. Thus, the principles of explanation that apply to all hybrids will not apply to every aspect of each subclass of hybrid. Rather than appeal to the fact that mules are hybrids, when explaining their sterility, as the opponents did, more relevant facts must come to the forefront.

As the passage on mules in the *GA* demonstrates, the principles employed to explain exceptional cases wouldn't always be exhausted by those mapped out in *GA* I and II. *GA* IV and V use a variety of principles and facts, including those listed below, many of which are not mentioned in *GA* I and II.

(F) Sexual Differentiation: Ability/inability (765b19 – already available in *GA* II).
Opposition (*antikeitai*) of male and female (766a3, 766b17)
Mastery (*kratein*) (766b16)
Blend (*summetria*; 767a14; prefigured in *GA* I at 723a30)
State of water (that is, hard and cold; 767a34–5)

(G) Heredity: said to be due to the same causes as sexual differentiation (767a36).
Re-uses: Mastery (*kratein*) (767b22, 768a4–6)
Opposite (*to antikeimenon*) (768a5)
Adds: 'Faculty' or 'faculties'[48] (*dunamis/dunameis*, 767b23, 36, 768a4–21)
'Movements' (*kinêseis*, 767b37)
'Departure from type' (*existatai*) (768a15)
Relapse (*luontai*) (768a17)
Nearer and further *kinêseis* (768a17)

(H) More specific phenomena: Deformity.
Re-uses: *kinêseis* (773a8)
Mastery (769b13; 772b32–5)
Summetria (772a18)
Classification by feet (770a37, 774b5–7)
Adds: Multiparity (770a11, b25)
Size of animal (771b33, 773b5, 774a22)
Amount of semen/generative fluid (771a25, 774a20, 775b15)
Dunameis in the semen (772a8)

[48] The translations in quotation marks will be questioned later on in this book.

Kinêseis in semen (773a8)
Perfection/imperfection of offspring (774b5, 13)
More rare still: Superfetation (occurs in only two kinds: human beings and horses)
Re-uses: Multiparity
Body Size
Amount of generative fluid
Mola uteri (only occurs in human beings)
Due to: amount of generative fluid

(I) Milk (in live-bearing quadrupeds); it is not a deformity.
Re-uses: Body size (777b3)
Perfection/imperfection of offspring (774b13, 777b10)
Residues
Adds: Gestational periods
Lifespan (777a33)
Blend (777b7)
Periods of time (777b19)

(J) Variations in parts which occur in some animals and not in others, explanations in terms of: Amount of fluid (779b27, 782a27).
Pneuma (781a25, 789b9)
Thickness/thinness of skin (785b2)
Exhalations (782b19)
Heat/coldness (783b9, 784b27)
Putrefaction (784b6)
Diet (786b2).[49]

Any worry that the investigation of the horse and ass species would involve entirely particular subject matter, which would not then count as generalizable knowledge (*Metaph.* 1003a13–14) can be appeased. Clearly, the factors Aristotle brings in to aid the discussion here *all* appear later on in the work and help to explain a variety of generative phenomena (the *GA* II references are followed by the *GA* IV and V ones): monoparity/multiparity (748a23 [770a8, 771a18–27]), more/less fertile

[49] In the introduction to his 1942 translation of *GA* Peck provides a lexicon of relevant terms, many of which line up with the more specific principles of explanation listed here (Aristotle 1942, xlv–lxvii). We need not assume that these terms originate in or are exclusive to Aristotle's biology. The idea of mastery is ubiquitous, for example, in several Hippocratic accounts of generation (e.g. *Genit., Nat. Puer.*). Aristotle himself suggests that the term was used by rival theorists Democritus and Empedocles (*GA* 764a10, 764b27). On Aristotle's use of Hippocratic terminology in the *GA* see van der Eijk (2007).

kinds (748a21, 748b8–19 [773b28]), amount of menstrual fluid (748a23 [771a25, 774a29, 775b15]), coldness of animal (748a25 [783b19, 784b27, 788a21]), gestational periods (748a30, 748b22 [777b13]), proportional blend of male and female residues (*summetria*, 748b2–8 [767a17, 772a18]) and body size (748b20 [777b3, 771b33, 773b5, 774a23]).[50]

In other works on natural philosophy Aristotle sometimes alerts his reader to the fact that his initial description of some occurrence or entity or quality is only a broad sketch in need of further elaboration. For instance, in the *de Caelo* his initial description of 'the heavy' and 'the light' await a more detailed analysis much later on in *Cael* IV. As he explains in *Cael.* I, 'we must first lay down what we mean by heavy and light, at present only so far as it is necessary for the purpose in hand, but later with more precision (*akribesteron*), when we come to investigate the real nature of the two' (*Cael.* 269b20–23). The *GA* also employs this methodology; topics that first come up in the early Books are not fully explained and elucidated until much later on. For instance, wind-eggs are first brought up in *GA* I at 730a33 and then are more fully elucidated in *GA* III (750b2–751a24). Also, sex difference is first set out in *GA* I (716a20–716b15) and then more fully explained in *GA* IV.1–2. In these chapters, Aristotle notes that he has given a general account of embryonic development but not an account of the development of male and female animals. Sex is not an essential attribute, but occurs accidentally (or *per se*) with reference to the body or matter (*Metaph.* 1.9). This means that Aristotle will not elucidate this occurrence directly in terms of essence and form. In the section on heredity (G) he begins by noting that the phenomena he plans to deal with are 'due to the same causes' as sexual differentiation, and he does indeed employ similar sets of explanations and references to such factors as *mastery* and *opposition*. He also employs several novel ideas, including 'faculty'

[50] Many of the concepts employed to explain sterility in hybrids at the end of *GA* II reappear to explain other types of deformity (including sterility) in *GA* IV. Clearly not all of these are *principles* of explanation. It will take a lot more work, some of which Aristotle undertakes later on in the treatise, to discern which, if any, of these observable facts and theoretical considerations are principles – that is, which are fundamentally explanatory and can be connected to essence. See Chapter 10.

(*dunamis* used in a new technical sense here; 767b17), 'movements' (present in the semen; 767b36), *'departure from type'* (*existatai*; 768a16) and 'relapse' (*luontai*; 768a16). Section (H) on monsters and deformation makes good use of many of the considerations introduced in *GA* III and IV so far – particularly multiparity (770a7, 773a36 and everywhere), mastery (770a7, 772b13) and *summetria* (772a18). As Aristotle explains, deformity is due to the same causes as the other specific phenomena explained so far in *GA* IV (767b3–11). Aristotle brings in all this more specific explanatory apparatus in *GA* III, IV and V. Meanwhile, the terms of discussion which were most useful when giving an abstract or general explanation are less useful and so are not mentioned as often.[51] The most commonly referred to general principle in *GA* IV and V is 'necessity' (e.g. 764b34, 766a21, 770b12, 776a26, 778a35, 782a24, 789a13, 789b13, 21), which becomes a more appropriate principle in all cases where teleological explanations are of limited use.[52] In the *GA* and elsewhere when a discussion begins to become more focused on one part of a general group of events, we often find that the most general account is unwieldy and therefore inadequate. For example, when Aristotle narrows down his discussion of change in *Ph.* VII.2 to alteration, and focuses on perception and perceptible qualitative change, he no longer uses the terms that were appropriate for local or quantitative change. In the *GA*, more general discussions which do not employ these more specific principles are unwieldy and inappropriate for analysis of more specific subject matter.

2.6 The female role: need there be any contradiction?

We might still feel uneasy about the female role in generation in the *GA*. Although it is clear that there is a general account and a

[51] One particular case in point is 'form', which occurs rarely after *GA* III.

[52] The more detailed analysis focuses on the material situation rather than teleology, for instance the food or the habitat and climatic conditions in which the animal lives. So, for instance, the shape of the uterus is often explained by the heat or coldness of the body of the animal in question (*GA* 753a31, 758a6, 750a13). In the case of the kestrel, an explanation is provided by reference to what the animal drinks (750a7). Various explanations of heredity, deformity and differences between animals within the same type that lack purpose (e.g. eye colour in humans) follow a similar pattern. See Chapter 10.

more detailed one, might it not be the case that the two conflict and are inconsistent with each other? Aristotle appears to argue so vociferously against the idea that the female parent contributes semen and can play any active role in generation that a reader might legitimately think that Aristotle backtracks once he reaches his discussion of heredity. Commentators remain divided. Balme feels that the distinction between 'general and specific discussions' is enough to declare that 'no contradiction need be found' in the *GA*.[53] At the opposite pole, committed problematizers contend that this text shows Aristotle's lack of resolve. For them, the interpreter is forced by the specifics clashing with the general to bite the bullet and accept that Aristotle either changed his mind or was never that committed to his initial views on the female role.[54] A midway position recognizes a tension while also hoping to avoid the need to see Aristotle as an entirely tentative thinker, without adherence to enduring doctrines. This interpretation views Aristotle as troubled, rather than purposefully pluralistic. It suggests that he wanted to achieve certain things when he set out to study any particular subject matter, but that he was not always able to follow through and that this must have led him to cut corners and equivocate. Here the *GA* is not seen as a harmonious work, planned and executed in every part; instead it supposedly contains sections in which Aristotle is stretched to his limit, where he appears to be thinking up his explanations as he goes along rather than slotting them into a pre-planned discussion.[55] Others characterize Aristotle's view on the female role as 'quite consistent' throughout the *GA*, suggesting that he did not modify his view or change his mind but that, when the earlier portions of the

[53] Balme in Aristotle (1972) 165.

[54] '[T]here are also plenty of signs, if we do not discount them, of tentativeness, of hesitation, of the pluralism and open-endedness of [Aristotle's] approach, of a readiness to backtrack, to qualify and modify even fundamental doctrines and principles. Of course many would be tempted, many have been, if not to discount those signs, at least to minimize their significance, in the name of the coherence and unity of Aristotle's thought', G.E.R. Lloyd (1996) 4.

[55] See van der Eijk (1999) who interprets *GA* I and II as the theory of the female role presented 'without qualification' and thus notes that in *GA* IV there is a 'more active role for the female than the sheer passivity of the first two Books suggest' (495). 'Aristote apporte des modifications ou précisions à sa théorie générale (exposée dans les livres I et II) concernant les rôles respectifs de mâle et de la femelle dans la génération', Van der Eijk (2007) 413.

GA are properly understood, they fit together with what comes later, which is presumed to be Aristotle's intention.[56]

I am in broad agreement with the final position. There is abundant evidence in the text that Aristotle had a plan: he shows no lack of resolve or tendency to equivocate when it comes to the female role. The conflict between descriptions of the female role appears most prominent when comparing Sections (B) and (G). However, when each is considered in terms of the different types of explanation employed in different parts of the work, the contrast is perfectly understandable and does not constitute any confusion or change of mind on the matter. In (B) Aristotle is concerned with explanations involving principles so general as to cover all existent things (e.g. activity and passivity) or all natural objects (e.g. matter and form) (e.g. *GA* 721b7–12, 724a14–18, 729a9–12, 729a22–34, 729b9–19, 733b17, 25–26). He needs to describe animal generation from this vantage point in order to impress upon various opponents that their outlook cannot account for the unity of individuals. The set of passages that infamously present the sexes as active and passive and to be associated with matter and form (crudely put) occur within that context.[57] When it comes to (G) the reader finds direct engagement in a much more particular study of aspects of generation and the phenomena associated with it. Aristotle not only uses more specific principles applicable more properly in this subject area, such as reference to the potentials (*dunameis*) in both generative residues, but also refers to necessity and material causes instead of reasons relating to outcomes (teleology).

For Aristotle each field of inquiry must begin with factors that are relevant to that study. When we inquire into reproduction and fertility in mules we cannot use principles appropriate to abstract logic. We must, instead, begin with direct considerations of the observable reproductive organs of horse and ass. Likewise, in *GA* IV, in order to explain heredity, Aristotle thinks that we must begin by considering that both male and female contributions to generation contain nutritive potentials that create and maintain all the

[56] For the idea that Aristotle is 'quite consistent' see Henry (2006a) 286. See also Gelber (2010).

[57] See Gelber (2010) for a clear description of how male and female are active and passive in terms of Aristotle's account of substantial change (205).

parts of the body. We cannot start by using the distinction between matter and form, since this is not appropriate for addressing the issue at hand.[58] The discussion is immersed in principles specific to generation and elaborates on how the process of generation takes place in a particular instance. In this context, he brings in terminology which he has not yet mentioned, and which did not play any part in his explanations in *GA* I and II. The *GA* endeavours to connect animal generation to broad metaphysical and natural concerns (as an instance of substantial change) in which context Aristotle gives a broad view of the roles of the sexes (B). However, one who reads the entire treatise is left in no doubt that this cannot be taken as any kind of comprehensive picture of maternal and paternal contributions to generation. The early Books do not in any way exclude the possibility of a much more detailed and positive account of the female role in generation – in fact, they require it. Thus the popular interpretive strategy of suppressing the content of *GA* III–IV is not only unnecessary but actually impairs our understanding of the text as a whole.

2.7 Notes on dialectic in the *Generation of Animals*

The difficulties encountered in the project of fitting *Analytics*-type procedures to biological practice has led some to suggest that the method at work is really the dialectical one (detailed most thoroughly in Aristotle's *Topics*).[59] Thus it seems important to address the question of whether Aristotle employed this methodology in the *GA*. The way in which Aristotelian dialectic is meant to operate is disputed. Some think that the method works by taking the appearances (*phainomena*) and attempting to bring them into line with one another. *Phainomena* is a somewhat ambiguous term which could indicate not only 'observations' or data, but also diverse points of view which 'describe the world as it appears

[58] We certainly ought not to start with an abstract general principle, such as the principle of non-contradiction. To do so would result in completely inappropriate conclusions. 'The problems, in such cases, will often need to be resolved, insofar as they can be resolved, not by a straightforward appeal to the laws of non-contradiction or excluded middle, but by an evaluation of the empirical data. They will require the acumen not of Aristotle the logician, but of Aristotle the field-worker', Lloyd (1996) 73.

[59] Owen (1961) and Kullmann (1974). See also Lloyd (1990) 33–4.

to, and as it is experienced by, observers who are members of our kind'.[60] There is some indication that Aristotle is epistemically optimistic, believing that everyone has access to the truth. '[F]or the best thing would be if everyone were seen to agree with what is said, or if not, to agree in a way, which will happen if they change their ground. For everyone has something of his own to contribute to truth' (*EE* 1216b28–31).[61] On this view Aristotle would have constructed his theories by concentrating on what others had to say on the matter – trusting these accounts and incorporating them all into his own theories.

In his *Metaphysics*, Aristotle offers that the love of learning and knowledge reaches fruition in the leisured class. This is why philosophy began in Egypt, where priests no longer had to toil for sustenance (*Metaph.* 981b24). The privileging of an elite point of view in Aristotle's works ought not to be exaggerated, however. He has cause to criticize the wise at many points, while also taking care not to dismiss the views of the non-elite. A case in point is the inclusion of women's points of view. Although some argue that Aristotle did not listen to women, because they would not have had the leisure necessary for knowledge, there is evidence that he took careful note of what some women told him. For instance, we find that he uses information coming from women to help determine gestational periods.[62] It seems quite likely that many other statements about vaginal discharges, pregnancy and birth came from women themselves, both midwives and mothers.[63] By listening to all these different

[60] Nussbaum (1982) 272.

[61] See also *EN* 1145b6-7 and *Metaph.* 993a30-b3: 'every one says something true about the nature of things, and while individually they contribute little or nothing to truth, by the union of all a considerable amount is amassed.'

[62] In antiquity it was common to think that human gestation lasted seven months. Proclus speculated that this was due to women only becoming aware of their state in the second month. See von Staden (1989) 298–9. Aristotle notes that in humans the length of gestation is variable, which would appear to have been a myth perpetuated by women to protect their knowledge of their children's paternity. See *HA* 584a35-b1, *GA* 772b5-11, 776a22 and the Hippocratic treatise 'On the Eight Month Child', Joly in Hippocrates (1970) and Lloyd (2003) 72–3 and 51.

[63] See, for example, *GA* 728a2-9, 739a37, *HA* 583a11-13, 584a25-584b25, 585a17-29; Aristotle must have ascertained most of this information from women. *HA* X, if authentic, finds Aristotle reporting direct accounts from women of their experiences of erotic dreams (*HA* 638a6: 'If women are telling the truth . . . '). For arguments for its authenticity see Chapter 3.3.

perspectives on an issue, it may be argued that Aristotle remains open to new pieces of information that could potentially change or modify his initial reasoning.[64]

With all this in mind, we can now consider reasons to resist viewing Aristotle as engaged in dialectical inquiry in the *GA*. For one thing, this would make him quite unscientific in his approach. So when he reports what shepherds say about the effect of wind's direction on conception, we do not like to suppose that he accepted these views uncritically.

Also, shepherds say that it makes a difference so far as the generation of males and females is concerned not only whether copulation occurs when the wind is in the north or in the south, but also whether the animals face north or south while they are copulating: such a small thing thrown in on one side or the other (so they say) acts as the cause of heat and cold, and these in turn act as the cause of generation (*GA* 767a9–13).

Since he qualifies these reports with phrases such as 'so they say', Aristotle does not appear to take the views as unquestionable premises or starting points of inquiry.[65] Here the shepherds' statements fit well with his theoretical account of the effect of hot and cold on generation. However, he is clearly wary of their origins. On closer inspection Aristotle very seldom accepts others' views uncritically. A brief survey of the places in which he refers to others' views and the context in which they occur helps to reinforce this point. The views of others fall roughly into two categories: the opinions of the wise (the *endoxa*) and the opinions of the many (common views). Both types can be found throughout the *Generation of Animals* and are noted in Table 2.2.

Most often in the *GA* Aristotle uses the views of others (*phainomena* in the broadest sense) as counterpoints to his own views. He concentrates on opinions (of the many and the wise) which seem incorrect or misguided to him and conflict with his preferred account. He almost never uses second-hand views to form his own theories in the *GA* – instead employing many theoretical

64 Some think that by concentrating on the dialectical method, feminist critiques of Aristotle's procedural rigidity and his dismissal of women's views can be mitigated. See, for instance, Modrak (1988) and Freeland (1994).

65 *HA* 493b14, 580a19-22. Cited in Lloyd (1987) 54 n.8.

Table 2.2 *Citations of others' views in the* GA

Commonly held views/myths/data from the uneducated[66]	The *Endoxa* (reputable opinions of the wise)
***GA* I–II**	
The observations of fishermen (720b32)	Pangenesis theorists (*tines*) (721b13, 727b8, 34)
Common views (721b6)	
Myths (e.g. from Herodotus) (736a10, 746b5–10)	Empedocles (722b8–16, 747a27, 34, 747bb18–25)
	Anaxagoras (723a1)
	Democritus (740a13, 36, 742b19–26, 747a27–b5)
	Unspecified: the 'like goes to like' theorists (741a2)
	Some *phusiologoi* (742a16)
	'Those people' (746a19)[67]
***GA* III**	
Observations of fishermen (756b2, 757b34)	Alcmeon of Croton (752b26)
'What is commonly supposed' (752b26)	Anaxagoras (756b16)
Myths (e.g. from Herodotus of Heraclea) (756a7, 757a2)	Unspecified: *phusikoi* (756b13)
	'Some say' (751a10, 755b8–13, 758a1, 759a11, 23)
***GA* IV**	
What 'people say' (769b14)	Anaxagoras (763b32)
	Empedocles (764a2–6, 12, 36, 764b6–20 769a18, 777a9–11)
	Democritus (764a6–11, 764b14, 769a18, 769b32–6)
	Leophanes (765a4, 25)
	Panspermia theorist (769a29)

66 Aristotle occasionally indicates that certain people are unreliable informants on particular grounds. Fishermen, for example, who are not interested in knowledge for its own sake, are not regarded as reliable (*GA* 756a32-5).

67 This is possibly a reference to Aristotle's medical contemporary, Diocles, as a dispute of this kind on foetal nourishment is documented to have occurred between the two thinkers. See Nutton (2004) 120, Diocles of Carystus (2000–1) vol. II, 40.

Table 2.2 (*cont.*)

Commonly held views/myths/data from the uneducated	The *Endoxa* (reputable opinions of the wise)
	Unspecified: other/certain of the *phusiologoi* (763b32, 769a8) 'those that suppose the female is hotter' (765b20–8) 'Some say' (771b28)
***GA* V**	
	Empedocles (779b17) Democritus (788b10–27, 789b3) 'Some people' (786b28, 788a11)

considerations as well as empirical facts.[68] Although Aristotle does not appear to build up theories out of others' views in the *GA*, certain patterns can be discerned. He tends to bring in the views of other intellectuals when he is establishing his own *general* theory as more comprehensive. Thus, he will frequently remark that other theorists are unable, by their methods, to account for as many obvious cases and phenomena as he is (for example, *GA* II.8 on mules and III.1 on bees). In contrast, when he is busy explaining the differences between different animals, and detailing the material factors that cause them to deviate from a general pattern, he is not engaging very much with the views of other theorists. This might, therefore, mark the distinctively new Aristotelian outlook: his biological research is meant to stretch to all types of living being, even the lowliest, rather than concentrating only on human beings. His rivals, being less able to abstract themselves from moral, political and medical matters (human concerns) need to broaden their analyses. Meanwhile, he does not value as much the minds of common folk, or casual observers, who he thinks are not at all focused in the correct manner and so are often inadequate – although the occasional dispassionate observation of fisherman, hunter or other tradesmen is noted as authoritative.[69] In these sections only the rare reference to generally held views is used either as support for his

[68] Karbowski (2009) 20–1.
[69] For example, fish copulation has been observed by fishermen (756a36).

view or in order to correct the common opinions and myths surrounding certain animals (for example, the 'stupid tale' that fish conceive by swallowing milt at *GA* 756b6).

When it comes to more specific phenomena such as heredity and deformity, Aristotle is most keen to bring up the views of experts, the *endoxa*, or reputable opinions (Anaxagoras, Empedocles, Democritus and Leophanes). In these contexts, then, he is probably working within already well-established theoretical frameworks, and possibly accepting and adapting many of the general themes and concepts employed by other thinkers. This might explain in part why he uses *summetria*, mastery and other concepts which we find in the Hippocratic corpus. Some of these explanatory principles were possibly also a part of the physiology of generation in such thinkers as Empedocles and Democritus, whom he so often cites (although this is merely speculative as we now have scant evidence of this in what is left to us of their writings). Even though he borrows explanatory terminology from some of his predecessors and contemporaries, he is attempting to employ his own unique overall methodology, whereby the more material explanations come after and fit together with a general metaphysical outlook. And he thereby adapts and changes the concepts he initially borrows. This brief survey can only lend the most tentative of conclusions concerning Aristotle's treatment of the views of others in the *GA*. He has to include others' views and explain why his, in contrast, make better sense of the phenomena at issue. But those particular problems have been highlighted and picked out by thinkers before him and he consciously engages in the same sorts of debate. However, we do not find the technique of taking (even well attested) views of others and building on them to create a theory in the *GA* – the Aristotelian dialectical method as commonly understood. Thus it seems that the depiction of the dialectical method as a robust scientific procedure used by Aristotle in the *GA* has been exaggerated.[70]

[70] As supported by Bolton (1987) and Karbowski (2009) 16–22. Föllinger (2005) sees more of a role for dialectic in the *GA*,arguing that Aristotle's method displays a mind at work. She describes this in terms of 'dialectical exercises' and maps out many structures within Aristotle's scientific writing, which seem to reveal repeated patterns not only of question and answer but also of analysis of both real and imagined opponents. See also Wardy (1990) who writes that the two versions we possess of *Physics* VII look to be 'differently stressed dialectical exercises dedicated to a common set of themes' (90).

PART II

RE-ASSESSING THE GENERATIVE ROLE OF THE FEMALE ANIMAL IN ARISTOTLE

Introduction

Part II closely analyses the theory of the female role set out in the *Generation of Animals*. In order to understand this theory one must first disentangle it from a prominent type of historiography about embryology, which has Aristotle standing on one side, denying female seed, and various other thinkers standing on the other affirming its existence. Although there is some truth in this contrast, simplified versions of it distort our view of the female role in Aristotle's theory. It is particularly distracting when Aristotle's 'one-seed' theory is seen as more sexist or less sympathetic to women. This is why the first section will focus on the theory within the context of his contemporaries and medical theorists in later antiquity, who discussed and responded to his views. This will also involve explaining why Aristotle adopts the theory that he does, and what he is arguing against. The second section will begin with the biomedical background to the link between the female role and nourishment and will end with the detailed physiology of nutrition so central to Aristotle's account.

Many think that we ought to be able to extract a consistent and comprehensive theory of reproduction from certain passages occurring in *GA* I and II. This chapter will challenge accounts, which, based on these passages alone, propose that the female contribution comprises a non-spermatic, non-specific or 'raw' material, which can be best characterized as either earth or nourishment and that the male contribution is the only spermatic component of generation, providing every characteristic to the offspring. In reading the text more carefully, ample evidence emerges that the first two Books do not contain a theory of this sort and that male and female contributions are comparable and complementary for Aristotle from the start; the female contributes active life-giving materials. The chapter as a whole serves to counteract the tendency to read only certain passages of *GA* I and II in an abstract manner. To do so, I will argue, robs the reader of the opportunity to appreciate a key insight into Aristotelian science found in this work – that soul, organization and life are not separable from the materials in which they occur.

3

MENSTRUAL BLOOD AND FEMALE SEMEN

3.1 Ancient seed theories: Aristotle as contrasted with Hippocrates and Galen

According to a well-worn narrative, Hippocratic doctors, and other ancient natural scientists, congregated around the idea that women produce semen (*sperma*). Aristotle, meanwhile, categorically denies, on several famous occasions, that females are capable of doing so (e.g. *GA* 727a28, 728a31, 729a22). In many accounts of the history of reproductive theory, Aristotle's view that, in a sense, women do not contribute semen invites suggestions that this robbed them of a more positive generative influence.[1] It is also common to regard the contrast between the ideas of Galen and Aristotle along similar lines. Galen writes that females produce semen as males do, and he represents his own view as strongly opposed to that of Aristotle. In the guise of Aristotle's opponent, Galen tends to be praised in a number of ways, most notably for his relative scientific sophistication and for his egalitarian attitudes towards the biological nature of women.[2] The idea that there is a significant contrast between two positions on the issue of female semen continued into the medieval and early-modern periods.[3] The influence of this historiography of

[1] Allen (1985) 97, Lloyd (1984) 43, Rousselle (1988) 30, Blundell (1995) 106–7, King (1998) 10, Cadden (1993) 117–19, Demaitre and Travill (1980) 414–15, Lloyd (1983) Part II.3, Jacquart and Thomasset (1988) 55–69 and Shildrick (1997) 31. Rouselle (1988) goes further, praising Hippocratic doctors for their supposed commitment to female equality, interest in female sexual satisfaction and attentiveness to what women say, while castigating Aristotle for failing on all three counts.

[2] Praise for Galen's scientific advances comes from Preus (1977) 81, Boylan (1984) 75 and Siraisi (1990) 80. Praise for his 'feminism': Jacquart and Thomasset (1988) 59, Cadden (1993) 108–9 and 117–19, Thomasen (1980) and as documented in MacLean (1980) 37–8. Those with more detailed knowledge of his writings are well aware of his sexism. See Flemming (2000) and Hankinson (2008) 2.

[3] See Jacquart and Thomasset (1988) 55–62, Aughterson (1995) 49 and Cadden (1993) 117–19.

embryology is widespread and remains apparent in many commentaries on the ancient works.[4]

The latent assumption behind the condemnation of Aristotle and praise of Hippocrates and Galen is that each thinker made a choice, either to acknowledge female significance or to try to demolish it. Added to this, Aristotle's 'decision' to pursue a one-seed theory has also been seen as ground for blame. His guilt is even dramatized by association with the literary *topos* of Aeschylus' *Eumenides*, where, in order to excuse matricide, it is claimed that a mother is not the true parent of her child.[5] Aristotle's real attitudes and the decisions he made are unavailable to us, thus the focus on blame for the sexism comes to rest instead on the theory he espouses. Even if blame is neutralized in this way, the claim that one ancient theory of generation is more sexist than others remains a source of distortion since it tends to simplify the views found in the texts. Aristotle's theory cannot be assimilated to that of Aeschylus' character in the *Eumenides*, in particular, for the simple reason that no ancient thinker can be. Nobody who studied animal generation ever regarded the female contribution to be wholly insignificant. Even if the female parent were merely a nurse or incubator to the male seed, she would, at the very least, have had to contribute a large quantity of nourishment, which, for almost all early thinkers, determined the outcome of generation to some extent.[6] Thus, it seems, Athena's assertion that a mother is in no sense related to her children finds no support in contemporary biomedical theory.[7]

[4] Giorgianni in Hippocrates (2006) talks of 'einer polarisierten Debatte' of female semen in antiquity (57).

[5] Tuana (1989) 151, (1993) 133 and (1994a) 192, G. Lloyd (1996) 43, Blundell (1995) 106, Giallongo (1981) 26–7 and Peradotto and Sullivan (1984) 2.

[6] See Chapter 4.

[7] In the context of the play, there is every reason to imagine that this speech is meant to be the unhinged pseudo-justifications of a tragic mother-killer, rather than a reasonable theory. See Hanson (1990) 314 n.27 and du Bois (1988) 122–4. The statement has been useful to the historiography of embryology, because it represents the extreme opposite of an egalitarian hypothesis. This historiography starts with Aristotle himself who attributes a male-only theory to Anaxagoras at *GA* 726b1. For a modern example see Needham (1959) 43–4. The hypothesis that there could be two completely equal seeds also does not exist, not even in modern embryology where the female's contribution of cellular cytoplasm and mitochondrial DNA outweighs the male's contribution of mere nuclear DNA.

It is important to disentangle Aristotle's theory from this history of interpretation because it misrepresents his views. It is particularly problematic when a commentator decides to adhere rigidly to the view that Aristotle had a one-seed theory and denied maternity, since there is so much within the text of the *GA* that tells against this. Thus, adherence to this view can cause readers to pass over or suppress these portions of the text. This type of commentary is born out of more modern concerns in general, in so far as it focuses on the relative sexism of accounts. But the fact that biomedical theories were used in the political arena to argue for female equality or as support for female oppression in later history does not mean that we can assume this was the case in the ancient context. Indeed, when we consider the ancient texts more closely, there is little evidence for this. Perhaps unsurprisingly, Aristotle and his biomedical opponents display no discernible differences between their gender biases. The female role in generation is never described as equal to that of the male; his is always superior. For this and other reasons it is not possible to contrast ancient thinkers as more or less misogynist in their reproductive theories, despite the ways in which their ideas of the female role differ. In order to get a better sense of Aristotle's actual view we must investigate why and how he rejects the ideas available in his predecessors and contemporaries concerning the female role without being distracted by a desire to blame or exonerate.

3.2 Aristotle's arguments for the differing functions of male and female

The basic difference between Aristotle and his opponents on the female role is as follows. In what I will call 'the parallel seed theory' (found mainly in the Hippocratic writing and taken up later by Galen) the female has (1) menstrual discharge (which may serve as 'food') and (2) a white fluid emitted at sexual climax. The male has (2); (2) is 'seed' or 'semen'. In Aristotle's theory (which I will call 'the differentiated seed theory') the female has (1) and the male has (2) exclusively. (1) counts as a type of 'semen'; it differs from (2) in the way in which it is emitted and by its degree of concoction (and only incidentally by its colour).

The female contribution is not emitted at the point of sexual climax but is released internally in the healthy female in preparation for receiving the male semen with which it will mingle (*GA* 730a33-b4). Any excess female contribution is evacuated from the body as menstrual discharge. Both theories are two-seed theories.

The key reason to think that Aristotle rejects a two-seed hypothesis and advocates a one-seed alternative is that he says that females do not emit semen. However, by looking at the context in which these statements occur, this assessment can be legitimately challenged. *GA* I and II develop preliminary ideas concerning generation by opposing the views of certain unnamed opponents: 'some people' (*tines*).

> It is thought that all things come to be from semen, and that semen comes from the parents. Because of this, it is the same inquiry to ask whether the female and the male both discharge semen or one of them only, and to ask whether the semen comes from the whole body or not. For it is reasonable, that if [it does] not [come] from the whole, then [it does] not [come] from both the parents. Since some people say that it is drawn from the whole of the body, we must first investigate how they can maintain this[8] (*GA* 721b6–14).

The opponent holds two doctrines: that the female emits semen like the male's (and thus 'parallel' to the male's) and that the semen is drawn from the whole body. The latter idea is often referred to in commentaries as 'pangenesis'.[9] Many have speculated as to the identity of 'some people' in this section of the *GA*. Similar remarks about the origins of semen are made in the Hippocratic treatises *On the Sacred Disease*, where it is said to 'come from all places in the body'[10] (*Morb. Sacr.* 2.11, L VI 364–7), and *Air, Waters, Places*, where it 'comes from all quarters of the body' (*Aër.* 14, 19–24, L II

[8] Δοκεῖ δὲ πάντα γίνεσθαι ἐκ σπέρματος, τὸ δὲ σπέρμα ἐκ τῶν γεννώντων. διὸ τοῦ αὐτοῦ λόγου ἐστί, πότερον καὶ τὸ θῆλυ καὶ τὸ ἄρρεν προΐενται ἄμφω ἢ θάτερον μόνον, καὶ πότερον ἀπὸ παντὸς ἀπέρχεται τοῦ σώματος ἢ οὐκ ἀπὸ παντός· εὔλογον γάρ, εἰ μὴ ἀπὸ παντός, μηδ' ἀπ' ἀμφοτέρων τῶν γεννώντων. διόπερ ἐπισκεπτέον, ἐπειδὴ φασί τινες ἀπὸ παντὸς ἀπιέναι τοῦ σώματος, περὶ τούτου πῶς ἔχει πρῶτον.

[9] See above at *GA* 721b9 and b13; and at 722a4–5: τοῦ ἀπιέναι ἀπὸ παντός; 722b15: ἀπέρχεται ἀπὸ παντός. Darwin (1868) used the term 'pangenesis' of his theory that reproductive gemules come from the parts of the body. Platt in Aristotle (1910) was the first to apply it to the theory of Aristotle's opponent in this section of the *GA*.

[10] ἔρχεται πάντοθεν τοῦ σώματος.

70–93).[11] Because there are aspects of these Hippocratic texts which do not fit with Aristotle's description,[12] some speculate that the target is Democritus or Anaxagoras.[13] It is effectively impossible, due to the fragmentary nature of either, to find any independent pangenesis theory there. I propose, then, to look to the texts of the Hippocratic corpus which loosely resemble the ones that Aristotle opposes, for some of the general points made against the latter apply also to the former.[14]

The principal reason Aristotle disputes the pangenetic hypothesis relates to his metaphysical views on the nature of generation. Sketching a theory similar to the one he opposes can help make this clearer; so let us now consider the theory that appears in the Hippocratic *On Generation*. The author first gives an account of the production of male semen. Vessels (*phlebes*), which extend throughout the body, hold the four essential bodily fluids or humours; these vessels are ultimately connected to the penis (*Genit.* 1.2–3, L VII 470–3).

> The semen is, as I say, secreted from the whole body[15]—from the hard parts as well as from the soft, and from the total humour. This humour has four forms: blood, bile, water and phlegm (*Genit.* 3.1, L VII 474–5).[16]

During sexual intercourse semen is drawn from the four humours.

> Friction on the penis and the movement of the whole man causes the fluid in the body to grow warm: becoming diffuse and agitated by the movement it produces a foam . . . what is secreted as a foam is the most potent and richest part of the humour (*Genit.* 1.2, L VII 471–2).

[11] The most thorough work on embryology in the Hippocratic corpus is the combination of *On Generation (Genit.)* and *On the Nature of the Child* (*Nat. Puer.*), which are thought to form one coherent work. *On Diseases Four* (*Morb.* 4) is also by the same author and furthers the themes taken up in *Genit.* and *Nat. Puer.* For Von Staden (1989) 'pangenetic theory dominates the Hippocratic Corpus' (290).

[12] For instance, no Hippocratic text argues that semen comes from nails and hair, which appears to form part of the theory Aristotle cites here. See Balme in Aristotle (1972) 141.

[13] For example, Democritus in Lesky (1950) ch. 8, De Ley (1980) 130–1 and Anaxagoras in Kember (1973).

[14] See van der Eijk (2007) for the view that Aristotle knows these particular texts and is addressing them here. However, only one of the proofs Aristotle mentions actually appears in the Hippocratic corpus. Mutilated offspring come from mutilated parents at *Genit.* 8; see Giorgianni in Hippocrates (2006) 20.

[15] ἀποκρίνεσθαι ἀπὸ παντὸς τοῦ σώματος.

[16] Translations follow Lonie in Hippocrates (1981).

Semen is secreted by women in the same manner and conception occurs when semen from the man and semen from the woman are retained in the uterus and mix together (*Genit.* 4.1–5.1, L VII 474–7). Both sexes have either male or female semen; male semen is stronger than female semen. The strongest or the most abundant semen conquers (*kratei*) the other, determining the outcome in terms of sex and inheritance (*Genit.* 5–7, L VII 476–81). Semen which comes from the individual parts of parents also battle against each other in order to establish resemblance in parts – here it seems that the determining factor is quantity rather than quality.

Semen is a product which comes from the whole body[17] of each parent, weak semen from the weak parts, and strong semen from the strong parts. The child must necessarily correspond. ... The child will resemble in the majority of its characteristics that parent who has contributed a greater quantity of semen to the resemblance—that is, semen from a greater number of bodily parts (*Genit.* 8.1–2, L VII 480–1).

Semen comes from the whole body in this theory, in so far as it comes from all of the fluid humours of male and female bodies and also from particular parts of their bodies.

Aristotle's main arguments against pangenesis and the parallel seed theory centre on the metaphysical requirements for animate individuality which emerge when he sets before his opponents the ridiculous consequences of their view.[18] Pangenesis, says Aristotle, cannot explain how the various parts of an embryonic body come together to form a whole animal. If miniature bodily parts arrive separately at the womb, how can these separate parts remain alive, since a part like an arm is useless and effectively 'dead' once separated from the organic whole (*GA* 722b4–5). Even if the parts could exist separately this would create a difficulty for any explanation of why they would recombine since each would be more like separately living animals than parts of one whole. Aristotle also questions the level of division envisaged by his opponents: when they speak of 'parts' do they mean arms and

[17] ἀπὸ παντὸς τοῦ σώματος.

[18] See Morsink (1982) 67–85, for a thorough analysis. See also Balme in Aristotle (1972) 140–51.

eyes ('non-uniform' parts in Aristotelian terms) or flesh and bone ('uniform' parts)? At either level, a certain amount of organization exists, until you reach an analysis into the four elements: earth, air, fire and water, at which point no animate existence remains (and also no possibility of resemblance to parental parts) (*GA* 722a34–722b3). In order to provide an adequate explanation, Aristotle thinks that generative investigations ought to concentrate on the agent which 'fashions' (*dêmiourgei*) the offspring rather than the parts that are fashioned (*GA* 722b2–3).

The *GA's* concern to emphasize an organizing or fashioning power in the generative process combines with a commitment to the difference between male and female functions which result in their having differentiated seed.[19] Only one parent can provide an overall organizing power, otherwise two consequences would arise: first, we would find that every couple produced two offspring at a time, one from each (*GA* 722b7–8).[20] Second, if females had the ability to complete the organization of offspring there would be no need for males; females would be able to generate on their own (*GA* 722b13–14). The impossibility of parthenogenesis is a serious challenge to the parallel seed theory; it indicates that both sexes are needed. This supports the hypothesis that the individual contributions of the sexes are differentiated.

Aristotle sets out his ideas more fully with reference to facts about generation in insects. He believes that female insects insert a part of themselves into the male in order to reproduce (*GA* 723b19–22). Because male animals do no emit semen in these animals, organization is instead achieved by a principle working from within the male's body. For those males that emit semen, it becomes the tool of this fashioning part. This is a step towards showing that the pangeneticists 'shouldn't have said that it is drawn from all the parts, but only from the fashioning part[21] as from the carpenter, but not from the matter' (*GA* 723b28–31). In order to

[19] See *GA* 721b6–10. These two positions are linked together in many other instances. See *GA* 721a33–721b2, 722b8–9, 724a8–9, 730a24–7.

[20] Aristotle also asks how, if the generative parts of both parents arrive in the uterus, offspring end up one sex rather than another and not as hermaphrodites (*GA* 723a32–723b3).

[21] ἀπὸ τοῦ δημιουργοῦντος.

reinforce that point, the *GA* argues for a certain account of the production of semen in the body. Semen cannot come from the diseased breakdown of the body, which is what Aristotle says it must for the pangeneticists. That account of seminal origins would make the seed something contrary to nature (*para phusin*) and what is contrary to nature cannot create a natural product (*GA* 725a1–3). Since offspring are natural, the semen must be formed from residues of the nutritive process, which are prepared to go *towards* all the parts, rather than coming from them (*GA* 725a21–23). Employing the potentiality/actuality distinction, which he thinks his opponents failed to hit upon, Aristotle posits that semen is potentially capable of being the parts of the body rather than actually having been them. In Aristotelian physiology blood is the usual product of digestion; semen is the residue of this process, further heated up (*GA* 726a26–28). Both blood and semen originate at the fashioning part (*demiourgountos*), that is, the heart, which is the centre of nutritive operations. Yes, Aristotle is saying, the new animal does come to be from semen and the semen comes from the parents – all the wise rightly agree on these points (*GA* 716a9–14). However, semen does not come from all parts of the body (as pangenesis would have it) but from the fashioning part (*demiourgountos*) and it is going towards and is potentially like all the parts of the living body.[22]

The fact that semen comes from the fashioning part explains the male parent's ability to organize offspring. This can seem to indicate that no such connection exists between the female and the fashioning power in generation. However, crucial similarities between male and female contributions remain in Aristotle's differentiated seed account. Although others could characterize his arguments here as a refutation of the two-seed theory in favour of the one-seed theory, there is good reason to believe that Aristotle does not reject but in fact endorses his own version of a 'differentiated' two-seed theory.[23] For, 'the menstrual fluid is

[22] Bos (2003) argues that for Aristotle what are potentially present in the semen are parts of the soul rather than parts of the body. However, this does not accord well with Aristotle's challenge to pangenesis, which focuses on the bodily parts.

[23] Balme in Aristotle (1972), Mayhew (2004) ch.3 and Henry (2007b) 256 n.14 all conclude that Aristotle thought of the female contribution as seminal.

[also] a residue, and it is the analogous thing in the females to the semen in males' (*GA* 727a2–4).

3.3 Aristotle on female semen

In this section, I will explain that Aristotle is a two-seed theorist because, for him, the female contributes semen to generation. This is evident from the very beginning of the *GA* where the term 'semen' refers to that which both parents contribute.[24] Semen (*sperma*) is a word which was used to indicate that 'from' (*ek*) or 'out of which' (*ex hou*) offspring come.

> As we said, the chief principles of generation may be set down as the female and the male: the male as possessing the principle of movement and generation and the female as [possessing the principle of] matter. One is most likely to be convinced of this by considering how semen comes to be and where it comes from. For while natural things come to be constituted from (*ek*) this, it is necessary not to forget how it [semen][25] happens to come to be from the female and the male. For such a part is secreted from the female and the male, and the secretion is in them and from (*ek*) them; because of this, female and male are the principles of generation (*GA* 716a5–14).

As Aristotle notes, natural things come to be 'out of' (*ek*) semen and semen comes to be 'from' (*ek* or *apo*) *both* parents. Later on in the treatise, however, after the section in which the reader is made aware of the organization brought about by the male animal, the term is temporarily restricted to that which comes exclusively from the male. A separate discussion on the topic of the female contribution follows (*GA* 1.19–20). The two generative residues must be differentiated; the female does not contribute semen *of the same sort* as the male (*GA* 727b6, 730a25–28), that is, a white fluid, sent forth[26] at sexual climax (*GA* 727b7). Because the term 'semen' is originally used to indicate the generative contribution of both parents, Aristotle must make an effort to establish a new

[24] '[A]ll animals come to be from semen and semen comes from the parents (πάντα γίνεσθαι ἐκ σπέρματος, τὸ δὲ σπέρμα ἐκ τῶν γεννώντων)' (*GA* 721b7). In his more general discussions, Aristotle takes it for granted that offspring must come to be from (*ek*) the father *and* the mother (*Metaph.* 1013a9, 1023b4). Balme in Aristotle (1972) 131.

[25] *Toutou* refers back to *to sperma* in the previous line.

[26] Προΐεται.

and exclusively male use for *sperma* or *gonê*.[27] When using these terms exclusively of the male contribution, ample indication is given that the male secretion is being spoken of by the addition of 'by the male' or 'of the male'. For example, deciding that an account of generation must concentrate on the male activity, Aristotle asks: 'How is the semen produced *by the male*?' (*GA* 729b2).[28]

Although the term is used exclusively of the male at this point, it also continues to be employed to indicate the contributions of both sexes. Despite an attempt to set out or re-label the female contribution to generation as not semen proper, Aristotle continues throughout the entire *GA* to treat it as a sort of semen and often to refer to it as semen or seminal. This is most evident in his account of the physiological basis of semen. Semen can be one of five alternatives within the body: parts according to nature, parts contrary to nature, residue, colliquescence or nutriment.[29] As explained earlier, Aristotle argues that according to pangenesis *sperma* is a colliquescence (*suntêgma*) – a waste product coming from the breakdown of bodily parts and the cause of disease.[30] His theory instead posits that semen is a residue, the leftovers of nutriment (*GA* 724b27). He supports this thesis in various ways. First, residues have receptacles in the body, whereas waste products roam around it (*GA* 725a34–725b4). Since semen has a receptacle, this suggests that it is a residue. Also, the loss of it makes an animal weak, indicating that it has a nutritive rather than a waste-like quality (*GA* 725b17–19). Its nutritive quality is also indicated by its absence during childhood, when all food is used for growth, and in old age and disease when the body is weak

[27] Both terms are used to refer to semen by Aristotle. Tress (1992) 314, n.14.

[28] τὸ σπέρμα τὸ ἀπὸ τοῦ ἄρρενος. Cf. *GA* 736b18.

[29] *GA* 724b24–6: τῶν κατὰ φύσιν ... ἢ τῶν παρὰ φύσιν, οἷον φῦμα, ἢ περίττωμα ἢ σύντηγμα ἢ τροφήν.

[30] On *suntêgma* see Coles (1995) 57–9 and Balme in Aristotle (1972) 146–7. This term does not occur in the Hippocratic works, which is another reason to remain cautious about finding the pangenesis theorist there. Whether Aristotle's opponent really thought this is questionable; it is no doubt a shrewd polemical move on Aristotle's part to associate the rival theory with such an unsavoury idea. Although there a few ancient authors who think that development results from a process of putrefaction or rot, labelling this *sêpedonê* rather than *suntêgma* (e.g. *Carn.* 3; Diogenes of Apollonia DK 64A32; Pre-Socratics [2010] 448–9).

(*GA* 725b19–25). Finally, fat or well-nourished animals do not have semen since all their residues have been used to fatten up their bodies (*GA* 726a3–7).

Within the initial section on why semen is a residue, Aristotle lists three receptacles for semen – the uterus, the male genitals and the breasts – clearly indicating that the female contributions of menses and milk are seminal (*GA* 725b3). As a residue of nourishment (*GA* 727a2) the female generative contribution shares all the features of semen listed above. As with the male contribution, female menstrual blood is absent in childhood and old age and in instances of disease (*GA* 727a5–10). Also, when saying that fat animals lack *sperma*, Aristotle indicates that this occurs in both women and men (*GA* 726a4). Indeed, the menstrual blood, which fat female animals fail to produce, is considered to be a sort of semen in the following passage.

> Menses are a residue as semen is. One may grasp, as proof of this, a number of facts concerning animals. For fat animals are less spermatic (*spermatika*) than thin ones, as we said before. This is because fat is a residue like semen (*to sperma*), being cooked blood, but it is not [cooked?] in the same way as semen. So, it is reasonable that when residue is consumed toward fat, the portion going toward the semen (*tên gonên*) is deficient (*GA* 727a31–727b1).

This is only one of many places where the size of the body or the diversion of residues in the female body indicates that the female contribution is seminal. Earlier, he states: 'people [whether male or female] in sound health, who put on rather a lot of flesh and get a bit fat, emit less semen (*sperma*) . . . than is normal' (*GA* 725b31–34). The same theory appears later on where the two seminal residues are differentiated: when too fat, males produce less *gonê*, females less menstrual fluid (*GA* 746b25–9). Aristotle also explains that the largeness of the bodies of female mules shows that nutritive residues have been diverted to semen (i.e. menses) (*GA* 748b20–2). And those (female) animals, such as the lioness, who use up the residues on the bodies of their offspring become infertile later on in life because 'the semen is diminishing as the prime of life abates' (*GA* 750b1). Furthermore, egg production is referred to as semen production at *GA* 718b15: small animals produce more semen, that is, more eggs.[31] Finally, evacuation of menses is explicitly equated with the emission of semen at *GA* 774a1.[32] Thus Aristotle often explicitly refers to the female contribution as semen or seminal.

Despite accusations to the contrary, Aristotle does not say that only the male has semen.[33] The female contribution must be semen by its original meaning, since it comes from (*ek* or *apo*) the female parent and offspring come to be from (*ek*) it. In fact, offspring come to be from it in a much more concrete manner than they do from male semen.[34] Also, the bodily process by which female menstrual blood is produced is almost identical to the process that brings male semen into being and both are affected by the same physiological factors. The connection between the female parent and her contribution and the subsequent offspring is identical to the same connection in male animals; both contributions have soul potentially and potentially resemble the parts they were originally 'going towards'. When he explains that (male) semen is a residue of nourishment that is undergoing change (*GA* 736b26) – he indicates that this is also the case with the menstrual fluid. Both male and female contributions are residues of blood, which is nourishment in its final form (*GA* 726b4, 737a18–20). The idea that loss of *sperma* brings on a certain weakness does not only apply to the male condition; Aristotle spends a long time explaining that the weakness of the female body is due to the loss of menstrual blood each month, indicating that it also contains vital powers (*GA* 727a16–26). When considering female fertility, Aristotle again treats the menstrual blood as seminal. He thinks that if a woman is fertile, then dye rubbed into her eyes will change the colour of her saliva. The test is designed to show whether the passages of the body are open, allowing

31 See also *HA* 582b28–583a11.

32 Shortly after this passage, Aristotle discusses milk production, which occurs in the chest because this is where semen is generated in the body. He then notes: 'this is the reason why the voice changes in both sexes when *they* begin to bear seed' (*GA* 776b16, my emphasis).

33 Dean-Jones (1994) 176–7, for instance, writes: '[Aristotle's] alternative theory of the female contribution to reproduction did away with the notion of female seed altogether . . . '

34 *GA* 725a11–13: '*Sperma* is, therefore, a certain part of a useful residue, and the final residue is most useful and so it is that [i.e. *sperma*] from which each of the parts comes directly into being.'

residue from the head to descend to the genitals and depends on the principle that, 'the head and the eyes are the most seminal'[35] (*GA* 747a14–15). After all, 'the nature of the semen (*tês gonês*) is similar to that of the brain' (*GA* 747a18), even if that semen is feminine.[36]

In some instances, Aristotle uses semen (*sperma*) exclusively of the female contribution. Indeed, since the term brings with it connotations of the material used to make up the foetal body, it is more suited to the Aristotelian female role (*GA* 724a18). For example, while arguing against pangenesis, it is noted that semen does not contain the parts of offspring in actuality but rather is able to *become* these parts (*GA* 723a16). Having established that the male contributes nothing material to the offspring, only the female contribution could be understood to become the material parts of its body. If the same semen can be *formed into* either a male or a female animal (*GA* 723a33), then semen in this passage must refer to the female contribution, which will come to constitute the sexed body of the offspring. Finally, in the formation of the foetus's physical parts, such as the bones and sinews, these are said to be formed 'out of the spermatic residue',[37] which must refer again to the female contribution.[38]

Commentators who wish to maintain that Aristotle never allowed for female semen and hold a one-seed interpretation of his embryology occasionally bump up against the type of passages I have here detailed. To maintain his consistency, as they see it, with regard to the female role, they try to deny that Aristotle ever refers to the female contribution as semen or seminal.[39] Even while noticing that 'the seminal secretion' is applied to both male and female residues at *GA* 746b26, one

35 σπερματικώτατός ἐστιν.

36 Support for the view that the female contribution is semen also comes from various other texts which detail the analogous roles of male and female generative residues in the body. For example, *HA* 544b15–33; *PA* 689a6–13.

37 ἐκ τῆς σπερματικῆς περιττώσεως; *GA* 744b29, 38.

38 If female menses do count as a sort of semen for Aristotle, then there is no difficulty with the numerous other passages in the corpus which say that semen becomes the animal (e.g. *Ph.* 190b4, 196a31, 199b7, *PA* 641b32, *Metaph.* 1049a2, 1058b23–4). Ross, among others, found this idea problematic, since he could not accept that female menses are a sort of semen (Aristotle 1924, vol. II, 255).

39 For example, Morsink (1982) says of *GA* 726b13–19, 'The semen referred to in the passage just quoted is the semen of the male and not that of the female. She does not discharge semen . . . ' (130).

commentator dismisses it by writing 'i.e. the semen of men and the catamenia of women; the latter have no semen according to A[ristotle].'[40] Others attempt to offer alternative translations of *sperma*, for instance as 'embryo', in these cases, rather than have to concede that the term applies to the female contribution as well as the male one.[41] These efforts, although seemingly quite innocuous, show how the thought that Aristotle had a one-seed thesis can serve to distort the way the *GA* is interpreted.

There is no good reason to reject clear indications within the text that Aristotle was perfectly happy, in some instances, to count the female contribution as seminal. If this is so, then it is worth considering evidence from other works that Aristotle adhered to a two-seed hypothesis, in particular the *Historia Animalium* X (or 'On Sterility'). Recent commentary on *HA* X argues that it ought to be included in the Aristotelian corpus as authentic, representing an earlier theory of the female contribution to generation. The theory found there, which is not presented in an overly theoretical manner but rather with reference to the diagnosis and treatment of female infertility, states that both male and female emit semen at sexual climax to a place in front of the uterus; the uterus then draws up the two semens (*HA* 634b29–39; 636a4–8).[42] Despite its difference from the *GA* theory, which does not allow the fluid that the female emits at climax to be seminal (*GA* 727b34–728a1),[43] both *HA* X and the *GA* reject pangenesis and both emphasize the necessity of the female contribution and its correct relationship to that of the male (e.g. *GA* 723a27–31, 767a16–27, *HA* 636b9). Whether the *HA* X text is indeed authentic is still being disputed.[44] However, from the point of view of fitting

40 Platt in Aristotle (1910) 746b26 n.2.

41 Adelmann (1966) 739; Morsink (1982) 99–100. Some even deny that male and female generative residues have anything in common. They are merely analogous according to Barthélemy Saint-Hilaire in Aristotle (1887) vol. 1, 91.

42 The *GA* can then be taken to be a modification of the earlier *HA* X two-seed thesis. Most commentators now agree that this work is not part of the *HA* but a separate (possibly medical) treatise. See especially Balme in Aristotle (1991) 26–30 and 487–489 note c. and van der Eijk (1999) 491, 498.

43 See van der Eijk (1999) for an alternate view of *HA* X. He argues that one need not read the text in this way but can allow that the female contribution remains unspecified and unconnected to sexual climax.

44 Dean-Jones (2012) has recently argued that chs. 1–5 are not authored by Aristotle and that chs. 6–7 are either by Aristotle or a later follower.

it with the doctrines found in the *GA*, there may be less of a problem if the *GA* is rightly acknowledged to contain a differentiated two-seed theory rather than adhering to a one-seed hypothesis.[45]

3.4 Contrasting views on the role of the female in generation

The contrast between Aristotle and his opponents remains. For Aristotle, the female's generative contribution is menstrual blood, for Galen and various Hippocratics, it is a white fluid emitted at sexual climax. Perhaps tellingly, when Galen contrasts his school with that of Aristotle over the issue of female semen, the debate is not focused around the relative sexism involved. Instead, he is concerned to emphasize his philosophical and epistemic authority. In order to get a clearer view of the contrast between the two it will be necessary to disentangle Aristotle's views from the way that they are represented by Galen and others, but a brief overview of biomedical theories in antiquity is first required.

Ancient seed theories can be divided into three key types: encephalogenetic, pangenetic and haematogenous.[46] Proponents of the encephalogenetic view include the Pythagoreans, Plato and Diocles, all of whom preceded Aristotle; they took the brain to be central to living functions and the origin of semen which passes down the spinal column to the genitalia during intercourse.[47] We have already met the pangenetic thesis in the Hippocratic corpus, which was also held more broadly and is associated with atomism. On this view, rubbing during sexual intercourse is similar to scratching,[48] in which tiny particles of the body are displaced; these tiny particles make their way to the genitals during sexual

[45] Balme in Aristotle (1991) 28. 'The *GA* theory . . . does not radically contradict the *HA* X theory, but refines it. It does not prevent Arist. from calling menses "female seed" [in the *GA*].' (489, note c). I would say that Aristotle does more than merely 'refine' the theory in *GA*, he would have had to 'modify' it.

[46] Von Staden (1989) 288–90. See also the distinction as made by van der Eijk in Diocles (2000–1) vol. 2, 86.

[47] Plato, *Tim.* 73b1-d2, Diocles (2000–1) vol. 1, 84–5.

[48] Scratching was also a euphemistic way to refer to sexual intercourse at this time. See Plato *Phlb.* 47a.

intercourse and are excreted at climax.[49] The third theory is the one Aristotle espoused, which has its possible origins in the ideas of Diogenes of Apollonia. As we have briefly seen, for Aristotle the semen is largely completed in the blood vessels before they reach the genitalia and before sexual intercourse takes place. After Aristotle, advances in anatomy were taken to support his view. Herophilus of Alexandria first discovered the female ovaries and, following the tendency to map female anatomy onto the male one, called them 'testes'. However, no radical change in the view of the female role in generation needed to follow from this discovery, for in all three of the above options, the testes did not play any role in the production of semen. Herophilus employed other anatomical observations in support of an haematogenous against a pangenetic theory. In a famous fragment, he declares to have seen that the nearer to the genitalia blood vessels get the less they have of blood and the more they contain a seed-like substance (von Staden 1989, 292). As a doctor and dedicated anatomist himself, Galen was much more aware of the structure of male and female reproductive organs than Aristotle, and certainly took into account Herophilus' observations. He integrates this into his theory that both men and women produce semen in the same manner and that their bodies operate in a parallel fashion, although the female is weaker and less capable.

Other differences in the characterization of female bodies and their healthful functions are just as important to consider when contrasting ancient biomedical theories as their stances on female seed. In fact, for many ancient medical writers, the theoretical intricacies of embryology were side lines to therapeutic recommendations and practice.[50] Views about the nature of the female body were more important in the setting of therapy, as they more directly dictated diagnosis and treatment. In this context, we can find a shift from considering men and women to be entirely different in terms of the structure, texture and functioning of their bodies in the Hippocratics writings (fifth and fourth centuries

[49] Lonie in Hippocrates (1981) 106, *Genit.* 1 and *Morb. Sacr.* 5.

[50] Soranus' views on the contributions of male and female seed to generation are obscured because his surviving work is a medical manual on conception and infant care (in his *Gynaecology*). His work on semen is lost.

BCE) to the idea of the two sexes as parallel, evident in Herophilus (335–280 BCE), Soranus (first and second centuries CE) and Galen (129–99 CE).[51] Galen, as physician at the Roman Imperial Court and self-professed philosopher, was keen to make his mark on many theoretical aspects of medicine in addition to medical practice. His book on reproduction and embryology, *De Semine*, focuses on humans and on the seed produced by both men and women and so lacks the zoological scope of Aristotle's *GA*. But what both works have in common is a desire to set their theories apart from others and strongly argue for the advantages of their own views. Galen was a vociferous arguer and in his *De Semine* Aristotle and his followers are the main targets of attack. Given the fame and authority of Aristotle, the better a person was at defeating his point of view, the cleverer and more authoritative he himself would appear. Thus we find that, although Galen shares the Aristotelian blood-based theory of semen, as well as commitments to teleology and hylomorphism, it is their differences that the writer plays up. Galen's own rhetoric initially suggested the contrast between one- and two-seed theories and their relative advantages, the traces of which remain in modern commentary on the history of embryology. In the following I will sketch out Galen's presentation of Aristotle and possible points of misrepresentation and set out the contrasts between the two thinkers on the female role in generation without the showcasing and posturing of the former. I will also explain why in terms of assessing which is more or less sexist or enlightened with respect to women, there is no significant distinction to be made.

Galen characterizes his disagreement with Aristotle as one about female semen, the existence of which he accuses Aristotle of denying strictly on the basis of logic, without observational evidence.

> Now the fact that the female animal has semen must be accepted on the evidence of the senses, as we said earlier, and the existence of what is clearly seen must not be overturned by argument (*Sem.* II.4, 5–6).[52]

[51] On the Hippocratic propensity to regard the female body as fundamentally different and suffering from different diseases see Bourbon's Introduction to Hippocrates (2008) LXI–LXVII, *Mul.* 1.62, *Loc. Hom.* 47 and Dean-Jones (1994). On the transition to viewing male and female as similar in Hellenistic medicine see von Staden (1989) 168, 297.

[52] Galen (1992) 172–3.

Part of Galen's critique focuses on the following passage from the *GA*, where Aristotle declares:

Now since what comes to be from females is as the semen from males, and it is not possible for two spermatic secretions to come to be at once, it is plain that the female does not contribute semen to generation. For if it (the female) had semen, it would not have menstrual fluid. Now because that one is present [in the female], the other one (semen) is not (*GA* 727a26–30).

In response Galen writes, 'let us not say that it is impossible for the female to accumulate both generative residues. For she is observed to accumulate them' (*Sem*. II.4, 20).[53] Galen's accusation that Aristotle failed to employ empirical methods is an exaggeration of the actual state of affairs. Although Aristotle did not have access to the evidence that Galen did,[54] he clearly engaged in observations in order to come to his conclusions. He attempted to provide empirical support for the supposition that menstrual blood is the analogue to male semen in female physiology and that no other seminal secretion exists in women.

One set of observations on which Aristotle and his parallel seed opponents agree is that the process by which males become fertile is similar to that by which females do and that fertility in both is affected by various factors, such as nutrition.[55] A large part of the Hippocratic *On the Diseases of Women* (*Mul*.), for instance, is taken up with comparing the beginning of menstruation (menarche) as an indication of sexual maturity and fertility with the advent of semen production in male bodies.[56] Even in the Hippocratic work *On Generation*, a treatise which almost exclusively deals with a theory of male and female seed, the author takes time to describe menarche.

[A]s both boys and girls grow, the vessels which extend in the boy's case to the penis and in the girl's to the womb, open out and become wider in the process of growth; a way is opened up through the narrow passages, and the humour, finding

53 Galen (1992) 176–7.

54 He did not see semen in the blood vessels near the ovaries (which he did not know about in any case), as Herophilus claimed to have done. Von Staden (1989) 292.

55 See *GA* 725b19–25, 727a5–10 on the timing of menarche and puberty and *GA* 726a3–7, 727a31-b1, 746b25–9 on situations in which semen and menstrual blood are absent, for example, when animals are overweight.

56 Mention of female semen is infrequent in these Books. See *Mul*. I.24 (L VIII 62–5) and also Dean-Jones (1992) 10–11.

sufficient space, can become agitated. That is why when they reach puberty, sperm can flow in the boy and the menses in the girl (*Genit.* 2.3, L VII 472–5).

The dispute between Aristotle and his opponents on the nature of female semen cannot be characterized in terms of the opponents observing phenomena that Aristotle refuses to acknowledge. Instead we see that he legitimately criticizes the idea that the female emits semen at climax on the ground that, in terms of the evidence, they would simply have too many seminal contributions.[57] If his parallel seed theorists want to have it that the female has seed emitted at climax just like the male seed, then they ought to give up on the idea that the menstrual fluid is an analogue to male semen.[58] The empirical evidence, however, would make that an extremely unattractive position to maintain.[59]

Another piece of empirical evidence Aristotle uses to propose that male and female contribute differently to generation is the morphological variation between their generative organs in most types of animal (*GA* I.3–16). The possession of different organs appears to indicate the possession of different functions:

Since they differ by ability (*dunamei*) and a certain function (*ergô*), and every function requires organs, and the organs for different faculties are parts of the body, then it is necessary in the case of those parts which serve for generation and copulation that these differ from one another. Accordingly, male differs from female (*GA* 716a23–8).

Because he thinks that this differentiation of function is in some sense obvious, Aristotle is keen to overthrow theories which cannot adequately accommodate it. Given all of this, one must not get too distracted by Galen's attempt to undermine Aristotle's scientific credentials.[60]

I now turn to other ways in which the contrast between parallel and differentiated seed theories has been misrepresented,

57 'It is not possible for [the female] to have two seminal secretions' (*GA* 727a26–7).

58 This is an issue that Aristotle himself may not have decided until writing the *GA*. See Balme in Aristotle (1991) 29 and van der Eijk (1999) 498 on *HA* X.

59 The coincidence of male ejaculation and female menarche is 'a sign that the semen is in the menstrual fluid' (σημεῖον δ' ὅτι ἐν τοῖς καταμηνίοις τὸ σπέρμα ἐστίν) and not in some other fluid secreted by the female (*GA* 728b22).

60 Aristotle also thinks that observations concerning generation in insects help to show that the sexes contribute differently (*GA* 723b19–31).

particularly by social historians. I will first consider Aristotle's nearer contemporaries before returning to Galen. Sometimes the parallel seed theory of the Hippocratics is praised for the attention it pays to female sexual pleasure and to the resemblance children bear to their female parents.[61] Certainly Aristotle's differentiated seed theory will have to approach both phenomena differently. However, Aristotle and the Hippocratic doctors were divided on such issues not due to any attempt to influence the standing of women in society, but as part of an argumentative arsenal they used to promote their own explanations.

Discussions of female sexual pleasure and hereditary resemblance appear in the *GA* as both are presented as proofs in support of the pangenetic stance (*GA* 721b15–24). In the Hippocratic corpus also the intensity of pleasure experienced during intercourse is said to show that the entire body produces semen, which presumably applies to both sexes.[62] Next, it is said that mutilated parents, for instance those who lack a part of the body, produce mutilated offspring, indicating that no semen came from that missing part.[63] Finally, the fact that offspring resemble parents in parts suggests that semen is drawn from each part of the parents and also that these possess their own peculiar (*idion*) semen (*GA* 721b13–28).[64] Aristotle does not think that these three pieces of evidence lend support to the pangenetic thesis. First of all, the intensity of sexual pleasure does not provide adequate proof of particulate semen because females do not have to experience it in order to conceive.

> Very often the female conceives although she has derived no pleasure from the act of coitus; and, on the contrary side, when the female derives as much pleasure as the male, and they both keep the same pace, the female does not bear (*GA* 727b8–11).

61 See, especially, Rouselle (1988).

62 *Genit.* 4. See also Democritus DK 68A142; Pre-Socratics (2010) 566–7.

63 See *Genit.* 8 and Hippocrates (2006) 20. Diseased parents produce the body of the diseased offspring because 'the seed comes from all quarters of the body' (*Morb. Sacr.* 5.7–15, L VI 368–71 and *Aër.* 14, 19–24, L II 70–93).

64 This idea is evident in the *Genit.* Aristotle appears to adopt something like it later on in the *GA*, where he attempts to explain variations in parts of the body with reference to the fact that 'each part of the body has its own peculiar (*idion*) heat' *GA* 786a21.

Some think that the above statement amounts to a disregard for female pleasure.[65] But Aristotle's remark hardly proves this. When he writes that female pleasure is lacking 'very often' during conception (*GA* 727b8), this is an exaggeration used for emphasis, rather than a call to ignore that pleasure. Elsewhere, he notes the frequency and naturalness of female sexual pleasure.

> Conception takes place even if the pleasure which women *usually* experience during sexual intercourse fails to occur, if the part concerned happens to be ready and the uterus had descended within. *Generally*, however, pleasure does occur, because when the secretion, which is usually accompanied by pleasure in man and woman alike, takes place, the mouth of the uterus has not closed, and in these conditions a better passage is afforded for the semen of the male (*GA* 739a29–34, my emphasis).

For Aristotle, some features of animals do not merely aid survival, but serve to make life 'better', for instance, the presence of testes and kidneys.[66] The female's pleasure is a feature like this. The frequent opening up of the mouth of the uterus is not necessary for conception but makes the situation better by aiding it.

With respect to female sexuality, one could argue that the parallel seed theory, in contrast to Aristotle's, has the disadvantage of assuming it to be just like male sexuality and thereby tying it to procreation. First of all, then, in many of these ancient texts, those who call the female contribution 'semen' impose a male framework onto the female. Consider the author of *On Generation*'s description of the pleasure experienced by women in sexual intercourse.

> Once intercourse has begun, [the woman] experiences pleasure throughout the whole time, until the man ejaculates . . . her pleasure terminates along with that of the man. . . . If, for example, you pour wine on a flame, first of all the flame flares up and increases for a short period . . . then it dies away. In the same way the woman's heat flares up in response to the man's semen, and then dies away. The pleasure experienced by the woman during intercourse is considerably less than the man's, although it lasts longer (*Genit* 4.1–2. L VII 474–7).

[65] Rouselle (1988) and McLaren (1985). These commentators argue that Aristotle's theory led to the neglect of women's rights to pleasure. The idea that someone has a right to sexual pleasure is quite modern (see Nussbaum 2000, 78) and probably would not have found a place in ancient Greece (see Omitowoju 2002).

[66] *GA* 717a15–21; *PA* 670b23. See more references at Aristotle (1972) 87.

The author thinks that women produce semen in exactly the same manner as men by ejaculating a white fluid at sexual climax.[67] He thereby focuses entirely on the sexual experiences of men and assumes that female sexuality matches these.[68] Arguably, this leads to a distorted picture of the female experience. It is supposed here that women's pleasure is produced by the sort of rubbing action that men enjoy,[69] and that the woman's pleasure always ceases at the same point as the man's, her 'flame' being extinguished by the incoming fluid. It is also noted that the female's pleasure is less intense than the male's. All of these points show a disregard for women's actual experiences.

Let's now turn to the second point: for the parallel seed theorist, female sexuality is tied to social purposes and so is not something that women may value or choose apart from these goals.[70] Some modern biological theories similarly suggest evolutionary advantages afforded by female sexual response. Feminist critics challenge the findings of this type of research on the grounds that attempts to discover such a link distort the evidence.[71] And, furthermore, this research seems to be motivated by a desire to straightjacket women to their social roles. Thus, when the parallel seed theorist makes the link between female sexuality and the goals of procreation it can in fact be used to undermine women's freedoms.[72]

Unlike some of his opponents, Aristotle is sometimes interested in pointing out that female pleasure is of a different type from male pleasure.

There are some who think that the female contributes semen during coition because women sometimes derive pleasure from it comparable to that of

67 Galen follows the Hippocratics on this matter (*Sem.* II.1, 25–6, Galen, 1992, 150–1).

68 No account is taken of activities other than heterosexual intercourse, which could be more intensely pleasurable for women, or the fact that many women, for anatomical reasons, seldom experience pleasure during typical sexual encounters with men (Hite 1976, Lloyd 2005). Of course, Aristotle is no better on this score when he asserts at *GA* 739a29 that women *usually* experience pleasure during intercourse.

69 τριβομένος. See Lonie in Hippocrates (1981) 106 for the idea that this theory is atomistic.

70 See, for instance, Galen *UP* XIV.2, *Sem.* II.4, 13, Galen (1992) 174–5.

71 This could be a case of confirmation bias. See especially E. Lloyd (1996) and (2005).

72 It is mere speculation to posit that this is what motivated two-seed theorists. Similarly, the scientists that E. Lloyd (2005) criticizes do not seem to be wholly conscious of the sorts of expectations which distort their readings of the data. We might argue, though, that tacit sexism in the modern context is much less forgivable. See Fine (2010).

> the male and also produce a fluid secretion. This fluid, however, is not seminal; it is peculiar to the part from which it comes in each several individual; there is a discharge from the uterus, which though it happens in some women, does not in others … Where it occurs, this discharge is sometimes on quite a different scale from the semen discharged by the male (*GA* 727b34–728a6).

At this point in the text at least, Aristotle appears to have considered evidence of women's different sexual nature.[73] He also notes that stimulation of a portion of the female genital anatomy other than the vagina is productive of their pleasure (*GA* 728a33–34).[74] Aristotle's theory, then, at least allows that the female body differs from the male one with respect to pleasure and so frees female sexuality to some extent from any necessary ties to procreation.

When it comes to sex bias, there is little to distinguish parallel and differentiated seed theories. Either of their views on sexual pleasure can be used to aid or harm actual women. If procreation is desirable, then the parallel seed hypothesis may endorse the occurrence of female sexual pleasure, a seeming bonus for women. However, Aristotle's theory could also be used in a similar manner, since he thinks that female sexual response to be important for 'opening up the uterus' in order to aid conception. Alternately, both can be used to promote a disregard for female sexual experience; Aristotle's differentiated seed thesis because female sexual pleasure is not necessary for procreation and the parallel seed thesis because in the end it disconnects female pleasure from female ejaculation. It does this in order specifically to account for the phenomenon noted above by Aristotle (i.e. of conception occurring despite the lack of female sexual response). The parallel seed theorist must concede that emission of female seed is not always pleasant or even perceptible.[75] Thus, the theories themselves do not dictate kinder attitudes towards women. Positive and negative attitudes

[73] Later in *GA*, female and male experiences are equated. In order to explain why female sexual desire slacked off after parturition, he likens the process to male ejaculation (*GA* 773b33–774a6).

[74] Aristotle is probably referring to the clitoris, see Dean-Jones (1992) 79.

[75] Van der Eijk (1999) 500 n.55.

can be accommodated by either theory, and either can be manipulated for social purposes.[76]

The disagreement about the role female sexuality plays in generation did not particularly interest Aristotle. That his view happens to allow for the separation of female sexual desire and pleasure from conception was not something he took any notice of; we certainly see no evidence of him using this idea to alter women's lives. He simply uses observations concerning female sexual response in order to support his thesis that females do not emit a separate seed as well as menstrual fluid. What is really at issue is not whether one ought to listen to women, but whether one can convince the reader that your theory is better able to explain accepted facts. This is certainly what seems to be going on in the brief refutation Aristotle gives of the opponents' contention that facts about hereditary resemblance support their thesis. Resemblance between parent and offspring can consist of similarity in voice or manner of movement, which are not parts of the body, strictly speaking. Also the propensity to develop grey hair or a beard can be inherited although a parent may not possess these at the time conception takes place and so the seed cannot come from these parts. Finally, offspring can resemble remoter ancestors, whose parts no longer even exist (*GA* 722a4–9). Aristotle is making a firm point here: the idea that parts of bodies create parts of offspring is too shallow an understanding of heredity. Offspring are a result of *organization*, the whole animal being more than a mere sum of its parts.

Returning now to the contrast between Aristotle and Galen, it is apparent that sex bias operates in a similar fashion in both. For Galen, although women contribute semen, this semen is cold, weak and infertile, the product of a colder animal. It is called

76 Soranus, who held that women could not conceive without sexual desire, proposed that in cases of pregnancy after rape the desire must be present but obscured (*Gyn*. I.37; Soranus 1956, 36). The idea that women had to feel not only sexual desire but also sexual pleasure in order to conceive was even used against them in the medieval period. As McLaren (1985) notes, at this time if pregnancy resulted from an alleged rape the accused could use this idea to exonerate himself, for the pregnancy proved the woman's pleasure and approval of his action. Unfortunately, similar ideas persist as came to light in the recent scandal over Todd Akin, a US Senate candidate in the 2012 elections, remarking that pregnancy rarely occurs in cases of 'legitimate rape' as 'the body has ways to try to shut that whole thing down' in such cases (Robinson 2012).

'nutriment' for the male semen and is incapable of generating anything on its own (*Sem.* II.4, 19; *UP* XIV.6). This allows Galen to avoid Aristotle's parthenogenetic problem: women are unable to generate on their own because of the cold state of their semen (*Sem.* II.4, 24–25, 31–32).[77]

For Galen, female infirmity explains not only what they contribute to generation but also their anatomy. In Galenic embryology, heat perfects the body, hot air (*pneuma*) being capable of pushing out and forming the parts (*Sem.* II.5, 28–36). The female's entire body is less formed and articulated than the male's, which is especially manifest in female generative parts, which are trapped in an unfinished state inside the abdomen. If their heat had been adequate, the female testes, uterus, vagina and other organs would have been pushed out of the body forming the superior male organs. In the context of this normative account, Galen treats female generative parts as failed organs, like moles' eyes that can never see (*Sem.* II.5, 52–60).[78]

[J]ust as the mole has imperfect eyes ... so too the woman is less perfect (*a-telestera*) than the man in respect of the generative parts ... this, though making the animal itself that was being formed less perfect (*a-telesteron*) than one that is complete in all respect, provided no small advantage for the race (*UP* XIV.6).[79]

In contrast, Aristotle compares the female organs, along with the male organs, not to non-functioning eyes but to eyes that can see.

There must be an organ both for the male and for the female; hence the male has his genitals and the female has the uterus. Nature gives each one its organ simultaneously with its ability, since it is better done thus. Hence each of these regions of the body gets formed simultaneously with the corresponding secretions and ability, just as the ability to see does not get perfected without eyes, nor the eye without the ability to see (*GA* 766a3–9).[80]

[77] Galen (1992) 174–9. See also Connell (2000) 417–8.

[78] de Lacy in Galen (1992) 184–93. Of course Galen cannot maintain this view for long and so he speaks of the uterus in particular as functional and unique in treatises that do not mention the embryology detailed above (e.g. *UP* XIV, *Ut. Diss.*).

[79] Helmreich in Galen (1907–09) vol. 2, 299, *ll.* 10–18.

[80] The fact that *HA* X contains a similar comparison between healthy uterus and healthy eye (*HA* 633b18–29) could be another reason to accept that text as authentically Aristotelian.

For Galen, the female animal is literally unfinished – *a-teles* (*Sem.* II.5, 60).[81] Aristotle, in contrast, very seldom uses this term to describe a female animal and, in most instances, considers an animal to be complete (*teleiôthen, teleion*) once it has become either sex (*GA* 715a21, 737b11–12).[82] We can see, then, that Aristotle's sexism differs from Galen's in so far as he regards the female organs as functional in their own right and not as failed male organs.[83] However, there is not very much to differentiate these thinkers' sexist attitudes.

Although Aristotle does not explicitly denigrate the female body anatomically, he does disparage it at many points due to its weakness. For instance, he writes: 'it is always very noticeable that the female is pale, and the blood-vessels are not prominent, and there is an obvious deficiency in physique as compared with males' (*GA* 727a23–26). Later he compares a woman's shape to that of an immature and a castrated man (*GA* 728a17–25). The idea that the female was less perfect than the male in the sense of not reaching the perfection of the kind (see Chapters 7 and 10) is present throughout the *GA*, most notably in *GA* IV where he describes the female state in terms of its weakness.

The process of completing female embryos is inferior to that of male ones ... for while still within the mother, the female takes longer to develop than the male does; though once birth has taken place everything reaches its completion sooner in females than in males – e.g. puberty, maturity, old age – because females are weaker and colder in their nature; and we should look upon the female state as being as it were a deformity, though one which occurs in the ordinary course of nature. While it is within the mother, then, it develops slowly on account of its coldness, since development is a sort of concoction, concoction is effected by heat, and if a thing is hotter its concoction is easy; when, however, it is free from the mother, on account of its weakness it quickly approaches its maturity and old age, since inferior things all reach their end more quickly (*GA* 775a12–21).

81 de Lacy in Galen (1992) 192–3.
82 Aristotle refers to female offspring as *atelês* in a tangential manner at *Pol.* 1335a17. It would seem likely that this is an extended sense of incomplete, like the extended sense of deformed used to describe the female at *GA* 737a26–7 and 775a15. The way in which Aristotle characterizes the female as a deformed or failed male will be explored in Chapter 8.
83 One of the reasons why scholars have not picked up on this difference may be due to the relative neglect of the later Books of the *GA*, where we find the above passage.

The abundance and the blood-like appearance of the menses indicate that females are colder than males and cannot fully heat up their seminal secretion (*GA* 728a18–22). The way in which Aristotle interprets the evidence is not in itself irrational: more concocted residues will be white (i.e. more full of *pneuma*) and scant (burned away), less concocted ones, bloodlike and abundant. However, it would be rather odd to conclude that sexism is not a factor here.[84] Instead, it is most likely that Aristotle's assumption that the female contribution must be inferior is firmly in place from the start. Galen's sexism is clearer than Aristotle's with respect to female anatomy only because he connects his anatomical theories more systematically to his physiology, providing a fuller explanation of how the infirmity of the female embryo results in a weakened adult body.[85] Although the idea that the female is a 'deformed', or less formed, male is apparent in his writings, Aristotle's theory along these lines is a lot less worked out than Galen's.[86] Both Aristotle and Galen assume that the female produces a weaker contribution to generation and that the female body suffers a related infirmity.[87]

A propensity to focus on the relative sexism of ancient reproductive theories is relatively modern. When the texts are considered closely in their historical context there is no evidence of any disagreement or opposing stances on such matters. There is not much to be gained, then, by contrasting Aristotle with Hippocrates and Galen in the way that has been sometimes undertaken. Instead, one must appreciate that most of the disputes in embryology focused around integrating the assumption of female inferiority into a persuasive account of how reproduction occurs, an account which fits best with the available empirical evidence. The main difference between Galen and Aristotle's theories is that for Galen males make a material contribution; the bodily semen goes to make up membranes, sinews, nerves,

[84] See Deslauriers (1998). [85] Connell (2000) 420–1.

[86] Galen is not such an inferior thinker that he does not recognize the importance of the functionality of female reproductive organs. Anatomically, the female is a reverse of the male anatomy, but is also functional. Galen *UP* XIV.6. See also Flemming (2000).

[87] Some ancient theorists thought that women were hotter than men, but they generally considered this to be a failure or disadvantage. See *GA* 765b20–8, *PA* 648a28–30, *Mul.* 1.1–2, L VIII 11–23.

vessels and bone.[88] Aristotle, in contrast, is adamant that the male contributes no material (*GA* 737a8–16), its role being like that of a carpenter. *GA* I and II, where Aristotle denies the presence of female semen, must be studied in the context of the views he opposes there. His dispute with pangenesis in *GA* I is primarily driven by his need to emphasize form and its importance in generation as part of his general drive to advocate formal and final causes. The opponents he cites tend to disregard the whole, describing detached pieces and parts as causes of the ordered animal that results. In this context, he is most keen to establish that form is better than, superior to and separable from matter.

[88] Blood, derived from the mother, composes the fleshy parts (*Sem.* I.5, 2; I.11, 1–2, Galen, 1992, 78–85, 102). Galen also believes that females contribute 'power' (*dunamis*) as well as matter to generation, and so also fashion the embryo, which he thinks helps to explain maternal resemblance (*Sem.* II.2, 15–16). Aristotle, I believe, allows that the female menstrual blood has powers which are capable of fashioning resembling parts, but he is unwilling to characterize this as the primary fashioning role. This will be discussed in more detail in Chapter 9.

4

MATTER

4.1 The female contribution as 'inert' or 'raw' matter

Famously, Aristotle declares very early on in the *GA* that the female contributes 'matter' to the generation of a new animal. This idea has struck feminist critics as not only dubious but also as deeply misogynistic. Such reactions, although understandable, should not cause the reader to abandon attempts to discover what Aristotle means by 'matter' in this context, and thereby gain insight into his metaphysics. For although he undoubtedly assigns a lesser role to females than to males in generation, the female matter is highly complex.[1] Attempts to get to grips with the female role as 'matter' run into a number of difficulties. The first problem is that there is no precise and definitive account of matter or of the matter/form relation in the Corpus; instead there are various different ones, in texts with widely divergent goals. For this reason, there have been countless ways of interpreting Aristotelian 'matter', even before the added complications of its operations in embryology. In a general context, the thought that the female matter must be 'inert' and thereby insignificant fits with a certain sort of craft analogy that Aristotle employs to describe matter and form. In the embryological context, the fact that the female matter is some kind of food for the offspring tempts readers to see it as 'raw' and so insignificant, like soil. In this section I will explain that matter for Aristotle is never inert; even in his crudest craft analogies, it is potent. Adding to this, the craft metaphor is of limited use for explaining materials in natural productions, particularly when it comes to embryology.

[1] The tendency to see Aristotle as more sexist than he is, as thinking that the female only provides 'inert matter' (Sissa 1992, 70); 'inert raw material' (Blundell 1995, 106); 'inert shapeless matter' (Grosz 1990, 155) or 'inert and cold matter' (Cavarero 1995, 70) can distort interpretations of his theory in the *GA*.

The female matter is also characterized as food; no food is inert, not even the rawest. However, the idea that the female contribution is 'raw' underlies many misrepresentations of Aristotle's theory, likening the female contribution to lifeless soil, which serves as nutriment for plants. I will show how wrong it is to think of the female contribution as 'raw', since it has been cooked by the female animal's soul.

One of the most important considerations when it comes to assessing Aristotle's theory of the female contribution as matter is the biological context in which this idea appears. The *Generation of Animals* is a mature work, coming at the end of a long research programme covering explanations of many complex happenings in living beings. One cannot hope simply to go back to a treatise such as the *Physics*, where the craft analogy is prominent, to clarify the nature of female matter. Aristotle's concern in his biological works is to explain some of the most difficult and seemingly mysterious natural events with reference to the explanatory tools and principles he feels are the most appropriate. But he must also adapt and describe these in much greater detail with reference to relevant bodily processes. The most crucial of these will be nutrition, a process whereby the parts of the body and their various capacities are created and maintained. Some critics are convinced that Aristotle adhered to the traditional view that the female acts like the earth or soil, and the male like a plant seed.[2] The idea that women are earth-like appears to justify their lesser role as the providers of food, which must be worked on in order to produce the living bodies of offspring.[3] This, I will argue, is a simplified view of the account of nutrition which underlies Aristotle's fuller and more compelling account of how the female animal contributes the materials necessary for living functions.

[2] E.g. Whitbeck (1973–4) 55.

[3] This type of image plays an imaginative role in various ancient texts before and during the time of Aristotle. Thus, when Aristotle or his biomedical contemporaries mention plants or earth in the context of animal generation, they are sometimes thought to be contributing to a more widespread literary degradation of female parenthood. See du Bois (1988).

4.2 The craft analogy and the metaphysics of animate existence

Aristotle introduces the matter/form distinction in his general work on natural science through the use of a craft analogy: matter is like timber, form like the shape of the chair (*Ph.* II). The Greek word Aristotle uses for matter (*hulê*) standardly meant 'timber' at the time, which would appear to support the view that all matter/form relations are like those that exist in crafts.[4] However, *hulê* swiftly becomes specialized terminology in Aristotle's philosophy and, thus, retains little if any of its original meaning. When it is assumed that Aristotle only equated the contribution of females in generation to timber, his theory can be immediately dismissed as implausible and misogynistic.[5] But to adopt this view is to disregard his more detailed account of matter.[6]

Aristotle's ideas about the metaphysics of living beings underlie his use of the term *hulê* (matter) in his biology. These must be viewed against a backdrop of the metaphysical questions which faced anyone at the time who attempted to explain the phenomena of life. A separation is commonly made between more imaginative accounts of life, such as those found in myth, literature and folklore, and the rational or 'scientific' ones that emerged a few hundred years before Aristotle's birth in authors who are generically referred to as the Pre-Socratic philosophers.[7] Explanations of the generation of new life found in the former variety usually involve references to the gods and their abilities to create living entities out of lifeless materials.[8] In conscious opposition to this

4 LSJ 1847, Grosz (1994) 5 and Allen (1985) 92–3.

5 'The craftsman analogy vividly combined the association of maleness with activity and form and the association of femaleness with passivity and matter, and in so doing displaced any notion that the female is the procreator' Horowitz (1976) 196.

6 I disagree with Dean-Jones (1994) 184 when she says that 'Aristotle generally used artefacts to illustrate this theory, but thought it applied just as well, if not better, to natural objects.'

7 I maintain the existence of such a distinction for simplicity's sake although it has been greatly problematized, especially in the work of Geoffrey Lloyd (e.g. 1983). I would not wish to deny significant overlaps.

8 Pandora and Prometheus were created by the gods from water and earth (Guthrie 1957, 30). Early cosmologies often focused on the order in which entities and gods were 'born' from one another, which can be taken to be quasi-rationalistic. See Kirk, Raven and Schofield in The Pre-Socratics (1983) ch.1.

viewpoint, the Pre-Socratics avoid reference to divine intervention. Thus we find that these thinkers bring in more abstract concepts ('elements', 'qualities', 'opposites') to elucidate natural processes, referring less and less to divine or human goals. The birth of animals, for many of these thinkers, was caused by the moist and hot nature of seeds.[9]

These new structures of explanation increasingly emphasized the material components of world construction. The 'elements' (*stoicheia*, which literally mean the letters of the alphabet) were now taken to be the primary existent objects in the world, out of which other more complex objects came to be and in terms of which these could be explained. Aristotle believed that elements for his predecessors and contemporaries were 'matter'. The precise nature of the elements in Pre-Socratic thought is hazy, especially since our main source for their metaphysics is Aristotle himself, who hoped to oppose his own new understanding to that which went before. However, it seems plausible that these elements possessed powers, often expressed in terms of opposing forces (or the 'opposites'), which had to be reconciled in order to produce cosmic order.

In addition to Pre-Socratic philosophers, Hippocratic medical writers were also an important part of the background to Aristotle's work in this area. Some Hippocratic authors took metaphysics to be a crucial preamble to any discussion of pathology (e.g. *Vict.* 1). Similar to the Pre-Socratic emphasis on a harmonious combination of elements and powers, many Hippocratic writers also held that life consists in a balance within animals. Whether the elements were fire and water (*Vict.*) or the bodily humours (*Genit.*), these were liable to assert themselves through the dynamic qualities or 'powers' of hot, cold, wet and dry.[10] To the modern reader 'matter' immediately suggests geometrical dimensions ('space'). This concept implies that matter is 'inert' because,

9 Aristotle on Thales *Metaph.* 983b20–28. Anaximander DK 2A11§6; Pre-Socratics (2010) 52–3. For a detailed account of Pre-Socratic theories of the generation of animals see Lesky (1950).

10 Aristotle was to share the view that life depended, in some sense, on the correct mixture of elemental powers. Of course, there is a significant difference between life consisting in such a balance as opposed to requiring it.

according to Newtonian physics, material objects are separable from their movements. In ancient science, in contrast, even such seemingly inert entities as lumps of earth or buckets of water contain some sort of 'inherent power of action, motion or resistance' which the modern term precludes. Already we can see that, even in the case of crafts, matter for Aristotle is not 'inert'.

Although Aristotle was, in important ways, a product of his cultural and intellectual milieu, his metaphysics in the area of animate existence differed from that of his immediate predecessors and contemporaries. By his own account, he was dissatisfied with many of the explanations other thinkers offered. This was due to their disregard for structure and organization and their propensity to be *too* heavily dependent on inanimate processes to explain animate existence. Some of Aristotle's most coherent and sustained complaints along these lines occur in his *Physics*, where he also introduces reference to the four causes as a superior and more comprehensive method of analysis than anything that came before. The work puts forward that formal and final causes ought to be prominent in any explanation of natural phenomena. In the strictly biological context, both the formal and final causes are what it is to be that sort of animal or plant (*Ph.* 198a26).[11] Previous thinkers, Aristotle argues, concentrated on the material cause, sometimes including aspects of formal and efficient causes, but always managing to ignore final causes. It is in explaining why he disfavours such views that Aristotle finds the craft analogy most useful. The analogy features frequently in the work of Pre-Socratic and Hippocratic thinkers (e.g. *Vict.* 1.13–24; L VI 488–96).[12] In comparing craft production with natural processes in the body, Aristotle complains, these authors failed to realize that crafts require craftsmen – that is, agents that arrange materials in order to produce serviceable results. If nature really does make things in a similar way to crafts, then natural objects will also require organizing principles, directed towards goals (*Ph.* 199a18–19).

Aristotle finds the craft analogy a simple way to illustrate that productive processes imply final causes, the very causes his

[11] See also *Ph.* 199a33; *de An.* 412a20.

[12] Plato, in his use of the craftsman (*demiourgos*) in the *Timaeus*, does recognize final causes but his teleology is differently focused. See Johansen (2004).

opponents prefer to ignore. At times he characterizes his break with the past as a new emphasis on form rather than his opponents' concern with matter.

In reading the ancients one might well suppose that the natural scientists only concern was with the material; for Empedocles and Democritus have remarkably little to say about form and essence (*Ph.* 194a19–22).[13]

In the biological context, he insists that the form, which consists not only in structural specifications of the animal or plant but also the functions and actions of its life, is more fundamental than the materials out of which it is made (*Ph.* 193b7, *PA* 640b28–9). By focusing on the material cause, he argues, instead of robust and satisfying explanations, we end up with uninformative lists of ingredients.

Part of Aristotle's point in the *Physics* is to suggest that the craft analogy has its limitations and can never fully elucidate nature. When other authors try to account for complex processes by merely referring to a craft process (e.g. foetal development is exactly like winnowing) they miss the important point that craft and nature *differ*. Objects with a 'nature' possess a source of change and rest within themselves, whereas the products of craft depend on external craftsmen (*Ph.* 192b8–19). But also, crucially, the material cause in each case differs. For crafts, the formal cause is the shape and function of the artefact, the efficient cause, the craftsman, the final cause, the completion, and the material cause, the ingredients. All four causes are clearly delineated, but the material cause is particularly distinct in the analogy; a craftsman equipped with an idea and capacity to put tools into action to bring it about, comes upon material which is separate and not directly related to the craft.[14] In nature, this is not the case: natural matter is present in the natural object. 'The crafts make their material . . . whereas in the products of nature the matter is there all along' (*Ph.* 194a34–194b8). Furthermore, the external

[13] Cf. *Ph.* 194a23–34; *Metaph.* 984a18–19.

[14] 'For what the house replaces by being made is the unordered relation of the materials to each other; and what passes away in the making of the statue, or any other shapely work, is the unshapeliness of the materials (ἥ τε γὰρ οἰκία γίγνεται ἐκ τοῦ μὴ συγκεῖσθαι ἀλλὰ διῃρῆσθαι ταδὶ ὡδί, καὶ ὁ ἀνδριὰς καὶ τῶν ἐσχηματισμένων τι ἐξ ἀσχημοσύνης)' (*Ph.* 188b18–20).

materials that crafts use require manipulation and adaptation (e.g. wood must be made into planks and earth into bricks), whereas the materials used for the production of natural objects have been specially prepared by nature for their purpose before the agent encounters them.

In the context of disagreements with other thinkers, Aristotle heavily privileges the role of form in natural explanation and, for this reason, some imagine that matter is insignificant in his theory of generation. The idea that the female contribution is 'raw material', exactly like the material used in crafts, appears relatively early on in commentary on Aristotle's embryology.[15] However, although Aristotle attempts to distance himself from those he characterizes as matter enthusiasts (or materialists), this does not mean that he thinks of *hulê* as inert.

4.3 The agrarian metaphor: the female principle as external food

For Aristotle, and many of his contemporaries, reproduction was crucially tied to digestion and nutrition: when an animal eats, it takes elements from its environment and transforms them into parts of a living body, precisely what it will have to do for another in order to reproduce.[16] This background explains why the matter provided by the female has been thought of as 'raw' by some. For the ancient reader the rawest of food would have been earth or soil, the type of nourishment taken in by the simplest of living beings – plants. If Aristotle thought that the female contribution was like the role of earth in nutrition, then this might suggest that he regarded it as indeterminate.[17] So how, if at all, does Aristotle's theory mimic the widely held mythical

[15] In his *Microcosmographia* of 1618, Helkiah Crooke states that '[the female] matter is made of the crude and raw remnants of her own aliment'. Quoted in Aughterson (1995) 54. Some feminist critics continue to emphasize the apparent separation of matter that the male semen works on in generation, describing it as 'raw'. See Blundell (1995) 106, Clack (1999) 30, Scheman (1993) 109 and 135, Shildrick (1997) 31, Sissa (1992) 70 and Tuana (1989) 150.

[16] Lonie in Hippocrates (1981) 67 and Coles (1995).

[17] Aristotle's was coined the 'flower pot' theory by Whitbeck (1973–4) and this is taken on board by many modern feminists, for instance, Sherwin (1991), 340 n.9.

association of mother and earth? He himself mentions this idea at the beginning of the *GA*, attributing it to others.

In cosmology they speak of the nature of the Earth as something female and call it 'mother', while they give to the heaven and the sun and anything else of that kind the title of 'generator' or 'father' (*GA* 716a15–17).[18]

Any endorsement of this view is unlikely once we consider Aristotle's thoughts on the role of the sun and earth in generation. For Aristotle, animal heat is different in kind to the heat produced by the sun (*GA* 737a7–8). The latter would be unable to arrange and limit its powers in order to produce viable offspring.[19] As for earth, Aristotle argues that inanimate elements (of which earth is one type) cannot mix together to start life without seeds that have been created by living beings (e.g. *Ph.* 199b8–9). So his views on the nature of the sun and soil do not map in any straightforward manner on to a characterization of them as male and female.[20]

Although Aristotle's views are not lifted from mythology, he does liken the nutrition that females provide for offspring *in utero* to the way that plants 'feed' off the earth, in a similar manner to other ancient biomedical writers.[21] The *GA* does not treat this as

[18] For the mythological *topos* of Gaia and Ouranos see Hesiod *Theog.* 117, 132–3, 154–5. Cf. Kirk, Raven and Schofield in The Pre-Socratics (1983) ch. 1 and Cornford (1912) 66–8. The metaphor is central to the traditional characterization of Zeus, the sky god, who is called *pater*, 'father'; Graf (1996). Meanwhile, the Greek term for mother, *hê mêtêr*, was often used of the earth. LJS 1129–30; Hippocrates *Nat. Puer.* 22–7, L VII 514–29.

[19] Although the sun can bring about generation of a certain kind, it is not specific enough to bring about the generation of entities possessing proper forms. See Chapter 7.4 on spontaneously generated animals.

[20] Many assign such a view to Aristotle. See, for example, Allen (1985) 95: 'Aristotle developed an association of the female with the earth and the male with fire and sun', and du Bois (1988) 126: 'The great scientific work of Aristotle . . . takes up metaphorical systems'. See also Elshtain (1981) 44. In contrast, Lloyd (1966) 85 argues that Aristotle's work on generation 'was far removed from the myth that derives all things from Sky and Earth'.

[21] Before the discovery of photosynthesis, most assumed that plants fed on earth or soil. For Aristotle, this is part of the reason that plants are upside-down, having their heads (and mouths) in the ground where they are best able to consume their proper nourishment (*PA* 650a21–22, 683b19–23). This should not seem that strange to us since the nourishment consumed by plants was thought to be liquid and to contain certain ingredients essential to the organism (*Nat. Puer.* 22.1, 27.1, L VII 528–9, 544–9). Similarly, modern horticultural science regards the type of soil and the minerals contained in the fluid it draws in to be crucial to proper plant development.

the only part the female plays in generation.[22] Instead, there are two stages or parts to the female's contribution:[23] (1) the contribution of seminal residue to the initial setting of the embryo (for Aristotle the menstrual blood is 'set' by the male semen at conception); (2) continual feeding (of blood or its equivalent) through the umbilical cord during gestation. Like the seed of a plant, the embryo is already a mixture of the male and female generative components (*GA* 724b14–18, 731a1–4). At this stage, it then becomes like a plant in the womb.

Once the foetus has been 'set' it behaves like seeds sown in the ground. The first principle is present in the seeds themselves too, and as soon as this, which at first was present potentially, has become distinct, a shoot and a root are thrown out from it, the root being the channel by which nourishment is obtained (*GA* 739b34–740a1).

The foetus's growth is supplied through the umbilicus in the same way that the plant's growth is supplied through its roots (*GA* 740b9–11, Cf. 745b24–6, 753b28).[24]

Aristotle likens foetuses to plants at the gestational stage (2), which does not show that the female serves exclusively as a container or 'feedbag' for him.[25] Only the female's subsequent and additional role of feeding the gestating offspring is sometimes described in this manner. Furthermore, the mother's role as nutritional need not be viewed negatively, if nourishment is thought of as potent.

Many of Aristotle's contemporaries held that earth or soil possessed dynamic qualities with the ability to affect plants growing in it. If powers are considered to be present in soil, then a plant sprouting in the wrong soil will be sick, die or be unable to

[22] See also *Genit.*

[23] I will be focusing here exclusively on generation in live-bearers, as this is the main emphasis in *GA* I and II. I will discuss Aristotle's explanations of egg-bearers and others in Chapters 7 and 10.

[24] See also *GA* 774b25–7: '[The sow] brings [many offspring] to completion because of its fine physical condition: it is like a rich soil which can provide plants with sufficient and indeed abundant nourishment'. This passage will be discussed in more detail in what follows.

[25] Keuls (1985) 144. Aristotle's female has also been characterized as exclusively a 'vessel' (Wilshire 1989, 93), 'container' (Sissa 1990, 162, Shildrick 1997, 162, du Bois 1988, 126, Bordo 1993, 90) or 'incubator' (Scheman 1993, 109; Tuana 1994a, 192), all of which are inaccurate.

germinate (*Morb.* 4. 33. 17–21, L VII 544–5). This picture of nutrition is then generalized to animals that are also affected by the 'powers' in food, a line of thought no doubt significant to medical practitioners, whose prescriptions often included dietary recommendations.[26] For some ancient theorists, the comparison of mother and earth is used to explain why embryos are healthy or become ill.

Nutrition (*hê trophê*) and growth depend on what arrives from the mother into her womb; and the health or disease of the child is relative to that of the mother. In just the same way, plants growing in the earth receive their nutriment from the earth and the condition of the plant depends on the condition of the earth in which it grows (*Nat. Puer.* 27.1; Hippocrates 1981, 18).

In certain medical texts, materials (or 'the elements') were considered to be powerful enough to determine the health of the plants growing in them and maybe even to determine what type of plants emerged from them. The comparison of the relation of plant and earth to that of embryo and mother in pregnancy might operate as in Table 4.1.

Table 4.1 *Ancient medical model likening foetus to plant*

seed/plant	*feeds on*	earth
passive		active/dynamic · determines the health of plants that grow in it. · determines what type of plants grow.
compared to:		
foetus	*feeds on*	**mother**
passive		active/dynamic · determines health of the child · determines other aspects of the child

[26] Indeed, throughout much of subsequent history, no clear line has been drawn between food and medicine.

4.3 The agrarian metaphor: the female principle as external food

Some commentators think that the medical model views maternity in a positive light, in so far as the female parent has an active or dynamic role in nutrifying offspring.

> By adopting an image like that of Aeschylus, seeing woman as the soil in which man plants his seed, one can still attribute to woman a richness and fertility. Such an image contains glimpses of the earlier view of the earth, Gaia, and her soil bursting forth in plenty. Although now the soil of woman must be planted to be fertile, still she continues to actively contribute to the growth of the seed planted within her. . . . woman remains a living, organic force in the production of life (Tuana 1994a, 193).

One might well imagine that Aristotle's view suffers by comparison. Rejecting the idea that a plant's (or embryo's) food is influential, he posits instead that, 'form' has the most influence over a living beings development. Perhaps, then for Aristotle, 'far from the fertile, flowering, burgeoning earth, the female principle is seen as inert wood. The matter of woman is not the living, nourishing soil, but the severed lifeless limb of a tree' (*ibid.*, 194). On this reading it would appear that when Aristotle compares plant to animal development he simply reverses the biomedical schema as in Table 4.2.

It is true that Aristotle does give at least part of the dynamic role in generation to the embryo itself, in exercising its formal

Table 4.2 *A postulated Aristotelian model likening foetus to plant*

seed/plant	*feeding on*	earth
active/form		passive/matter
dynamic		inert
determines type of plant		resists formation, causes disease and deformity
The supposed comparison to animal generation would be as follows:		
foetus	*feeding on*	**mother**
F1: active		M1: passive
F2: form		M2: matter
F3: dynamic		*M3: inert*
F4: determines type/form		*M4: resists type, causes disease and deformity*

capacities. However, the contrasting schemas misrepresent Aristotle and his opponents in several ways. The medical doctrine actually regards the female animal's nutritional role in a negative manner – rather than the mother determining positive aspects of growth (as perhaps good soil does in the horticultural case) it was most often evoked to explain bad generative outcomes (e.g. *Nat. Puer.* 27.1). Although Aristotle will attempt to clarify both good and bad influences, it is mostly the good ones that interest him in his account of embryonic nutrition.[27] More importantly, he does not compare plant and animal growth in the simplistic manner set out above. Although he resists the idea that soil powerfully influences plant development, he does not similarly downgrade the role of the nourishment provided by the mother in animal embryology. This is because animals are blooded, whereas plants do not have internal nutriment which means that only animals transform their food into living materials. The next section will set out how Aristotle characterizes nutrition and how this can explain his ideas about what the female contributes both (1) initially and (2) during gestation. While the female contribution is M1 and M2, it is not at either stage *M3* or *M4* (so in fact the italicized portion of the above schematization is inaccurate). In the case of the initial contribution, this cannot be characterized as M1, *M3* or *M4*, but is F3 and F4 along with the male contribution.

4.4 Nutrition: bodily mixture versus bodily structure and function

As previously mentioned, ancient science tended to associate nutrition and reproduction.[28] Generally, it was thought that whatever process brings about the initial construction of the animal will also explain how they continue this existence through subsequent nutrition. Often this process is described as one of mixing or blending. For the author of the Hippocratic treatise *On the*

[27] As with levels of fertility, the diet of a pregnant animal may produce positive and negative effects on embryonic development. Aristotle recommends exercise and a moderate diet for pregnant women (*Pol.* 1335b11–19).

[28] For Aristotle, the nutritive and the generative faculty of soul are one and the same (*de An.* 415a23–27).

Nature of the Child, for instance, growth of the foetal body occurs due to the addition of like parts to an initial mixture (of male and female semen).[29]

As the flesh grows it is formed into distinct members by breath. Each thing in it goes to its similar –the dense to the dense, the rare to the rare, and the fluid to the fluid. Each settles in its appropriate place, corresponding to the part from which it came and to which it is akin. . . . Suppose you were to tie a bladder on to the end of a pipe, and insert through the pipe earth, sand, and fine filings of lead. Now pour in water, and blow through the pipe. First of all the ingredients will be thoroughly mixed up with the water, but after you have blown for a time, the lead will move towards the lead, the sand towards the sand, and the earth towards the earth ... Now the semen, and the flesh, is separated by precisely the same process, with like joining to like (*Nat. Puer.* 18.1; Hippocrates 1981, 9).

In this account, the construction of the foetal body occurs due to the rearrangement of (very small) juxtaposed parts.

One of Aristotle's predecessors, Anaxagoras, also used a juxtaposition model of nutrition and generation. He is thought to have maintained that there are invisible parts of all things existing in everything. Nutrition occurs when 'the flesh already present is joined by flesh that comes from the nourishment' (*GA* 723a11–12). On this picture, when an animal or plant is nourished, parts of the world are merely rearranged, altering their juxtaposition.[30] The theory of nutrition found in the Hippocratic *On Diseases* IV similarly posits that the substances present in external food are separated off and attached to the body, because that which is similar goes towards what is similar to it.

When someone has eaten or drunk, the body draws to itself, from the stomach, the humour which has been mentioned, and the sources draw off through the veins from the stomach, the same drawing from the same, and these are distributed to the body. Therefore, in plants the same humour draws to itself the same humour that is present in the earth (*Morb.* 4.33.17–21, L VII 544–5).

[29] The Greek term commonly employed for sexual intercourse in the context of generation is *mignumi* which means 'to mix with' and is properly used of the combination of two liquids (LJS 1092). Aristotle almost always uses the word in the latter sense, mostly employing *sunduazomai* and *ocheuomai* for sexual intercourse. See also *Genit.* 5, which describes the initial construction of the foetus in terms of a mixing of male and female semens.

[30] DK 59A46; Pre-Socratics (2010) 288–9.

The juxtaposition (sometimes termed 'mechanical mixture') model of nutrition had a strong following before Aristotle,[31] and it is one that he is particularly keen to refute, advocating instead a different model of the mixing involved in nutrition: blending or irreversible change.

Although the cruder 'mechanical' model of construction, growth and nutrition was prominent in early science, a more sophisticated idea of mixture involving the maintenance of a proper blend or due proportion of elements was also becoming popular. Aristotle was not the first to offer this alternative account. Empedocles, a near predecessor and frequently named authority on reproductive matters in the *GA*, made advances before him.[32] For Empedocles, birth and death are inappropriate terms because they are not instances of coming to be and annihilation but of mixing and separating (DK 31B8, 9, 11).[33] He explains that the differences we discern between types of animal are due to the mixture of elements in their bodies.[34] Empedocles described the mixing together of elements using terms that indicate a fitting together or fastening of separate parts.[35] But he was not simply indicating a mechanism; the term also conveys the sense of appropriateness. There are clear signs in his writings that 'mixture' had a privileged position as proper blend rather than mere juxtaposition.

In addition to Empedocles, some Hippocratic works lay great emphasis on a 'proper blend' and insist that it is this normative concept that determines health (*Aër.*).[36] In line with these

[31] Plato's *Timaeus* presents a juxtaposition model to account for nutrition when the character Timaeus notes that blood consists of cut up bits of food (80D-E).

[32] Aristotle does not appear to acknowledge this, but instead criticizes Empedocles for holding to the older and simpler juxtaposition model (*GC* 334a26–30, Cf. Empedocles DK 31A43; Pre-Socratics [2010] 418–19). See especially Sedley (2007) for the view that Empedocles was a 'creationist'.

[33] Pre-Socratics (2010) 346–9. *GC* 314b7–8: 'Empedocles also is speaking to this effect, when he says that "there is no origin of anything, but only a mixture and separation of things which have been mixed (φύσις οὐδενός ἐστιν, ἀλλὰ μόνον μίξις τε διάλλαξίς τε μιγέντων)".' *Metaph.* 1015a1–4 repeats this quote. See Furth (1988) 89–93.

[34] 'The various species of animals were distinguished by the quality of the mixture (*kraseis*) in them' (DK 31A72 and The Pre-Socratics [2010] 382–5).

[35] *Gumphein* – to fasten with bolts or nails (DK 31B33); *Kollêin* – to glue, cement together (DK 31B34 and B96). Pre-Socratics (2010) 378–81.

[36] The author of *On Regimen* I, for instance, has it that humans are a mixture of hot and wet elements that combine and balance with food taken in. In this account 'balance' is

accounts, Aristotle holds that the 'bodily blend' (*tên tou sômatos krasin*) of an animal plays an important part in any account of health, the function of the heart and brain, and life-span.[37] Proper blend is also connected to habitat, so that animals living in water, for example, have a particular bodily blend (*krasis*) that corresponds to this environment (*PA* 669a11, *HA* 589b22, 590a15). Since the blend is the result of the nutritive process, this might mean that food affects the animal, as the Hippocratic model suggested. Although Aristotle acknowledged that food can affect body size (*HA* 606a11–25), health and fertility (*GA* 767a28–32), he did not think that food retains its own powers after animals digest it. A fuller account of nutrition in his philosophy will make this clearer.

First of all, Aristotle clearly distinguishes between mixture as juxtaposition of particles, on the one hand, and mixture as transformation on the other.[38] In the former case, one can easily extract the original ingredients through separation; in the latter a sort of change has occurred that disallows reversal, a change we would refer to as 'chemical' (*GC* I.10). For Aristotle 'chemical change' makes better sense of the process of nutrition and this supports the association of proper blend with soul and form. At first blush, it may seem that the evident expansion of the body which occurs when the body grows (and is nourished) supports a juxtaposition model. After all, it looks as if eating results in the attachment of external matter to an already existing structure, by adding similar parts. However, nutrition is not only about growth, but also more importantly ensures the maintenance of life, even when a body diminishes in size (*GC* 322a24). The idea that 'the body is

crucial. Many other Pre-Socratic natural philosophers and doctors held similar views. See Tracy (1969).

37 He describes health as the correct proportions in the mixture at *Ph.* 246b3–6, *Top.* 139b21, 145b8. Coldness of the brain is needed in order to counteract the heat of the heart and thereby establish a good blend – *eukraton* (*PA* 652b27). Longevity, meanwhile, is assigned to those animals that have a similar 'blend' to the air surrounding them (*GA* 777b7–8). See Tracy (1969) for a comprehensive survey of the concepts of proper blend (*eukrasis, krasis*) and balance (*summetros, metriotes, isonomia*) in Aristotle's predecessors. Tracy argues that these ideas deeply influenced Aristotle's thought in biology and in ethics.

38 There are actually three possible types of mixture: juxtaposition, solutions and compounds. See McKirahan (1994) 262–4, who argues that Empedocles was aware of all three.

nourished just so long as it is kept alive' (*ibid.*) places emphasis on the dynamic action of *living* rather than on physical bulk. As for those thinkers who already focus on 'proper blend' rather than juxtaposition of tiny parts in their accounts of animal nutrition, Aristotle encourages them to consider the normative force of the concept 'proper blend'. What makes the blend *proper*, or *good*, are not the materials themselves but the fact that this arrangement of materials fits to a pre-ordained purposive structure. Nutrition 'provides for the being of the entire animal and its parts' (*GA* 744b35), meaning that the process of nutrition keeps the animal alive as a functioning structured whole – which is why it is good.

Aristotle's explanations most often proceed from the top down – from form and function to the matter that will facilitate these goals. The form or essence of each animal is what it is and is the key to any explanation of its living functions. Most matter with which a living being interacts is significant only in terms of why it is useful to its goals. Thus, in *Ph.* II.9 he explains that matter is 'hypothetically necessary' – materials are the ingredients needed if specific organic structures are to come into existence and remain alive.[39] In contrast to this outlook, the Hippocratic author of *On the Nature of the Child*, assuming the existence of a stable structure upon which to build, failed to explain either its original organization or why it remains appropriately structured throughout life. For Aristotle, in contrast, a body maintains its organization when it grows through the capacities of the soul, which ensure that the shape rather than the matter expands (*GC* I.5).[40] Aristotle's concentration on form goes beyond shape and structure to cover the functional and dynamic aspects of living beings. For Aristotle, the soul maintains 'the being' (*to ti einai, ousia, tode ti*) of a living substance. Even in the case of evident outward growth in space, mere addition of matter to a fixed shape does not adequately

[39] This picture of an essentialist and teleological Aristotle is too simplistic, of course. However, it is a good way into understanding his views on nutrition. For problems with Aristotle's essentialism see Charles (2000) and Chapter 10.

[40] 'For nature is an origin more than matter. Even Empedocles occasionally stumbles upon this lead by the truth itself, and is forced to say that the substantial being and the nature is the account, e.g. when he says what bone is. He does not say that it is some one of the elements, or two or three, or all of them, but rather that it is an account of their mixture' (*PA* 642a16–22).

describe the process. Food is not potentially the form of 'flesh' in terms of the proportion of the elements in it that will be stuck onto it, but is potentially the *entelecheia*, the animate activities, of flesh.[41] When the soul transforms materials before growth can occur, it does so by making them into *living* parts.

> [T]he principle of growth present in that which grows … lays hold of the added food which is potentially flesh, and turns it into actual (*entelecheia*) flesh (*GC* 322a10–13).

What is actual (*entelecheia*) about the flesh is not its inert structure but its activity as a functioning part of a complete animal. The animal's nutritive soul changes food into the living body.[42]

For Aristotle, nutrition and generation result from the activity of the soul or the nutritive portion of soul.[43] So, in contrast to the Hippocratic stance, 'food' does not have a power of its own; it is only food in relation to being eaten by an animal and worked on by the nutritive soul (*to threptikon*).

> Food has a power which is other than the power to increase the bulk of what is fed by it; for it is due to the ensouled thing having bulk that food can produce growth, but is it due to it being a particular thing and a substance that it is food (*de An.* 416b12–14).[44]

The particular substance referred to here is the plant or animal. Thus, the animal makes food into a living part of itself. External food, the inanimate stuff in the environment, does not feed the animal. Instead, the animal feeds itself by changing external food into the type of matter it needs for its life, full of dynamic potentials (*PA* 650a5–8). For Aristotle, nutrition is determined not by what you eat but what you do with your food.

[41] See Philoponus' commentary on *GC* (1897) 115 at 321b35 for an astute clarification of the difference: 'If the food is added to flesh by changing to flesh, that would be generation of flesh and addition, and not growth (εἰ μὲν μεταβάλλουσα ἡ τροφὴ εἰς σάρκα οὕτω προσετίθετο τῇ σαρκί, γένεσις ἂν ἦν σαρκὸς καὶ πρόσθεσις, οὐκ αὔξησις).'

[42] What it is to be flesh is to be living flesh. *PA* 640b34–641a6; *Metaph.* 1040b5–10; *GA* 722b4–5, 33–723a1.

[43] The nutritive soul (*to threptikon*, introduced at *de An.* 413b7) is the most widely distributed soul faculty. It is that 'in virtue of which we are said to have life' (*de An.* 415a24–5). Animals also have sentient soul (*to aisthetikon*) while (most) humans (along with gods) possess intellectual soul (*nous*).

[44] ἔστι δ' ἕτερον τροφῇ καὶ αὐξητικῷ εἶναι· ᾗ μὲν γὰρ ποσόν τι τὸ ἔμψυχον, αὐξητικόν, ᾗ δὲ τόδε τι καὶ οὐσία, τροφή.

Within the zoological corpus, including in the *GA*, Aristotle displays an increasing interest in the materials which are involved in living processes. So, for instance, the *HA* discussions of animal diversity often begin by detailing matter or materials. It seems possible that, in this context, Aristotle is prepared to think of materials as explanatory of animal behaviour, for example in what they desire and seek. He writes, 'because in their constitution when they are being formed, they partake to some extent of matter such as provides their food, each animal likes what accords with its nature' (*HA* 590a9–11).[45] There is nothing particularly problematic for his top-down focus in this, though. Although certain animals are described with reference to a particular material, for example those that are 'watery' tend to partake of watery food,[46] it is not water itself that makes them what they are. In the very beginning it is the parents' souls that determine their original constitution and not the external environment.

It is also consistent for Aristotle to acknowledge that external food can affect animal bodies in ways that are not essential to it. For instance, this food often retains some of its elemental powers when initially entering the body.[47] Earth, which is the prime component of food, being dry and cold (*GC* I.5) can counteract the natural heat and fluidity of the body. Normally, the animals change the dry and cold into the wet and hot, making a substantial change to the external food.[48] What had no proper form takes on the 'form' of a part of the organic body, such as flesh.

The whole, however, has become greater by the accession of something which is called food, the 'contrary' of flesh, and by the change of this food into the same form as that of the flesh (*GC* 321b35–322a2).[49]

45 Animals seek out different types of food because 'their food differs chiefly according to the matter out of which they are constituted (αἱ δὲ τροφαὶ διαφέρουσι μάλιστα κατὰ τὴν ὕλην ἐξ οἵας συνεστήκασιν)', *HA* 589a6–7.

46 See *Resp.* and *HA* VIII.

47 Although most consumables were once parts of living beings, it does not make sense to view them as animate. Once either dead or separated from the entire living being, food stuffs, such as fruits and bits of meat, are parts of the original animal or plant only hynomymously (*de An.* 412b20–22, *Mete.* 389b28–390a2, 10–14).

48 See also Boylan (1982) 99.

49 Μεῖζον μέντοι τὸ ὅλον γέγονε προσελθόντος μέν τινος, ὃ καλεῖται τροφὴ καὶ ἐναντίον, μεταβάλλοντος δὲ εἰς τὸ αὐτὸ εἶδος.

However, sometimes the body is cooled by the incoming food and occasionally the soul's functions will be impaired by the subsequent change of temperature (*Somn.* 457b17–19.). But it seems that most of the time food adversely affects the body only to a very limited extent, as newly brought in and not yet converted into that living body. These slight changes only affect the body as an alteration (*alloiôseôs*), leaving its substance intact.

In other works Aristotle suggests that food occasionally affects body size. Again, this is not significant. Aristotle distinguishes between body size (*megethos)* and form making clear that form remains the same when an animal changes size due to growth (*GC* I.5).[50] However, he also thinks that body size is part of what it is to be a certain animal type, part of its essence (*de An.* 416a15–18, *GA* 745a5–9).[51] Certainly, a significant difference in body size would indicate that form was not properly present or operative, for dwarfishness and giantism are clear deformities. Thus, when Aristotle notes that nutrition can alter body size he must mean to indicate that these variations are within the bounds of normal differences within the kind. Although food eaten gives some indication of why inessential variations occur, the mere accident of bodily mixture could never make one type of animal change into another, as Empedocles' theory implies (*GC* 314b7–8). In cases of increased size, the food produces results that cannot tell us anything crucial about the animal's real nature.

In general, Aristotle's explanations of nutrition require reference to the overall goals and dynamic functions of living beings. External food does not have the power to determine that animal's form and structure. Let us return, then, to consider the passages in which Aristotle compares the female's feeding role to that of soil in plant nutrition and set out what view we can fairly ascribe to him on the basis of these. One such passage appears in *GA* IV.

Most of the time [the female pig] brings forth many offspring, and brings their nourishment to completion, because of the rich condition (*eubosian*) of its body.

50 *Megethos* is in the category of quantity, it does not affect the form or substance (*ousia*) of the animal (*Cat.* IV).

51 This idea will be discussed in more detail in relation to multiparity in Chapter 10. Aristotle also indicates that the size of parts of animals is fixed by nature. See especially *PA* 685b12, 683a18–19, *IA* 708a9–20 and *Pol.* 1309b18–35.

For it is like the rich earth which is sufficient and abundant nourishment for plants (774b25–27).[52]

The word used of the body of the female pig is *eubosian* which means 'good pasture'. The analogy compares two relations: one of the foetus to the womb in maternal nutrition and the other of plant to earth. Thus, Aristotle is not comparing the female contribution to the role of earth in *animal* nutrition, which would be extremely minimal, but to earth in *plant* nutrition, and when it comes to plants, earth is influential, to some extent at least, on their development. Plants, being simple, do not have many organs and are composed mainly of earth (*GA* 761b13–14, *Resp.* 477a28), which is why earth serves as their food.[53] Earth is raw to an animal but not to a plant, because the sun has partially prepared or digested the warm soil for it. Food for an animal embryo must be much more specialized in order to facilitate the types of function plants do not enjoy, such as locomotion and perception. The female animal purifies the food which she passes on to offspring, employing her nutritive soul, whereas there is no soul involved in the preparation of soil for a plant. Thus the actions of the female animal's soul, which are not at the mercy of the powers of any external food (such as earth), are crucial to the proper development of offspring. The sow's soul is strong and so is able to nourish its offspring unlike the passive soil. Passive soil can, however, if it is rich, nourish a plant.

In several other places Aristotle connects the female role with plant nutrition. The first occurs in the context of putting forward the ideal state in his *Politics*. Each citizen must be a good person, and this therefore requires eugenics. Conception itself will be regulated in order to ensure that children have the best minds and bodies and will only occur when husband and wife are healthy and in the prime of their mental faculties (*Pol.* 1335a12–30,

52 πολυτοκεῖ τὰ πλεῖστα, ἐκτρέφει δ' εἰς τέλος διὰ τὴν τοῦ σώματος εὐβοσίαν· ἔχει γὰρ ὡς πίειρα γῆ φυτοῖς ἱκανὴν καὶ δαψιλῆ τροφήν.

53 *GC* 335a12–14 and Sprague (1991) 222. Plants do not have stomachs and intestines and so do not digest their own food (*PA* 650a21–5). Their relative coldness also explains why the nature of the soil, weather and the general surrounding environment has more direct effect on plants than on animals. This might also explain why Aristotle's student Theophrastus (author of the *Inquiry into Plants*) is more overtly sceptical than Aristotle about the usefulness of teleological explanations (see Theophratus *Metaph.* 10b).

1335b30–5). Aristotle also recommends that women maintain a healthy regimen while pregnant: 'for children before birth are evidently affected by the mother just as growing plants are by the earth' (*Pol.* 1335b12–19).[54] The point of comparison to plant development seems to be very simply that the state of the woman during pregnancy will affect how well she nourishes her child through the umbilical cord (stage 2). This does not add anything more to the picture we already have of Aristotle's views on the correct type of nutriment, well prepared by the soul of the mother, during gestation. Another passage within the *GA* explains that in the case of hybrid animals the female contribution influences the shape of the body of future generations of offspring, which is linked metaphorically to the effect that soil can have on plants grown in it (*GA* 738b27–36). What the passage does not say is that external food, such as earth, can determine an animal's structure. The female animal's contribution to generation in Aristotle is more complex than the role of the earth in plant development. To better understand this complexity, we must consider both (1) the initial contribution and (2) the subsequent feeding role of the female in more detail. In live-bearing animals, both (1) and (2) are blood or a blood-like fluid created in the female body by her nutritive soul.

4.5 Blood: the female principle as internal food

Aristotle does not only use the term for nutriment – *trophê* – of what we put on our plates, but also of materials contained within our bodies. For Aristotle, as a blood-like stuff, the female contribution is *internal* nutriment and so is more influential than any external sort of food. Most generally, Aristotle indicates in several places that blood (or its analogue) must be present in order for something to be alive.

> The blood in animals throbs in the blood-vessels all over the body at once; and blood is the only fluid which remains throughout the whole body, and throughout life so long as it lasts (*HA* 521a6–9).

[54] Aristotle also says that women ought not to do anything too mentally taxing while pregnant. Scientists continue to speculate about the possible connections between healthy pregnancy and mental infirmity. See Peterson (2012).

All living beings need blood or its analogue (*PA* 648a20) because it is nourishment and all parts of the body require it.[55] In several places, Aristotle indicates that there are two types of nourishment, one that is growth promoting and the better sustaining variety, which is more properly called nourishment or food (*GA* 744b34–6, *de An.* 416b11–14, *GC* 322a23–7). Blood is the most finished of internal nourishment, and so maintains the being of the parts. The nutritive soul, which refines raw food into blood is then also able to constitute the initial body of an animal in instances of generation (*de An.* 416b14–16). In these cases, blood is that 'out of which the parts come to be' and that 'from which the parts directly come to be'.[56] Although the parts 'come to be out of' blood, as explained in the previous section, the addition of material bulk is not what is primarily meant by this.[57]

Importantly, the particular constitution of blood is relative to the parts of the type of animal that is being kept alive by it. The internal food has to be the right variety to do this work. Each animal requires its blood to possess particular qualities for its functions, character traits and manner of life.

> The nature of the blood affects both the temperament and the sensory faculties of animals in many ways. This is indeed what might reasonably be expected, seeing that the blood is the material of the whole body.[58] For nourishment supplies the material, and the blood is the ultimate nourishment.[59] It makes then a considerable difference whether the blood be hot or cold, thin or thick, turbid or pure (*PA* 651a12–19).

Variations in blood consistency and temperature occur between different types of animal and also between the sexes and for the old and young (*PA* 647b31–6; *HA* 521a22-b3).[60] The quality of

[55] See also *PA* 678a6–9.

[56] ἐξ οὗ γίνεται τὰ μόρια; *Juv.* 469a1; ἐξ ἧς ἤδη γίνεται τὰ μόρια τοῖς ζῴοις; *Resp.* 474b4–5. Cf. *GA* 726b5–6. 'The most useful of the residues is that which is produced last, that from which each of the parts of the body comes to be' (*GA* 725a11). In the *GA* blood is described as matter for the bodies of animals (τὸ δ' αἷμα ὅτι ἐστὶν ὕλη τοῖς σώμασιν; 751b2).

[57] There are some parts of the body that are formed as material bulk from the blood. For instance, concoction of blood results in fat (*HA* 521a18); lard and suet are cooked blood (*PA* II.5).

[58] ὕλη γάρ ἐστι παντὸς τοῦ σώματος. [59] ἡ ἐσχάτη τροφή.

[60] Although a proper blend in the blood is necessary for the life and health of any animal, Aristotle also thinks that some animals are better than others, and the best (animals/

blood also varies from one part to another within the same animal (*PA* 647b34–6). 'The finest and purest blood is contained in the head, while the thickest and most turbid is that in the lower parts' (*Somn*. 458a14–15). Such variations in the blood correspond, for example, to certain animal activities and abilities.

Particular consistencies of blood are required, in most instances, for the operation of the emotions.[61] Fearfulness and the behaviour associated with it often correspond to a fluidity and coldness of the relevant body parts (*Resp*. 479b23). As Aristotle explains, 'a great excess of fluid makes animals timorous, for fear chills the body . . . so that in animals whose heart contains such a fluid blend (*krasin*) the way is prepared (*proŏdopoiêtai*) for the operation of this emotion' (*PA* 650b27–30).[62] Similarly, anger often corresponds to heating in the parts: 'such animals as have thick and abundant fibres in their blood are of the more earthy nature, and of a choleric temperament and liable to bursts of passion, for anger is productive of heat' (*PA* 650b34–6). The ability or propensity to remain afraid or angry is also aided by bodily constitution in many instances. Once the soul has initiated the changes in the bodily parts which display emotions, often materials out of which these parts have been specially constructed retain the necessary warm or cold state which accompanies that emotion.[63] '[O]utbursts of temper or fear, when they have once produced an impulse, do not cease . . . ' (*Mem*. 453a26–7). Some blood types, constructed for such lives, are especially suited to certain emotions and their bodies aid the retention of these emotions for longer.

The close association between mental faculties and the blood within the body for Aristotle would appear to have much in

people) have blood that is neither too thick nor too thin, neither too abundant nor too scanty (*HA* 520b21, 30).

61 I will continue to qualify any pronouncement about the necessity of bodily conditions for psychological states and processes in this section as it is unclear whether Aristotle was a non-reductive materialist or a spiritualist. The latter characterization allows that material preconditions are not necessary and may not be present in many instances. See Chapter 6.3–4 and Caston (2007).

62 At *Rh*. 1389b31 Aristotle says that because elderly people are cold, the way is prepared (*proŏdopoiêke*) for cowardliness in them. In this case, the material cause of fear is not a hypothetical necessity for the species but rather a consequence of the external environment and the fragility of the human constitution. See Chapter 10.

63 'Solids, when they have been made hot, give off more heat than fluids' (*PA* 650b36–51a1). It is, presumably, the same in the case of cold.

common with his contemporaries' emphasis on bodily mixture. In order to explain intellect and thought, many ancient thinkers concentrate on giving an account of the blend of the blood. For instance, in the Hippocratic *Vict.* I (where the healthiest body is said to be a combination of the finest [*leptotaton*] water and the purest fire, *Vict.* I.32, L VI 506–10),[64] intelligence is explained in terms of the composition of the body and the physical soul (*Vict.* I.35, L VI 512–22). For Empedocles also 'it is especially with the blood that [animals] think; for in the blood above all other parts the elements are blended (*kekrasthai*)'.[65] In places, Aristotle appears to agree with the idea that a good blend in the blood often corresponds to advanced cognition. For instance, he emphasizes that blood is finest (*leptotaton*) and purest in humans, the most intelligent of all animals (*Sens.* 444a10; *HA* 521a3; *GA* 744a29–31). However, for him, thin, cold blood could never on its own make animals intelligent (*PA* 648a3–8); intelligent species mostly require that their matter be such. He does not think that a certain blood consistency itself constitutes thought.[66]

Unlike his opponents, Aristotle focuses on the form of an animal and its overall goals as a living being and so does think the composition of appropriate blood types occurs by chance. For him, form or soul creates blood of the right variety to fulfil the functions the animal requires for its appropriate cognitive activities and emotional responses.[67] The internal nourishment, including blood, of each animal is the result of nutritive processes. Recall that Aristotle described the mixing occurring in nutrition as a blending in which previous constituents cannot be extracted. One of the most familiar irreversible transformations to the ancient

[64] Males are warmer and females colder, and the various ages of humans are classified by their heat and wetness (*Vict.* I.33–4, L VI 510–12).

[65] This is according to the account of Theophrastus. See Empedocles DK 31A86§10; Pre-Socratics (2010) 400.

[66] A particular consistency of blood can be a necessary but not a sufficient condition for intelligence, according to Aristotle. See his criticisms of Parmenides at *Metaph.* 1009b13–16 and van der Eijk (1997). In other places (e.g. *de An.* III.5) Aristotle denies that thinking requires any material underpinning whatsoever. For further discussion see Chapter 6.3.

[67] As Aristotle explains in the *Nicomachean Ethics* II, the emotions are passive potentials to respond appropriately in a given situation. Human beings can train these to fall in line with their higher cognitive capacities (*EN* VI.13).

mind was that effected by cooking; for Aristotle, nutrition is a sort of cooking (*pepsis*).[68] The production of blood, the internal nourishment (*trophê; Somn.* 456a35, *Juv.* 469a1), requires that food be cooked three times in three different places from its entry into the body to its eventual transformation into internal nutriment. First, it is cooked in the stomach, and next at the liver. The heart brings about the final change of the food into blood, the 'ultimate' form of nourishment (*Somn.* 456b2; *PA* 650a34–5; *Resp.* 480a2–9).[69]

As well as that very general account, Aristotle also describes in a more detailed manner the processes by which an animal takes raw food and turns it into internal blood. This account occurs mainly in the *Parva Naturalia*, an adjunct to his general work on the soul (*de Anima*). There, he explains that at first after eating, the earthy substance of food goes straight into the system, still dry and cold. The animal's soul must then 'separate' off the purer and thinner part and warm it up. By separating the more corporeal from the purer blood, the nutritive soul of an animal ensures the correct degree of thinness and thickness of the blood (*Somn.* 458a12). This same soul capacity also maintains a certain proportion of fluid to fibrous blood. The more body-like and earthy part occurs mostly in animals that display brave (or rash) feelings and the thinner part is most often required for those that can think and in those that regularly feel fear. The relative heat of the blood must also be put in place at some stage after the food has entered the body and been changed into blood. For Aristotle, blood is not essentially hot: heat has been added to it (*PA* 649b24–8). This means that the variations in heat of different blood types depend on the nutritive soul having heated the blood to a certain degree. The composition of blood may make it prone to stay heated or cooled, but it is the soul that does the heating and

[68] This term is often translated by the obsolete English word 'concoction'. For more on *pepsis* see Lloyd (1996) ch. 4.

[69] Blood is formed by the cooking of ichor (*HA* 521a18). Ichor is, according to some texts, the watery part of blood (*PA* 651a17). This part of blood is either not yet concocted or corrupted (*HA* 521b2). It is thought of as the better, light and watery part of blood or, alternately, as phlegm or diseased blood (*Mete.* 384a32–3, *Pr.* 891a16–17). The other part of the blood is the earthy fibrous portion (*is, inas*) (*PA* 650b14–19). *Morb.* 4 says that by eating something bloody, the body is able to draw off the blood-substance to itself from the food (38, L VII 554–7). Unlike the Hippocratic author, for Aristotle, blood is something only living, functioning bodies have. See footnote 47.

cooling and the soul that controls the blood's original consistency and temperature.[70]

We have seen that Aristotle does not think food which enters the body, food which a creature consumes, has much influence over that animal.[71] Instead, he regards the animal's soul as key to the processes of nutrition, transforming external food into useful internal food, which underlies many physiological and psychological functions.[72] We can now relate these findings to embryology and specifically the female contributions of (1) the blood-like menses and (2) the subsequent feeding of the new animal through the umbilicus. The female's initial contribution has undergone a series of transformations in order to make it the complex and appropriate material basis for many of the soul's various needs. In fact, the portion that goes to constitute the embryo's initial body is not even ordinary blood, but a more pure and concocted portion of the female's menstrual flow (*GA* 744b13–14).[73]

When the embryo is first 'set', it is as yet undifferentiated. At this stage, its blood comes directly from the mother, since at first it does not have a heart with which to concoct its own blood.[74] Indeed, it does not possess nutritive soul in actuality (although it has it potentially) and so cannot yet nourish itself. The next significant stage of development occurs once the new animal has its first part, the heart. Through the heart, it can then feed itself and make itself grow (*GA* 735a14–17). Being initially established in this manner, it counts in sense as an individual (*ti on*; *GA* 735a23).

> Once the foetus which has been formed is separate and distinct from both the parents, it must manage for itself, just like a son who has set up a house of his own independently of his father (*GA* 740a6–8).

70 See also Johansen (2012): 'the capacity of nutrition works for the sake of the capacity of perception' (143).

71 However, Aristotle sometimes holds that food and climate have some effect on emotions and even intelligence. See *Pol.* 1272b24–34.

72 See also Leunissen (2010) 88–9.

73 Although not as cooked up as the male residue, it is still more so than blood in other areas of the body. Cooper (1988) 16 and Boylan (1982) 111. As a sort of final nourishment (i.e. blood), according to the account in the *de An.*, it is 'in some sense active in growth and nutrition'. See Johansen (2012) 136. Cf. *de An.* 416b3–15.

74 The embryo contains blood (ἔναιμος) from the outset (*GA* 740b3; *Resp.* 480a7–9).

Once it possesses actualized nutritive soul, the average embryo is not as independent as this simile suggests. Although its heart attempts to control what comes into its undeveloped body, it is still far from being self-sufficient. Aristotle knows full well that at this point the new living being is completely dependent on the female parent (*GA* 740a24–8).[75] Unlike a mature living being, the embryo cannot yet turn raw food into materials necessary for the construction, development and maintenance of its living body. Thankfully, nature does not require it to have these abilities yet because the mother – or rather the mother's soul – has already done this work for it. The menstrual residue continues to flow after conception and thereby feed the new animal or animals (2). The food that the female provides in this way is not just any food – it is certainly not the sort of thing we find on our dinner plates. This *internal* food is the female's blood which has already been digested and converted into the type of dynamic matter required for that specific animal life.

In mammals the offspring's dependence on the actions of the maternal soul extends beyond birth. For at least the first few weeks or months of life, milk provides the offspring with fully digested food, coming from the potent blood of the mother.

> Clearly, milk has the same character as the secretion out of which the animal is formed . . . the material which supplies nourishment and the material out of which nature forms and fashions the animal are one and the same. And this material, in the case of blooded animals, is the bloodlike liquid, since milk is concocted, not decomposed, blood (*GA* 777a4–8).

Initially after birth, a new mammal cannot survive without this additional nutritive aid (*GA* 777a22–6.). For Aristotle, young animals *in utero* are in some sense similar to plants. The soil which plants ingest as nourishment has been worked up by heat in the environment (*PA* 650a21–3). Thus, in the case of plant nutrition, there are powers and potentials in their external food, designed to aid them, being prepared and digested for them (*PA* 655b35–6). And the same can be said for the embryo *in utero*.

[75] The new animal will either die (e.g. in the case of the mule, *GA* 748b29–31) or be deformed (e.g. in the case of the 'many-toed' 774b10–17) if the female parent cannot provide adequate nutrition at this stage.

Table 4.3 *Comparison of the Aristotelian and ancient medical models likening foetus to plant*

Medical model:	Foetus	feeds on	mother
	passive		active, dynamic · determines the health of plants that grow in it. · determines what type of plants grow.
Aristotle:	**Foetus (like plant)**	feeding on	**mother (like earth)**
	F1: active *F2: form* *F3: dynamic* *F4: determines type/form*		*M1: passive* *M2: matter* *M3: inert* *M4: resists type, causes disease, deformity*

Because the embryo does not have a fully developed nutritive soul and cannot yet convert external foods, the mother provides the sort of nourishment the embryo requires before and even for some time after birth.[76]

Recall that, for some interpretors, Aristotle reversed the usual associations between the mother and the earth in which a plant grows (see Table 4.3). When viewed in much more detail, Aristotle's account of embryonic nutrition does not match up with this. The foetus is not active, at least not at first, and the menstrual blood which the female initially and subsequently contributes is not entirely passive, instead containing dynamic potentialities that maintain the living parts of the body. The form – the specific type of animal and its living functions – is also not fully possessed by the embryo to begin with. After all, it is only potentially a human being or a cat or an elephant. Its

[76] Of course, this model applies mainly to live-bearers and even more strongly to mammals. In these cases, it is striking, though, that the natural relationship between mother and child is envisaged as positive and supportive rather than parasitic and antagonistic. In Hippocratic medicine, the foetus must fight the mother for release during childbirth (Hanson 1991, 88). For modern medicine there is also a 'maternal-fetal conflict'.

form is hardly present at all at the very beginning of its development and then gradually begins to come into being; meanwhile the mother's soul nutrifies and develops this form until such time as the embryo can do this for itself. Also, it would seem that *F*3 and *M*3 are somewhat wrong-headed; both the embryo and the maternal contributions are dynamic in so far as they are living materials, imbued with the powers and potentialities found in blood and its analogue. Finally, it appears that *M*4 is almost never the correct way to characterize the female contribution of concocted blood; her nutritive soul, working through her stomach, liver and heart has made sure that the matter given to the embryo from conception through to the end of exclusive nursing is fully worked up and not recalcitrant. The schema thus must be re-described in Table 4.4.

It is clear, then, that the female contribution is very far from being 'raw' material. Unlike soil, which we encounter externally, the material employed for the purpose of animal generation has

Table 4.4 *Revised Aristotelian model likening foetus to plant*

(1) first stage: point of conception		
Father's contribution	Foetus feeding on	mother
P1: active	F1: passive	M1: passive
P2: form	F2: potential form	M2: matter
P3: dynamic potential		M3*: dynamic potentials (in blood)
P4: determines type/form		M4*: determines type/form

(2) second stage: after appearance of foetal heart	
Foetus feeding on	mother
F1: active	M1: passive with active elements
F2: potential form	M2: matter (full of potentials for form)
F3: dynamic potential	M3*: dynamic potentials (in blood)
F4: determines type/form	M4*: determines type/form

been digested by the soul of the female animal, a process which results in a material that is highly specialized and full of dynamic potential. Although the blood that the mother contributes ceases to be part of her own bodily system when it enters the embryo, the qualities that made it able to maintain her life are what make it indispensable for the development and maintenance of the new living being. The coldness of the nature of the female makes it unable to concoct semen thoroughly (*GA* 738a13–14, 34–35), but the menstrual blood is well on its way to becoming a living body. The difference between the generative residues of the male and female is thus merely one of degree, undermining the idea of any mutually exclusive opposition between their roles.[77]

The connection between blood and generation possibly suggested itself to Aristotle through the empirical coincidence of regular menstrual bleeding (in human females) and fertility, that is, the capacity to become pregnant and give birth. Empirically, the male and female contributions to generation Aristotle identifies are very different in colour, consistency and physiology. A more straightforward hypothesis would have been to posit that they are not the same sort of thing. But he did not choose to do so; instead both are blood-like. Semen, for Aristotle, must be blood-like because blood is produced in the heart, the first principle, and contains all the parts of the living body potentially.[78] The connection between the haematogenous thesis and this physiology is much more obvious in the female than the male, whose discharge does not appear to be blood-like. For his parallel seed opponents, the male generative contribution is their focus – a paradigm that the female contribution is required to fit to. And so they have to assume that the female also contributes a white fluid emitted at sexual climax, despite scant empirical evidence for this occurrence.[79] Aristotle, on the other hand, takes the blood-like

[77] Deslauriers (1998) 149: 'understood as seminal residues at different stages of concoction the difference between the material contribution of the female and the formal contribution of the male seems slight'.

[78] This is why the heart is the first part to form in the embryo (*GA* 735a22, 738b17, 740a4–5, 743b25–6, 766b1). The heart is the source of blood production (776b10–13).

[79] The encephalogenetic theory, for instance, has it that semen is different from and superior to (nutritive?) blood, originating in the spinal column and coming from the brain. See von Staden (1989) 288–90 and Chapter 3.4.

female contribution as his starting point; the male contribution, for him, must be like it. Only by extrapolation or other theoretical means could Aristotle connect the male role with the crucial capacities of blood. Aristotle's haematogenous thesis can be viewed, then, as more focused on the evident contribution of females, as internal bloody nutriment which has the potential for form. This is what all semen is, whether male or female semen.

4.6 The metaphysics of female matter: the female principle as the body

Now that we have understood that Aristotle's female contribution is blood-like, it may seem difficult to understand in what sense it is matter to the male's contribution of form. Most commentators acquainted with Aristotelian metaphysics acknowledge that the matter which the female contributes in generation is not 'simple'.[80] But disagreement persists about how to cash out this complexity. One account of the female contribution emerges from an understanding of the matter/form dichotomy as levels of compositional relations.[81] On this view, uniform parts of the body, such as flesh and bone, are more formed than the mere elements, because they consist of a certain mixed proportion of those elements. Then, the non-uniform parts, such as hands and faces, represent the next level of formation, since they are constructed out of the uniform parts.[82] One can imagine up to six possible levels, as detailed in Table 4.5.

The female contribution, on this view, is generally taken to be a uniform part like blood or flesh (level 4). Its complexity, then, would consist in its being constructed from the four elements and thereby standing at two removes from 'prime', that is, completely indeterminate, matter. Against this reading, it should be urged that Aristotle does not espouse a compositional account of the material which the

80 Thus, 'the Matter with which we are concerned in the generation of animals is far from being "uninformed"'. Peck in Aristotle (1942) xii-xiii. Gelber (2010) 198: 'the matter in the context of biological reproduction has a quite complex form'.

81 '[T]he catamenial mass is not *wholly* formless, but to the contrary – relative to the primitive materials or "simple bodies" – represents an extremely rich and specialised potentiality, even as much as a plant, perhaps.' Furth (1988) 123 n.3.

82 Dean-Jones (1994) 186, Furth (1988) 76–83, Preus (1975) 96 and Freeland (1987).

Table 4.5 *Levels of formation of matter in Aristotle*

(1) Prime matter (entirely indeterminate).
↓
(2) Hot Cold Wet Dry (material potentials).
↓
(3) The elements: water (cold+wet); earth (cold+dry); air (cold+wet); fire (hot+dry).
↓
(4) Uniform parts: flesh, blood, etc. (as ratios of the elements in them, e.g. 2/3 water, 1/3 earth).
↓
(5) Non-uniform parts: hand, face, etc. (bone+blood+flesh).
↓
(6) Shape of the entire body.

female contributes. Instead, in the context of generation, he has in mind a dynamic and teleological model in which the material body and its parts are understood in relation to their active roles within living bodies.

Although Aristotle clearly had an interest in some contexts in speaking of parts in the body in terms of structural composition,[83] there are a number of reasons why attempting to characterize the female role in generation in this manner limits our understanding. First of all, the idea that matter waits around for further levels of construction to happen to it obscures Aristotle's insistence that the matter of each thing is what is appropriate to it.[84] For him, different types of matter are required by different forms.[85] The carpenter cannot make a bed out of anything but timber (*GA* 743a25–6). The form 'axe' will require wood and iron, the form human being, female menstrual blood (*Metaph.* 1044a34–1044b2). In embryology, Aristotle insists that the female matter

[83] *Mete.* 388a13–19; *HA* 486a5–14; *PA* 640b18–22; *GA* 715a8–11.
[84] See also Gelber (2010) 201–2 on this point.
[85] See Freeland (1987). 'There is a matter proper to each thing (οἰκεία ἑκάστου)' *Metaph.* 1044a17, see also *Ph.* 194b9 and *PA* 646a34–5. Aristotle thinks that it is part of the craft to know about the appropriate materials needed *Ph.* 194b5: the helmsman must have knowledge of useful timber and how it should be finished. Aristotle's student Theophrastus takes this seriously in his discussion of the uses of different types of timber in *HP* V.

must be matter of a particular type (*hê toiautê; GA* 733b32), relating to a given final outcome.

The second point against the static reading is that parts of the body for Aristotle are best explained not only in terms of their structural composition but also in terms of how they serve the needs of the whole animal. This is why the bottom-up analysis in the *Meteorology*, which inspires the compositional reading of animal parts, is qualified by the consideration that all the parts of the body are determined by their definition (*logos*) and function (*ergon*). A good example of this is flesh, the complexity of which is not the ratio of the elements in it but the function it serves in the animal.[86] This is also the case with the blood; its materiality must be understood in relation to the goals of the whole animal – the blood maintains the living body. Although structured and organized to some extent, neither uniform nor non-uniform parts can exist independently as a 'this' (*todê ti*), an animal, which uses these parts.

A static description of the female contribution, standing apart from this teleological background, is inadequate. None of the levels of composition listed above is sufficient to characterize it. Much disputed throughout history, 'prime matter' (level 1) is imagined to exist below the level of qualities that the elements possess, being entirely indeterminate.[87] The concept is attractive to the modern mind as it is closer to our idea of matter as: empty, without power or quality, in effect 'inert', and has often been identified with the female role in generation.[88] For scholars who study the *GA* in any detail, however, it is patently obvious that the female contribution cannot be prime matter.[89] In any case, it is

86 'All uniform bodies consist of the elements described, as matter, but their essence (*ousia*) is determined by their definition ... What a thing is is always determined by its function (*ergon*): a thing really is itself when it can perform its function; an eye, for instance, when it can see. When a thing cannot do so it is that thing only in name, like a dead eye or one made of stone ... The same, then, is true of flesh, except that its function is less clear than that of, e.g. the tongue' (*Mete.* 389b29, 390a10–15).

87 See Weisheipl (1965).

88 Aristotle's female is taken to contribute indeterminate 'prime matter' by Irigaray (1985b) 162, Sissa (1992) 70 and Saïd (1983) 117. Henry (2006a) 258 n.3.

89 Matter for human beings must be peculiar (*oikôs*) to them (*Metaph.* 1044a33–b2). The most specific matter is referred to as either the nearest (*eschatê*) or the first matter (*protê hulê*) (*Ph.* 192a31–2). The female matter must be *protê* in this sense and translating this term as 'prime' matter is a mistake.

widely thought that Aristotle did not allow for such a concept, which was invented by later thinkers.[90]

Next, the female contribution is not, on Aristotle's view, a non-uniform part (level 3). For he insists that seminal residues cannot contain completed parts of the body (a crucial move against pangenesis at *GA* 723a15–23). No one would want to say that the matter the female contributes is formed in such a way that it is already like a hand or a face. What about uniform parts (level 2)? Although the female contribution is a blood-like stuff, it is hardly adequate to say that menstrual blood, like any other blood, is merely composed out of a certain ratio of the elements in it (*Mete.* 384a16–17). This would serve only as a necessary description of it. Although it may be the case that blood has to be fluid in order to serve its function, its fluidity is not what makes it what it is, but rather its relation to the goals of the process of generation.

Animal generation begins with materials possessing various potentials or capacities, which will gradually be actualized. Form is the goal (*telos*) of development and existent things, 'substances', are complete adult animals and plants (*Metaph.* 1032a19–20). Only a complete substance can have a proper form. Before completion takes place, embryos are primarily material. However, this matter which is the potential animal cannot be static, partially formed material parts anticipating further formal structures.[91] Of course it is possible, given this, to describe the female contribution as a uniform part, and as the combination of elemental powers in a certain proportion. However, such a description is not informative of its functional role in the process of generation. Even non-generative menstrual blood is best described as something that can potentially become the living body of an animal, otherwise it is just a waste-product. Indeed, Aristotle does not tend to characterize the female principle as a uniform part (level 4) but rather as 'the body' (level 6) of the offspring.

The female always provides the matter and the male the productive/fashioning part (*to dêmiourgoun*). For we said this is the ability (*tên dunamin*) of each of

[90] Charlton in Aristotle (1970) 129–45, Furth (1988) 221–2, Jones (1974), Gill (1989b) and Williams in Aristotle (1982) 211–19.

[91] This is why I reject the idea that, according to Aristotle, the female animal 'provides matter in the right state to receive a higher form' (Dean-Jones 1994, 186).

them, and that is what it is to be female and male. Thus, the female must provide *the body* (*sôma*) and bulk, and the male must not. For the tools necessarily are not present in the things that come to be or in the maker (*GA* 738b20–6, *my emphasis*).[92]

In this passage the term for body, *sôma*, initially indicates the bulkiness of the female menses. However, the passage goes on to associate the female contribution with the outward shape or structure of the new animal's completed body.[93]

Thus the body (*sôma*) is from the female, and the soul is from the male. For the soul is what it is to be a certain kind of body. Due to these reasons, when a male and female of different types/species copulate . . . the first generation resembles both parents equally . . . but as time goes on, the animals that result resemble the form (*morphê*) of the female, just as happens when seeds are introduced into a strange land. The land provides the material and the body (*sôma*) for the seeds (*GA* 738b26–36).

Over several generations, hybrid animals come to resemble their female ancestors more in terms of their bodily form. If the female determines, in some sense, the shape or structure of the body of an animal, she also determines what its soul is – since the outward appearance and general structure of any animal's body reflects inner psychic capacities and functions. If the female ancestor was a dog, and her descendants are mostly dog-like in appearance and activities then it is through the influence of her contribution that the essence of dogness has been preserved.[94] As surprising as this might seem, given Aristotle's insistence that only the male contributes form to generation, this inference fits with two key

92 The point of this statement is to show how it is that the male does not contribute any matter to generation (*GA* 729b18–22). Since the female provides the body and the bulk (*to sôma kai ogkon*), the male does not have to. However, it would be wrong to infer from this that the female cannot contribute towards the fashioning part (*to dêmiourgoun*), according to Aristotle (*GA* 738b21). This inference is drawn by, for instance, Dean-Jones (1994) 179: 'the male and female contribution to conception was totally dissimilar', and Preus (1975) 52: 'matter and movement are strictly segregated to female and male respectively, the female bringing about no change, and the male providing no material'. Peck translates the *kai* at *GA* 738b23 as 'i.e.' and thus reduces the female role to merely providing bulk. Whether there is scope for a fashioning role within the matter (or body) which the female contributes requires more thorough analysis. See Chapter 9.

93 Unlike the modern understanding of matter as inert bulk, Aristotelian matter can encompass (in his biology at least) a fully developed organic body (*PA* 642a12–14; *de An.* 412a28–413a9).

94 Se M4* in Table 4.4.

aspects of his account. The first concerns generative physiology, the second the developmental focus of his embryology.

In terms of generative physiology, by now it is clear that the female animal's menstrual blood is a bodily fluid prepared by her nutritive soul in such a way as to be able to constitute and maintain her body – a living body of a particular kind. Thus, it must of course be able to constitute and maintain a body very much like her own, that is, that of her offspring. Second, the *conceptus*, a combination of the generative contributions of both parents which possesses the potential to become a new animal, develops gradually. In this context, one cannot say that the female contributes the body *on her own* or that the male contributes the form or soul *on his own*. The body and soul of the new living being are intertwined and co-dependent.[95] There is never a clear point when the female hands over a structured living body or when the male ensouls such a body.[96] The materiality of the female contribution is a dynamic specific potential present in the blood rather than a static structure, at a lower level than a more complex structure. Although the female matter does not possess form, it can be understood with reference to form; it anticipates or facilitates the advent of future form in the offspring.[97]

4.7 Craft analogies revisited

The previous two sections show clearly how the simple craft analogies, such as the bed and timber and the sphere and wax, are unable to accommodate the complexities of female matter. In

[95] Matter and form are the potentiality and actuality, respectively, of substance (*Metaph.* 1045b18–21). In animals, matter is the body, form the soul and then there are also different levels of the actualization of soul (*de An.* 412b27–413a31).

[96] The fact that the body must be alive in order to count as the proper matter of the animal (*de An* II.1) has long been seen as an insurmountable challenge for any attempt to literally separate the two, as seems required if one is assigned to the female contribution and the other to the male. See Ackrill (1972–3), Sorabji (1974) and Sheilds (1988) on body and soul in Aristotle.

[97] The specialized potentialities in the menstrual blood that make it capable of becoming a new living animal might seem to us to be somehow 'formal' but for Aristotle the female contribution's potentiality is a trait of the matter or material. Balme (1987c) and (1990) uses 'formal' for anything that is descriptive. In English, the term 'formal' is misleading in this context. Descriptions are not always 'forms' in the robust Aristotelian sense.

light of this, one could try to adapt the analogy.[98] Aristotle himself elaborates the image when attempting to account for the effect of material factors on hereditary resemblance. In *GA* IV he explains that matter can sometimes resist form, as tools can become blunted by what they cut.

The reason why the 'movements' (*kinêseis*)[99] relapse is that the agent gets acted upon by the patient, as the thing which cuts gets blunted by what it cuts and the hot thing is cooled by what it heats, and generally, the external first mover gets moved in return somehow . . . (*GA* 768b16–20).

The passage does not help very much, however, in understanding the female role.[100] Timber may have a grain that serves the craft by being easily shaped into a bed, but this differs from the sense in which the female matter has properties that cause it to be readily formed into a specific animal. The grains of timber *qua* timber have to do with the life cycles of the tree of which it was once part; they do not have any relation in themselves to the craft. By contrast, the properties of female matter that make it able to become offspring belong to it as menstrual blood, the exclusive function of which is to be matter for offspring of a particular kind (*GA* 738b2–3).[101]

There may be another way to modify the craft analogy, which Aristotle himself did not suggest in the *GA*, appearing instead in his *Metaphysics*. There he says that certain crafts prepare materials for other crafts: for instance, the brick-making craft produces refined materials for the house-building craft to use. The female principle could be likened to a lower level craft such

98 This idea was initially suggested to me by Nick Denyer.

99 *Kinêsis* is more properly translated as 'change' and its plural, *kinêseis*, as 'sources of change' as I will explain in Chapter 5.1.

100 One difficulty with pressing too hard on this passage as key to understanding the active resistance of the female principle to being formed is that it refers to the difficulty that both male and female *kinêseis* have in imparting their character to the matter, which explains how offspring end up resembling ancestors on *both* sides. Since the male contribution cannot have any material characteristics, and cannot resist formation, a better explanation of relapse, which applies to both contributions, must be found. This is achieved later on in the work. See Chapter 9 for a more detailed analysis.

101 The presence of the menstrual blood is teleological only in an indirect or secondary way since it accumulates as a side effect of material interactions in the body (See *GA* 738a33). For a more thorough account of Aristotle's indirect teleology see Chapter 10 and Leunissen (2010) ch.4.

as brick-making which provides very specific materials. Perhaps females provide the ability to refine materials, producing something like the parts of flat-packed Scandinavian furniture. The male role would then be like that of providing the action of screwing in the bolts and following the sheet of (cryptic) instructions. The male role as master craftsman (*architektôn*) or overseer of other lesser crafts only brings about the final stage of construction, an alteration more slight than the simpler image implies.[102] If Aristotle were to extend the craft analogy along these lines, he would have to admit that the female is not entirely excluded from the productive, active role, since it would represent a sort of craft in itself. One might worry, however, that when adapted in this manner, the analogy misrepresents Aristotle's idea of the female role; again, she cannot provide polished and refined static parts but rather the living potentialities present in her blood.

Ultimately, it seems that the carpenter or architect analogy of generation is often made out to be much more significant than it actually is for Aristotle. There are a number of other images in the *GA* which are not taken so seriously and there are no grounds to distinguish the carpenter as more representative of the process of generation than these. At one point Aristotle compares semen to paint left over on a palette when a painting is finished (*GA* 725a27).[103] This analogy is clearly not shorthand for Aristotle's theory as it suggests that nutrition is like the addition of material parts – the palette's colours are the building blocks of the two-dimensional painting. Although this is better than an image of lego blocks, in that the colours might still need to be blended by the artist, it is still unrepresentative of Aristotle's fuller account of nutrition. As we have seen, for Aristotle the concocted male and female generative residues are not mere passive ingredients – both have the capacity to organize and maintain animate existence and so possess an element of the artist within

[102] For Aristotle's description of a hierarchy of crafts, see especially *Metaph.* 981a30-b6, b31, *EN* 1094a9–17, *MM* 1198a33-b3 and *Ph.* 194a34-b2.

[103] Although representational art could be classified as a *techne* it is certainly quite different from his usual example of carpentry. The comparison is similar to Empedocles' idea that the creation of trees, human beings, beast, birds and fish is like an artist mixing pigments (DK 31B23–4; Pre-Socratics [2010] 356–9).

them. Aristotle does not mean us take this image too much to heart; instead, as with all craft examples, he is only supporting the point he is making in the particular context. Here, he uses it in order to sum up the notion that semen is a residue rather than a colliquescence (*GA* 725a24).

Another analogy which Aristotle employs is that of the automaton or automatic puppet, which helps to explain how offspring come to be by the agency of the male parent (*GA* 733b33). Natural objects generally possess an internal source of change, according to Aristotle. He must, therefore, make clear how embryos, although natural, have an *external* source of change in the beginning.

> How is it not possible that [the new animal] is generated by the agency of something external? For in a way it is possible, but in another way it is not. The semen . . . has in it the movement (*kinêsis*) which the mover set in motion. It is possible that A moves B, and B moves C, and that this works like the miraculous automatic puppets. For while at rest their parts somehow exist potentially; and when something external moves the first part, then immediately the next part comes to be in actuality . . . It [the semen] moves not by being in contact anywhere now, but by having been so (*GA* 734b5–15).[104]

Although the comparison is meant to explain the male role, it brings to light crucial characteristics of the female one. After all, it is the female contribution that is being likened to the automatic puppet, poised to react in complex and specific ways to a stimulus.

> As the parts are potentially present in the matter, when the principle/start of movement (*kinesis*) is there, one thing follows after another as in the miraculous automatic puppets (*GA* 741b8–9).

This comparison has not been thought of as shorthand for Aristotle's theory in the same way that the craft analogy has. But if it were to have more prominence in commentaries, then this would provide a more balanced perspective on the female contribution.[105] Like an automatic puppet, the female matter is

[104] It may be that 'movement' is not the best way to translate *kinêsis*. For my argument to this effect see Chapter 5.1.

[105] The image has received some attention recently, especially from Henry (2005) and Berryman (2003). However, neither regards this as significant for our understanding of the female role.

complex and difficult to make.[106] More generally, the image also conveys this idea: matter in a particular situation has a very specific potentiality to change or react in a certain manner (e.g. *Metaph.* 1049a5–12). The very specific potential of the female contribution is like this because it is potentially the same in character as the female's living body (*GA* 738b4).

Matter and form in the biological context cannot be perfectly elucidated by any comparison.[107] For Aristotle, animate existence has its own set of metaphysical constraints that do not allow for further simplification.[108] Each analogy highlights different aspects of generation, and leaves out others. Certainly the craft analogy is limited because it leaves out the more positive aspects of the female role. Once he leaves behind the craft analogy in his discussion of natural generation, Aristotle still considers the female role to be material. However, the fact that he is not thinking in terms of crafts makes it easier to see how he can maintain this. The female contribution is not precisely like timber since it has undergone the nutritive process, imbuing it with the potentialities of a living body and, further, with the remarkable ability to become a new living creature of the same kind as itself. The female contribution is a set of complex potentialities to become an organized functioning body and is material in this sense and no other.

[106] In antiquity it was thought to require great skill to construct automata. 'The study of automaton-making has been considered by our predecessors worthy of acceptance, both because of the complexity of the craftsmanship involved and because of the striking nature of the spectacle', Hero of Alexandria, *On Automaton Making* (Murphy in Hero of Alexandria 1995, 9). Farquharson's notes to *MA* 701b4, b5, b7 in Aristotle (1912) speculate about what the toys were that Aristotle refers to here.

[107] Aristotle criticizes others' use of metaphor, complaining that they have not made anything clear by it. For instance, Aristotle is unimpressed by Empedocles' remark that the sea is the 'sweat of the earth' (*Mete.* 357a24).

[108] As we shall see in Chapter 4.6 matter in the biological context also comes to represent any incomplete or unfinished substance, even when this is a living being. Although matter in crafts, such as bronze, cannot move itself (*Metaph.* 984a23–4; 1071b30–2), in nature 'sometimes matter moves itself' (*Metaph.* 1034b3–5).

PART III

RE-ASSESSING THE GENERATIVE ROLE OF THE MALE ANIMAL IN ARISTOTLE

Introduction

Male and female principles are jointly necessary for generation according to Aristotle. Since they co-depend, a full understanding of the female role requires a clear view of that of the male. On this issue several broad and influential interpretations from the past create challenges for the modern reader. In particular, sometimes the female role is misrepresented as a consequence of certain understandings of the male role which do not accord either with the text or with Aristotle's philosophy more generally. In Chapter 5, I first set out how Aristotle describes the male role in generation as the principle of change. This account will then be placed in the context of Aristotelian physics and metaphysics. The second part of this section explains how male and female principles work together to ensure generation. Chapter 6 takes up the challenge of disentangling Aristotle's views on the male in generation from numerous interpretations and the uses to which they have been put. These interpretations tend to two extremes – those which posit that the male role is somehow divine and others which characterize it in materialist terms. The second perspective on the male generative role varies its focus. For some *pneuma* in the semen (or the inborn *pneuma*) is key while for others it contains a sort of special 'vital heat'. Still others take its role as efficient cause to be quite literally a motion or movement, and think that, for Aristotle, the male role is mechanical or machine-like. I explain what inspires these interpretations, how they stray from what Aristotle means to convey and how this can mar our understanding of the female role.

5
THE MALE AS EFFICIENT AND FORMAL CAUSE OF GENERATION

5.1 The male as generator and source of change

In this section it will be urged that the male, according to Aristotle, does not straightforwardly contribute or 'impart' soul or any particular portion of soul to offspring. The male is the agent in what is, for Aristotle, a joint actualization of the prior potential to change or develop into a new animal. I will conclude by considering what else can rightly be said of the male contribution – whether it could be characterized as living or actually alive and also whether it makes sense to say that, for Aristotle, the male is the only true parent, as some allege.

In *GA* I and II, male and female roles in generation are counterpoised within a teleological framework. The male role is an efficient cause in an instance of substantial change, where a new substance (i.e. animal) comes to be. The male semen must 'set' the female contribution (*GA* 729b31, 730b3, 731a16, 733b21, 737a21, 739a8), a process also described in terms of the male moving or changing (*kinei*) the female principle (*GA* 737a22). Only after this occurrence does the offspring begin to differentiate and assume its proper shape (*GA* 730a30, 733b20–2). Here efficient and formal causes are tied together – the male is like the carpenter who is active in shaping and forming matter, working to a model or form.

In terms of his metaphysics, Aristotle's favoured way of characterizing the male's role is to say that it is *hê archê [tês] kinêseos*, which is usually translated as 'the principle of movement'.[1] This phrase remains vague in English, a problem exacerbated by the fact that both *archê* and *kinêsis*[2] can be

[1] This is Peck's translation of the term at *GA* 715a6, 726b22, 740b26, 741b8 in Aristotle (1942). Balme in Aristotle (1972) translates 'the beginning of the movement' at 715a6 and the 'source of movement' at 726b22.

[2] This is the nominative form of the noun. In the phrase *hê archê kinêseôs*, it occurs in its genitive form meaning 'of *kinêsis*'.

translated in several different ways. *Archê* can mean beginning or start, as well as rule, domination and 'principle' (*Metaph.* 1012b35–13a23). In Aristotelian physics, *archai* (in the plural) is often used to indicate the primary ingredients or elements of the world (*Ph.* I). As for *kinêsis*, 'movement' is only one of its many meaning, and one which can be misleading in the context. The phrase is also Aristotle's preferred term for efficient cause in other texts (e.g. *Ph.* 194b29–32, 198a25–8).

In his *Physics* Aristotle explains that *kinêsis* occurs in a subject when it changes between two conditions, which are opposites or contraries (*tois enantiois*) (*Ph.* V.1 224b29–31). These opposing factors exist in three different categories – quality, quantity and place (225a34, 226a24–5). A quality is 'all affections of substances in respect of which bodies become different when they change, e.g. hot and cold, heaviness and lightness' (*Metaph.* 1020b9). Examples of opposite quantities include the more and the less and the great and the small (*Metaph.* 1020a21). A transition in place is the one sort of *kinêsis* that coincides in certain circumstances with the English terms 'motion' and 'movement'. Finally, *kinêsis* can indicate an unqualified change (*Ph.* 225a13–15), which occurs when a substance comes to be where there was previously no such object. Thus, it can refer to a 'substantial change'.

Despite its various Aristotelian meanings, many interpretations of the male role as *kinêsis* take their lead from the modern connotations evoked by the words 'movement' and 'motion'.[3] In one instance, the *kinêsis* present in generation is equated metaphorically with spatial movements around the body, which can draw the body's parts, like a pencil on tracing paper.[4] Another commentator writes, '[i]t is tempting to view Aristotle's talk about the same movement as referring to one continuous, unbroken motion or flow of energy' (Morsink 1982, 108). Yet others

[3] Some commentators more cautiously urge that *kinêsis* should not be translated 'motion' (Gotthelf 1976, 238 n.20).

[4] Coles (1995) 74: 'To begin with, parts can be sources or originators of movement, it seems, because blood flows around each and every part. This starting-point provides the clue for the reproduction of shape in movement. The analogy, where Aristotle provides none, is presumably in the simple use of tracing-paper to follow an outline: a superimposed sheet of transparent paper over a picture allows the reproduction of shape in the movement of the pencil around the outline.'

conjecture that *kinêsis* labels the movement (of the male) during copulation.[5] These somewhat outlandish suggestions can be quite easily put aside. A more compelling idea is that the male generative *kinêsis* consists in tiny spatial motions or movements in the male semen which fashion or shape the embryo.[6] It has also been attractive to regard these 'movements in the semen' as the motion of tiny particles which bring about heat: 'sperma is a fluid emitted by males which has the motion of the heat of the *pneuma* which it contains'.[7] There are two key problems with taking the male *kinêsis* to be a change in location or a spatial movement. The first is that the modern idea that heat, which can act on something else, is brought about by the motion of miniscule particles, is entirely un-Aristotelian. The second is that Aristotle here talks of a single *kinêsis* not a set of plural *kinêseis*. I will detail each of these difficulties in turn. First, then, Aristotle does not think static objects contain particles that move about spatially, disagreeing on this point with the atomists. If a *kinêsis* occurs, then the Aristotelian characterizes this as the subject undergoing a transition or change (*Ph.* 225b20).[8] If heat is the quality, then it will be becoming warmer. This is a basic analysis: heat cannot be further reduced to spatial motion.[9] Motion or movement does not occur unless the change is a change in location or place. The only spatial *kinêseis* Aristotle recognizes in generation is growth outwards (*GC* 320a21). Thus, the male role as *kinêsis* does not look to be spatial movements or even one spatial motion for Aristotle.

5 Sissa (1992) 68, Coles (1995) 68 and 70–1 and Morsink (1982) 109.

6 Furth (1988) 118–9, Cooper (1988) 19, 30, Tress (1992) 333 n.41. In *GA* IV.3 on heredity, when Aristotle says that both the male and the female principles have hereditary *movements*, Witt (1985) 54 imagines patterns that can meld together. In Lewis (1988) the movements from the male are taken in a completely mechanistic manner, as efficient spatial movements, part of a causal chain (60).

7 Bolton (1987) 163. See also Balme in Aristotle (1992) 161: '[Aristotle] does not explain this in physical terms, but judging from 737a18–19 we may guess that he conceived it as a conveying of movements superimposed upon the heat's own movements perhaps as a liquid conveys waves'. See also Dean-Jones (1994) 195. For some, soul is to be identified with these motions. See Balme (1987a) 282 and Witt (1985) 52.

8 *Kinêsis* often indicates that a process is underway. Broadie (1982) 94: 'change is to be regarded as essentially a changing'.

9 Burnyeat (2002) 82–3.

The second issue is even more important for our understanding of the role of the male animal in the *GA*. Aristotle's initial description of it is as *hê archê kinêseos* – the principle of *kinêsis* in the *singular*. Later on in the treatise, *plural kinêseis* in the seminal residues of both parents account for resemblance of offsprings' parts to the parts of parents and ancestors (e.g. *GA* 768a12–22). Almost all modern commentators on the *GA* do not clearly differentiate between these two uses of the term and instead assume that the singular *kinêsis* is shorthand for the plural *kinêseis*, which they think differentiates the offspring and gives it form.[10] This is not, however, the most plausible way to understand what Aristotle is trying to explain. For one thing, if he had meant to say that the male role is manifested through 'movements' in the plural why does he characterize it as the principle of *kinêsis* in the singular? It seems likely that this actually has a different import for him. The male begins the change; in that sense it is the *archê*, principle or beginning, of *kinêsis* – the one substantial change that takes place. This is a different consideration from the *kinêseis*, the sources of change (to be discussed in more detail later on in this book), which exist in the generative residues of not only the male but also of the female.

Another reason to resist the idea that the male *kinêsis* is the same as the plural *kinêseis* (present in generative residues) is that it has weak textual support. Cooper refers to *GA* 738b12–15, which supposedly says that movements in the male semen fashion offspring. Yet this text does not establish the interpretation he favours as it mentions only one *kinêsis*. Three other cited texts (*GA* 738b20–3, 771b21–4, 722b31–3) do not mention plural *kinêseis* (or singular *kinêsis* for that matter) at all; they note instead that the male fashions (*demiourgei*) the offspring. It is part of this interpretation to suppose that this implies that the male performs this action through movements in his semen.[11] Gelber joins in the

[10] Cooper (1988) 19, Balme (1987b) 282 and Witt (1985) 52 all speak of 'movements' (in the plural) in the semen. Cooper (1988) 19, who does not allow that the female residue has movements, states 'Aristotle repeatedly emphasises that only the male, through the *movements* in his semen, is capable in any way at all of fashioning'.

[11] Balme also describes the single initiating male *kinêsis* as the plural 'movements', which fashion the offspring. He writes '[n]ow what the sire transmits is in fact soul, which is therefore to be identified with the movements' (1987b) 282. See also Witt (1985) 52 on Balme. Coles (1995) acknowledges that there is a difference between the initial *kinesis*

conviction that, for Aristotle, '[i]n both [craft and natural] cases form is conveyed by the way of certain motions or *kinêseis*' (2010, 187). The use of the plural is unjustified for those passages she cites where *kinêsis* occurs only in the singular.[12] The passages cited where there are plural *kinêseis* all occur in the context of heredity (See e.g. *GA* 767b36, 770b26) except for one. The following is the only passage in the first two Books of the treatise where *kinêseis* (plural) occurs; I will quote it in full before explaining why it does not provide evidence of demiurgic motions in the male contribution.[13]

The male does not emit semen at all in some animals, and where he does this is no part of the resulting embryo; just so no material part comes from the carpenter to the material, i.e. the wood in which he works, nor does any part of the carpenter's art exist within what he makes, but the shape and the form come to be from his agency through *kinêsis* in the matter. It is [the carpenter] soul and his knowledge, in which the form resides, that moves his hands (*kinousi*) or any other part of him with a motion of some definite kind, a motion varying with the varying nature of the object made.[14] In like manner, in the male of those animals which emit semen, nature uses the semen as a tool and as possessing motion in actuality, just as tools are used in the products of any art, for in them lies in a certain sense the motion of the art. Such, then, is the way in which these males contribute to generation. But when the male does not emit semen, but the female inserts some part of herself into the male, this is parallel to a case in which a man should carry the material to the workman. For by reason of weakness in such males nature is not able to do anything by any secondary means, but the *movements* (*hai kinêseis*) imparted to the material are scarcely strong enough when nature itself watches over them. Thus here nature resembles a modeler in clay rather than a carpenter, for she does not touch the work she is forming by means of tools, but with her own hands (*GA* 730b10–31, my emphasis).

Here Aristotle is detailing the exceptional case of insects. Normally semen acts as a medium between parent and offspring.

and other movements but equates these with spatial motions around the body (74). '[M]otive *pneuma* also requires an extra condition, namely pulsation, if it is to leave the heart and begin its movement around the parts of the body' (64); 'as it courses its way around the various parts of the body . . . [it follows] the contours of the various parts' (77). Coles appears to wrongly assume circulation of the blood in Aristotle.

12 For example *GA* 740b25–34 where *kinêsis* only occurs in the singular, of which Gelber (2010) writes 'Aristotle tells us in this passage that the *kinêseis* [plural] . . . are used by nutritive soul . . . ' (207).

13 Cooper (1990) also cites this passage (19).

14 A scribe adds in at this point: 'his hands move the tools and the tools move the matter'. See Aristotle (1942) 120.

In insects, males have no generative residue, so the male role must be brought about directly from the fashioning part inside the animal. Aristotle uses the craft analogy to try to explain how this takes place. It is as if the offspring were a lump of clay, which someone brings to an immobile craftsman who puts his hands out and moulds that clay. When Aristotle uses *kinêseis* of the movements of the craftsman's hands, he is using it of the spatial motion of those hands; this is appropriate to describe the image. However, this does not indicate that in the actual case of animal generation (rather than the analogous case of the potter), spatial motion is meant. The male animal does not move his hands (or any other part of his body) to form the offspring; no spatial motion is required or implied.[15] Thus it is wrong to equate *kinêseis* in this passage, where it refers to the craftsman's action, with the technical use of the term later on the in the treatise.[16] The male role as *hê archê kinêseos* ought not to be taken, then, as a set of spatial motions.[17]

Spatial motion is also not an ideal way to think of the male role as a singular *kinêsis* in the context of generation. One commentator recognizes this and suggests that 'the *kinêseis* in the spermatic fluids are probably much more like the changes in a chemical reaction than locomotive movements ... '.[18] The modern idea of a chemical reaction, whereby atoms within the molecules of materials recombine to form a new material, may not be so promising a way to conceptualize Aristotle's male role either, though. The closest he comes to chemical change is in his description of the type of mixture whereby the original ingredients cannot be extracted – as happens in cooking. Indeed, Aristotle recognizes that such changes are very often accompanied by heat; thus one might wish to explore the relation between *kinêsis* and heat in generation. Is the *archê kinêseos* a qualitative change in heat, such

[15] Menn (2002) also argues that there is nothing analogous to the movement of the artisan's limbs in cases of natural generation (137).

[16] See also Henry's rejection of *kinêseis* as demiurgic motions in the semen (2006b) 436–42.

[17] Later on in this chapter I will argue that, at least in certain situations, the soul does not require a physical mechanism. See Chapter 6.3.

[18] Gelber (2010) 208. I will disregard for the moment that she assumes plural *kinêseis*. She adds that, '[c]oncoction endows the blood with *kinêseis* – a series of heatings and coolings, similar to a chemical process'. See also Gotthelf's (1976) reluctance to translate *kinêsis* 'motion' (238 n.20).

that the body (*sôma*) of the male semen is undergoing change from hot to hotter and conveys this to the matter?[19] Qualitative *kinêsis* from cold to hot and from hot to cold certainly do occur during embryonic development (*GA* 740b32).[20] Aristotle describes the development of the non-uniform parts, such as hands and faces, as the effect of a *kinêsis* towards hot. For instance, the brain and the eyes must be cooked into their final form, which is difficult to achieve due to their coldness. Only a very strong *kinêsis*, the heating action of the heart, will bring them to completion (*GA* 744b3–4).[21] However, this type of *kinêsis* never constitutes a full or adequate account of the formation of parts of living bodies.

> Hot and cold and the *kinêseis* set up by them are, since solidification is due to hot and cold, sufficient to produce all parts of this sort, that is to say, all uniform parts like flesh, bone, hair, sinew and the like . . . But no one would suppose that this was the case with the non-uniform bodies which they in turn compose (for example, head, hand or foot), for though cold and heat and their *kinêseis* will account for the production of bronze or silver, they will not account for the production of a saw or a cup or a box. Here human craftsmanship is the cause, while in the other cases it is nature or some other cause (*Mete.* 390b3–14).

Explanation of the formation of bodily parts requires reference ultimately to 'nature or some other cause', the analogue of craftsmanship in craft productions. The heating *kinêseis* are more like the necessary material conditions which normally underlie generation: necessary but never sufficient.[22] The formation of the embryo is not due, therefore, to the heating action of the male semen only. Instead the male, being the formal cause of generation, provides the primary transition which makes an animal exist where there was not one before. In fact, the male would appear to merely begin this transition – an animal will not be completed until birth, or possibly not even until maturity. For

[19] *GA* 736b35. See Bolton (1987) 163. Code (1987) 57 understands the actual movement(s?) of the male contribution as heat: 'The vital heat in pneuma causes the generative motions in the menstrual fluid when, as a result of physical contact, this heat is transferred from the semen to the menstrual fluid.'

[20] Hot and cold play a part in setting the uniform parts (*GA* 743a5).

[21] Another passage where Aristotle describes a change in temperature as *kinêsis* occurs at *Somn.* 457b12–13: 'the stomach is hot when empty, but the filling of it chills it because of the *kinêsis*'.

[22] See Cohen (1989) and also Chapter 6.4.

this reason, *archê* is more likely to mean 'beginning' than 'principle' and the *kinêsis* indicates the process of substantial change which is about to start. So far, then, it has been established that the male role as *hê archê kineseôs* is not a spatial motion or a qualitative change in temperature and it is not initially characterized by Aristotle as comprising a set of motions or changes. Instead, the male provides *hê archê kineseôs*, the beginning of the process of substantial change which must occur if a new substance, that is a new animal, is to come into existence.

5.2 Male and female as jointly actualized potentials

The most straightforward way to understand the male roleas *hê archê kineseôs* is to think of it as an agent. Animal generation is an instance of substantial change. From a general standpoint, every such instance is a process and requires an *agent* to bring about the change. It also requires a *patient* – something with the capacity to change (*Ph.* III.1–3).

It is necessary that there be a generator and that from which it generates, and even if these be in one thing,[23] they must differ in kind and there will be a different account of each. In those in which these abilities are separate, the bodies of the active one and the passive one naturally differ. Therefore, if the male is like the mover and actor, and the female, *qua* female, is like the passive recipient of that action, then the female would contribute to the semen of the male not semen but matter (*GA* 729a24–31).[24]

Both male and female generative residues exist in a state of potentiality. In generation, the male principle is the active agent and the female the passive one in an interactive process.

When that which is active and that which is passive come into contact, in the way in which they are active and passive (by way I mean manner, place and time of contact), then immediately the one acts and the other is acted upon (*GA* 740b22–5).[25]

[23] This would be the case of plants and hermaphroditic animals.

[24] ἀνάγκη γὰρ εἶναι τὸ γεννῶν καὶ ἐξ οὗ, καὶ ταῦτ' [Peck: ταῦτ' vulg. τοῦτ'] ἂν καὶ ἓν ᾗ, τῷ γε εἴδει διαφέρειν καὶ τῷ τὸν λόγον αὐτῶν εἶναι ἕτερον, ἐν δὲ τοῖς κεχωρισμένας ἔχουσι τὰς δυνάμεις καὶ τὰ σώματα καὶ τὴν φύσιν ἑτέραν εἶναι τοῦ τε ποιοῦντος καὶ τοῦ πάσχοντος. εἰ οὖν τὸ ἄρρεν ἐστὶν ὡς κινοῦν καὶ ποιοῦν, τὸ δὲ θῆλυ,ᾗ θῆλυ, ὡς παθητικόν, εἰς τὴν τοῦ ἄρρενος γονὴν τὸ θῆλυ ἂν συμβάλλοιτο οὐ γονὴν ἀλλ' ὕλην.

[25] καὶ ὅτι τὸ ποιητικὸν καὶ τὸ παθητικόν, ὅταν θίγωσιν, ὃν τρόπον ἐστὶ τὸ μὲν ποιητικὸν τὸ δὲ παθητικόν (τὸν δὲ τρόπον λέγω τὸ ὣς καὶ οὗ καὶ ὅτε), εὐθὺς τὸ μὲν ποιεῖ τὸ δὲ πάσχει.

One thing that Aristotle is not saying is that the male animal is active and the female animal passive.[26] It is the male and female principles of generation that merit these descriptions. Another possible confusion stems from imagining that only the female is potential. Instead male and female principles represent two different kinds of potency in Aristotle's general understanding of change: that of being able to act and that of being able to be acted upon (*Metaph.* 1046a19–26). Crucially, both are potentialities.[27]

One must also take into account the co-dependency of male and female. In the *Physics* Aristotle explains that the active agent is dependent on meeting with its passive counterpart.

> [C]hange occurs in the changeable object. It is the fulfillment of this potentiality by the action of that which has the power of causing change; and the actuality of that which has the power of causing change is the same as the actuality of the changeable; for it must be the fulfillment of both . . . there is a single actuality of both alike, just as one to two and two to one are the same interval (*Ph.* 202a13–19).

This means that the male contribution is, in a sense, just as dependent for its actualization on the female as the female is on it.[28] Further, the active and passive components in the process are both potentially able to bring about the same result. There is one single outcome of two joint potentialities. Aristotle explains that the generation of animals is like house-building in so far as the location of the activity is in the houses being built in the one case (*GA* 730b8), and in the new animal in the other.[29] In natural

26 It is common for Aristotelian commentators to view the female animal herself as passive in his theory. See also Barthélemy Saint-Hilaire in Aristotle (1887) I.109 who writes, 'l'agent est le père; le patient est la mère.' However, the passive role of the female is a specific point about the status of the female contribution as potential and not a denigration of it to utter passivity. Certainly, it is not a comment on the passivity of womankind in general. Allen (1985) 411 and Cantarella (1987) 60.

27 Aristotle does not often refer to the male contribution as potential before conception, focusing instead on the potential state of the female contribution (e.g. *GA* 737a23, 738b4, 740b20, 741b8). This may be because the male contribution is too hot to *become* any functional living part of an animal; instead, it must dissolve and evaporate (*GA* 737a12). Thus the male cannot be potentially alive in the same way that the female contribution can. Bos (2009) disputes this reading.

28 Both are necessary conditions and neither are on their own sufficient for substantial change.

29 'Where, then, the result is something apart from the exercise, the actuality is in the thing being made . . . and in generation, the change is in the thing that is being changed' (*Metaph.* 1050a30–4).

generation the actuality of the new living being contains within itself a source of change, which is how it takes over its own development at some quite early stage (*GA* 735a14–23). In any case, actuality is not properly applied to the male principle or the male semen, which is a detached bodily part or residue. Only a complete functional animal is the actualization of form.

This general account of the male role in terms of active agency raises a number of questions. The first is how the male and female roles as active and passive potentials manifest themselves in embryological development. Does the female contribute certain potentials and the male others, which are actualized as the embryo develops? In other words, does the female contribute some soul capacities and the male others, as some commentators allege? Or is such division of tasks improperly representative of Aristotle's view? Indeed the question of whether the male contributes differently from the female with reference to the soul of the offspring is a pressing one. And if the male does contribute differently, would this make it somehow alive, as opposed to the potentiality represented by the female matter? The next two sub-sections address these questions.

5.3 Male and female interaction and gradual foetal development

In *GA* II.3 Aristotle states that before the male semen performs its action, the female portion lacks the principle of soul[30] (*GA* 737a30). However, when faced with accounting for female-only wind-eggs in *GA* II.5, this idea is qualified. Because these eggs never become complete animals, but instead rot, the female contribution is said to provide potential nutritive soul. Aristotle concludes: 'the male always completes generation—it implants sentient soul' (*GA* 741b6). One line of interpretation makes much of this 'new' characterization of the contribution of the two sexes, viewing it as a refinement of the matter/form dichotomy.

[T]he meaning of the statement that 'the male supplies the Form' can only be that the male supplies that part of the Form known as sentient Soul: everything else,

[30] τὴν τῆς ψυχῆς ἀρχήν.

> including nutritive Soul, can be, and is, supplied by the female ... the only difference between them is that the male's contribution possesses also *sentient* Soul potentially (Peck in Aristotle 1942, xiii–xiv).[31]

The idea here is that, faced with wind-eggs, which clearly have some soul-like content or potential, Aristotle decided to divide the male and female contributions to generation in terms of soul parts, and specifically in terms of nutritive and sentient soul parts. The male supplies the sentient soul to the female nutritive soul.[32] There are two ways, however, in which this description is misleading. First of all, it presupposes that nutritive and sentient souls are in some quite robust way separable. Second, it imagines, in its most simplified form, that a portion of soul can be handed over or supplied by the male parent.[33] Any close analysis of Aristotle's account of gradual foetal development renders both assumptions suspect. In light of this, one must attempt to spell out more carefully how both parents' contributions relate to soul in Aristotelian embryology.

Before the passage in question, Aristotle explicitly undertakes to examine the relation between soul and the generative residues, assessing what soul faculties ought rightly to be attributed to (1) the female contribution, (2) the male contribution and (3) the embryo.

> [Is soul] in the semen and in the foetation to begin with, and if so where [does] it come from? No one, of course, would maintain that the foetation (*to kuêma*) is quite without soul, completely devoid of life in any sense, for the semens (*ta spermata*) and the foetations (*ta kuêmata*) of animals are just as much alive as plants are, and up to a point they are fertile. Thus it is clear that they possess nutritive soul ... It is while they develop (*proïonta*) that they acquire sentient soul as well (*GA* 736a32-b2).[34]

[31] See also Witt (1985) 52; Albertus Magnus, *De Natura et Origine Animae* in Allen (1985) 367.

[32] Henry (2006a) 283 and Gelber (2010) 199 accept this interpretation.

[33] Peck's idea that form in generation for Aristotle is the sentient soul supplied by the male has proven influential. See, for instance, Lange (1983), Horowitz (1976) and Tuana (1989). Nielsen (2008) also distinguishes male and female roles in generation in Aristotle in terms of the ability to 'impart sensitive soul' (402–3).

[34] πότερον ἐνυπάρχει τῷ σπέρματι καὶ τῷ κυήματι ἢ οὔ, καὶ πόθεν. οὔτε γὰρ ὡς ἄψυχον ἂν θείη τις τὸ κύημα κατὰ πάντα τρόπον ἐστερημένον ζωῆς· οὐδὲν γὰρ ἧττον τά τε σπέρματα καὶ τὰ κυήματα τῶν ζῴων ζῇ τῶν φυτῶν, καὶ γόνιμα μέχρι τινός ἐστιν. ὅτι μὲν οὖν τὴν θρεπτικὴν ἔχουσι ψυχήν, φανερόν ..., προϊόντα δὲ καὶ τὴν αἰσθητικήν.

The passage is a difficult one to interpret, since foetation (*kuêma*) is a term used of both (1) and (3) (e.g. *GA* 738b34).[35] (1) and (2) have soul of a kind because both, coming from a living body, serve a function, just as blood or hands do when they are parts of a living animal. It is clear, however, that neither male nor female residue could have actual soul, since neither are individual substances.[36]

The new animal as a genuine individual substance is the focus of this passage. If (1) and (2) meet, then this entity will begin to develop; it is only then that it can acquire sentient soul. In a sense, (1), (2) and (3) are all items that potentially develop into the new animal.[37] The *proïonta* (*GA* 736b1) refer back to the semens (possibly both 1 and 2) and the foetations (either 1 or 3) in the earlier sentence (736a34). Clearly, however, neither (1) nor (2) on their own can develop soul capacities. It is only in relation to the new individual that both can be said to have soul and both have it in the same way: as potentially nutritive.[38] Once they meet together and generation is successful, this potential is actualized; the embryonic heart forms nutritive soul first so that the embryo can begin to nourish itself and grow (735a14–23).

Separating out the sentient soul appears to be impossible. After conception, sentient soul is there already (*GA* 736b15–16). Animals have to have sentient soul immediately, for 'it is impossible for any part of the body whatever (hand, face, etc.) to exist unless sentient soul is present in it, whether actually or potentially'

35 The female contribution on its own is referred to as a *kuêma* in discussing wind-eggs (*GA* 750b10, 13); it is always the bodily mass which becomes the new animal. See Peck in Aristotle (1942) lxii–lxiii.

36 It is a mistake to ever construe Aristotle's male semen as literally alive, as in, for example, Duprat (1899) 307: 'il faut que le sperme soit vivant (it is necessary that the semen be alive)'.

37 It makes sense to say that male semen and female menstrual blood are potentially able to develop soul capacities, in so far as we can trace the embryo back before conception. Modern bio-ethicists, in their discussions of the moral standing of embryos and generative cells (in the context of abortion and stem cell research), often wonder about the way in which it is best to characterize the female ovum and male spermatozoon as potentially the offspring. '[O]ne could even say the same about the egg alone [i.e. that it is potentially the person], treating the presence of sperm as part of the environment necessary for further development . . . ' (Singer and Dawson 1988, 96). Quinn (1984), in contrast, does not wish to characterize the female ovum as potentially a person until the sperm has triggered development.

38 As Gelber (2010) points out, although the female contributes potential nutritive soul it is not a plant; wind-eggs do not actually have nutritive soul (200).

(741a10–12). Furthermore, it is clear that the embryo's sentient soul will emerge gradually in a process that cannot be separated from its nutritive actions and capacities. After the heart is in place, other parts develop successively from the mother's blood and this process eventually fulfils the first actualization of the senses – the possession of sense organs. Where a certain activity requires certain organs, then it does not enter into actuality until such time as the organs are there to operate. So there can be no locomotion without feet and no seeing without eyes (*GA* 736b23–4). Second potentiality consists in the potential for activity in the sense organs. Second actualization – the functioning of these sentient parts – occurs around the time an embryo starts to sleep. Being known to be 'asleep' means that the living being has the potential to become 'awake' and actively use its senses (*GA* 778b32–779a12).[39] Levels of awareness exist in the animal embryo itself, once it has become established.

(1) 1st potentiality of sentient soul = existence in embryo of: heart
(2) 2nd potentiality of sentient soul/ 1st actuality of sentient soul = existence in embryo of: sense organs (e.g. eyes, skin).
(3) 2nd actuality of sentient soul = the exercise of the senses (e.g. seeing, touching).

The second actualization of the senses, the experience and operation of the senses, is always dependent on the first actualization of sentient soul being in place.

Even in fully grown animals, nutrition and sensation are not strictly separable. The agency of the nutritive soul produces blood throughout life, blood which serves the needs of the sentient soul by maintaining the proper consistency for its particular sense faculties to operate (See Chapter 4). Furthermore, the nutritive soul is responsible for producing and maintaining the sentient parts of the body, that is, the first actualisation of sentient soul. It

[39] Plants cannot sleep because anything that sleeps has to be able to wake up (*GA* 779a2–4). Animal foetuses display their ability at all times to jump to the second actuality of the senses by waking up in dissections (or vivisections?) (*GA* 779a9, *Somn.* 454a15, b8). Chicks can be awoken and heard to peep in their shell (*HA* 561b27). While it sleeps, the developed foetus does not exercise the senses only because it is busy nourishing and growing: the operations of nutritive soul bring on sleep because the body expends its heating ability on the digestion of food so that no heat is left for the senses (*Somn.* 457a17–21).

makes little sense to separate the sentient portion from the rest of soul in the embryo either, particularly before the actualization of the senses. The male contribution is the source of the development of sentient soul; but it first needs to work together with the female matter in order to bring about the actualization of nutritive soul. Indeed the joint actualization of the potential of both male and female generative residues results in the gradual development of nutritive and sentient capacities located in the offspring.[40] Thus, the male contribution is not solely responsible for any portion of the soul. Rather, the female contribution, being poised to receive the impetus conveyed to it by the male, must 'share in' (*metaschê*) the change that occurs.[41] There can be no straightforward sense, then, in which the male 'supplies' the sentient soul.[42]

Given his views about the interconnectedness and inseparability of nutritive and sensory capacities, one may wonder why Aristotle associates only the male role with the sentient faculty of soul at *GA* 741b6. First of all, he is looking directly at the phenomena of wind-egg production. Faced with a complex female component of generation with differentiated parts, he simply means to remind the reader that the male is necessary as the efficient cause of generation; without the male the foetation will never develop into an animal at all, instead remaining an undeveloped wind-egg (which eventually rots, *GA* 741a23). Second, Aristotle takes it that soul in animals comes with the capacity for sensation and perception (*de An.* 434a30). The study of animal life often focuses on their ability to sense, something which makes them a superior sort of substance.[43] The male ensures that sentient soul is established – that an animal rather than a plant will come to

[40] See also Johansen (2012): 'It is the foetus' own nutritive soul that propels the development of the sense-organs' (140).

[41] *GA* 737a33. Platt translates *metaschê* as implying that the change or movement is 'imparted to' the female contribution rather than that the female 'shares in' it. The verb *metechô* cannot have the meaning 'impart to' (LJS 1120).

[42] Summaries along these lines tend to take the craft analogy too seriously. For example, Lewis (1994) 249 writes: '[Aristotle] describes how the male parent, acting as a craftsman and using heat and motions in the *pneuma* as his tools or instruments, imposes successive layers of form or soul – nutritive, sensitive, locomotive soul – on the material that comes initially from the female.'

[43] Sensation is a primitive sort of distinguishing capacity of which thought is the most advanced variety. The best or highest substance is God, which is the activity of thought thinking itself (*Metaph.* 1074b30–5).

be – and, in this way, is better than the female. Neither point need imply, however, that the male contributes sentient soul all on its own.[44]

5.4 The male role as actual

In *GA* II Aristotle explains that just as a carpenter brings shape and form through the *kinêsis* of his tools, so also there is a sense in which semen has within it the same kind of actual change.

[H]is [i.e. the carpenter's] hands move (*kinei*) his tools and his tools move (*kinei*) the material. In a similar way to this, nature acting in the male semen-emitting animals uses the semen as a tool, as something that has *kinêsis* in actuality (*energeia*) (*GA* 730b17–21).

It might seem puzzling that the male contribution is here described as being in actuality (*energeia*), when both male and female contributions are joint potentials poised to meet and bring about the actuality of the offspring (Chapter 5.2). In order to better get to grips with the sense in which male semen can be thought of as actual, we must turn to Aristotle's explanations of changes brought about by external agents in his *Physics*.

In many instances of change, the agent of the change must itself be changing. This is part of Aristotle's idea that there can be no action at a distance – so, for instance, what moves something else in space must itself be in motion (*Ph.* 202a2–5) and what heats must itself be getting hot (*Ph.* 256b31). An important exception to this general picture is the case of incorporeal changers[45] – of which soul is one type (*de An.* 405b33–406a4, 408b15). As Aristotle explains in his *Physics*, the thing which changes another, the efficient cause, could either be changed itself in reaction or could be an unchanged source of the change. If the former, then a chain of causes must always end with something that is a source of change in itself (*Ph.* 256a14–22). Each animal has a source of change and rest in itself (*Ph.* 200b10), its nature, form or soul

[44] When taken out of context, it is somewhat misleading to claim that 'the father alone provides the offspring's sensory soul' (Henry 2006a, 283).

[45] 'The manner in which the changer changes the changed thing is not always the same: whereas one kind of changer can only impart change by being itself changed, another kind can do so though remaining itself unchanged' (*GC* 323a12–14).

(*Ph.* 192b21–2). In animal generation, the father's soul serves as the source of change in another as well. The male parent's nutritive soul is unchanged and the source of change. Meanwhile, the material parts, first of all his demiurgic part, the heart or its analogue, are instruments of this soul. The heart conveys the change to the male semen, which means that the male semen is in a process of change itself, while also conveying this to the menstrual blood.[46] Thus, in instances of animal generation there must not only be an unchanged changer (i.e. the soul of the father) but a changed changer which the father's soul uses as a tool, that is, the male semen.[47] This is why the semen is said to be actually changing – because in order for it to convey the change that comes from the father's soul, it must itself be changing in that direction, just as the tool that heats another object must itself be heating up.

We must be cautious here about assigning anything like a special life force to the male semen on the basis of this account.[48] Although there is a sense in which the tool of the male animal's soul is in actual *kinêsis* – a state of changing – it does not thereby possess form or soul. The *kinêsis* here is for a *substantial change* into a living being and it would be wrong for us to regard the male semen as a living thing. The process of substantial change is more like teaching or building than heating. When we say that the tool of a heating agent is itself changing, it is actually becoming hot. This is not the case, though, when it comes to the tools employed in the

[46] The craft analogy is useful to explain this: the carpenter's hands move the tools which push around the materials (*GA* 730b21–31). However, it is not the case that the changes involved are local motions which alter parts. Instead, male semen conveys a single transition: the change from potential to actual substance. I am in general agreement with Henry (2006b) that *kinêsis* in the male semen is not motion of semen as it fashions parts of the offspring or makes changes in the menstrual blood (439–42).

[47] 'For there must be three things – the thing changed, the source of change, and the instrument of change. Now the changed thing must be in a process of change . . . the instrument of change must both change something else and be itself in a process of change (for it changes together with the changed, with which it is in contact and continuous, as is clear in the case of things that move other things locally, in which case the two things must up to a certain point be in contact); and the cause of change – that is to say, that which causes change in such a manner that it is not merely the instrument of change – must be unchanged' (*Ph.* 256b13–20).

[48] For example, de Beauvoir (1953) 8. While the female principle was passive for Aristotle, 'the male principle contributed force, activity, movement, life'. See also Blundell (1995) 106.

processes of teaching and building. Since 'that which teaches and that which builds need not itself be taught or built', similarly we may imply that that which makes the embryo come to life is not itself being brought to life.[49]

Another idea found in Aristotle's critics, related to the idea that male semen is alive or has special life force, is that only the male animal is the true parent (*gennan*) of offspring.[50] This hypothesis is unsound for a number of reasons. First of all, although *gennan* is sometimes used exclusively of the male, it is also used exclusively of the female.[51] This is not an anomaly or a mark of clumsiness on Aristotle's part, for the relationship between parent and child he refers to in many passages is a relation which the female parent also bears to offspring. Often where only the male is mentioned, there is no theoretical bar on the parental relation described applying just as well to the female's connection to its offspring.[52] Consider, for example, the idea expressed in *GA* I that semen has *megalên dunamin* – great ability or power, which explains why offspring resemble their generators (*tois gennêsasin*) (*GA* 726b12–14). The fact that the parts of offspring also come to resemble female parents would be easier to understand if *tois gennêsasin* here referred to female as well as male parents. This is legitimately part of Aristotle's haematogenous account: both male and female contributions are residues of blood, which is the final form of nourishment (*GA* 726b3–11), containing qualities peculiar to the

49 The male's nutritive soul is the source of change and in terms of the change taking place, it is unchanged. The tool of the male parent's soul is not actual (it is not the new living being) but is actually changing (in a sense) and bringing about change (*Ph.* 256b27–257a7). As Aristotle explains: 'it is the changeable that is changed; and this changed potentially, not in actuality, and the potential is in the process of fulfilment, and the process of change is an incomplete fulfilment of the changeable. The source of change on the other hand is already in actuality: e.g. it is that which is hot that produces heat, and in general that which produces the form possesses it. Consequently, the same thing in respect of the same thing will be at the same time both hot and not hot' (*Ph.* 257b6–11).

50 Sissa (1992) 77: '*Gennan* is literally the exclusive province of the father'. See also Keuls (1985) 145, Wilshire (1989) 93 and Lange (1983) 11.

51 For example, *tês gennôsês* at *GA* 741b38 where he argues that the new animal's *pneuma* is not derived from the mother. See also *GA* 731a11, 742a3, 768a6–9, 20, *Pol.* 1335b19.

52 Until recently, the male pronoun has been employed quite generally in English to indicate persons of either gender, for instance in the phrases 'when a judge decides his case' and 'an agent attends to his reasons for action'. I suggest that Aristotle's use of the masculine pronoun for the parent in Greek is similar to this.

animal possessing them.[53] Why, then, does Aristotle tend to speak of only one parent? A possible reason is that one parent producing one offspring is a simpler way to refer to reproduction, which is why it occurs in many of Aristotle's more general works.[54] Another reason he sometimes excludes the female is because, although it is a generator, the capacity to generate comes in degrees. As we will see, not all animals are as good at it as others. And so, the female, on a certain Aristotelian way of sketching matters, is not as good at it as the male. Only the male can initiate the substantial change which brings about the actualization of nutritive and sentient soul in the embryo, and this makes him a generator in a more complete and stricter sense.[55] Despite this, in those kinds where males and females exist separately, there is no single generator. The two parents together generate: superior and inferior are both generators of offspring.

53 Cf. *GA* 737a18–20. Semen 'is a residue of nourishment that is undergoing change (*metaballousês tês trophês*)' (*GA* 736b26). The idea that for Aristotle both parents contribute semen has been argued for in Chapter 3.3.

54 *anthrôpos anthrôpon genna* – 'a human being generates a human being'; *Ph.* 193b12, 198a26, 202a11, *PA* 640a25, *Metaph.* 1033b32, 1070a28. Aristotle never writes that *andros andron genna* (a man generates a man) or *andros anthrôpon genna* (a man generates a human being). The idea that reproduction is a one to one relation is also common in modern biology because it is easier to work with. Some feminists have criticized this usage as inaccurate. See, for example, Fox Keller (1989) 43: 'In much of the discourse of evolutionary theory, it is commonplace to speak of the "reproduction of an organism" – as if reproduction is something an individual organism does; as if an organism makes copies of itself, by itself. Strictly speaking, of course, such language is appropriate only to asexually reproducing populations . . . '

55 This is why the following sentiments in Aristotelian embryology apply most properly to the male: something which is potentially human must be made actually so by the agency of something which is actually human (*GA* 734a30–31, 734b22). The parent's form exists in actuality before that which it generates can exist (*Metaph.* 1032a25, 1049b29–30, 1034b17–18).

6

INTERPRETATIONS OF ARISTOTLE ON THE MALE ROLE

6.1 The principle of life (*zôtikên archên*)

Because the male parent initiates a new life, and a new soul, how it operates in embryology promises to inform us about Aristotle's theory of soul. In the previous section I have assumed a top-down approach, as detailed in Chapter 4 on the female role: the father's soul uses semen as a tool to bring about substantial change resulting in a new living being. I have also described how important the composition of the body and the internal nutriment are in underpinning many soul activities in the *GA*. However, it seemed clear that the capacities contained in the blood in particular were specific potentials for form, originating from the nutritive souls of the parents. Many interpretations of Aristotle wish to deny the vitalism inherent in this stance and some do so by reference to the male role in generation. As mentioned earlier, it is especially important to set out Aristotle's male role since misunderstandings on this score can result in a more negative account of the female than was intended. In this section I will argue that the *GA* account of the male role makes best sense if Aristotle is understood as some kind of vitalist. Opposing interpretations will be rejected, including Aristotle as hylozoist, theological vitalist, reductive and non-reductive materialist, emergentist and extreme hylomorphist.

Any philosopher of biology endeavours to address the following questions: 'What is it that makes X alive?' or 'How are living things to be distinguished from the non-living?'[1] Some speculate that changes in ancient approaches to this problem of life led to advances in rational science.[2] According to this account, before

[1] The very existence of life might be thought to create difficulties for our knowledge of the content of the world. See Zuby (2000), Weber (2003).

[2] Cornford (1912). See also Hammond (1895).

'science', myth posited that life comes from the gods in a series of births; the gods, then, provide the answer to the question of animate origins, deciding what to make alive.[3] In this story, as an explicit rejection of the earlier religious account, the Ionian Pre-Socratic thinkers (Thales, Anaximander and Anaximenes) held a hylozoic hypothesis that all, or some significant portion, of matter is alive. Thorough-going hylozoists, by taking everything to be alive, make explanations of animation available without recourse to divine intervention.[4] For them, there is no differentiation between the living and the non-living. Life is present at all levels of organization, making it unnecessary to posit any divine origin for them. For those hylozoists who only made *some* sorts of special matter alive, they explain that what is alive contains within it these special living materials.[5]

Hylozoism has been attributed to Aristotle on the basis of certain readings of *GA* II.3 where it is said that male semen contains 'the so-called hot', which makes it able to start life (*GA* 736b35). Before assessing this in light of a fuller quotation, the context must be detailed. Aristotle is considering the male semen as a material made-up of water and *pneuma*,[6] which is caught in tiny bubbles giving the fluid its white appearance and thick texture (*GA* 735a30–736a3). He proceeds to ask what happens to this bodily part (*sôma*) of the semen, which leads to the question of how male semen relates to the soul of offspring (*GA* 736a24–32). He notes that nutritive soul can be present in semen and foetuses; however, this capacity of soul is not separable from the physical body and so cannot enter from 'outside' during conception. Similarly, sentient soul depends on body and so is the actualization of prior potential in the appropriate material; only rational soul can be said to 'enter in from outside' (*GA* 736b27). He then attempts to clarify how the male semen is able to realize the higher soul faculties in offspring.

[3] The Pre-Socratics (1983) 7.

[4] Cornford (1912) believed that the hylozoism of the early Pre-Socratics rested on a rationalization of pre-religious magical practices (89, 123).

[5] For example, air for Anaxagoras (DK 59A117) and Diogenes of Apollonia (DK 64B4). Pre-Socratics (2010) 306–7, 438–9.

[6] Glossed at this point as 'hot air' (θερμὸς ἀήρ), *GA* 736a1.

6.1 The principle of life (*zôtikên archên*)

(1) Now, it is likely that every faculty of soul has to do with a body, different from, and more divine than, the so-called four elements. Souls differ from one another as being more noble and ignoble, so also the nature of this body differs. (2) For in all cases there is in semen that which makes it fertile, the so-called 'hot'. This is not fire or any other such power, but that which is enclosed in the semen and in the foamy *pneuma*, i.e. the nature in the *pneuma*, analogous to the element of the stars. (3) Because of this, fire generates no animal, and we do not see anything taking shape either in wet or dry fire. But the heat of the sun and that of animals [does generate], not only through semen, but also any chance residue different in its nature; this also has the principle of life (*zôtikên archên*). From these things it is clear that the principle of heat in animals is neither fire nor does it come from fire.

(4) The body of the semen, in which the principle of soul (*to tês psuchikês archês*) travels, on the one hand being separate from body, i.e. for those animals that enclose something divine [within them] (the so-called intellect is such), and on the other hand being inseparable: this body of the semen dissolves and evaporates (*pneumatoutai*), having a moist and watery nature (*GA* 736b30–737a17).

Some argue that sections (1) and (2) indicate that semen contains physical soul – the presence of which makes something living. The physical soul, on this view, 'the finest purest material', is usually thought to be the *pneuma*.[7] One key flaw in this reading is that no material passes from father to offspring, as Aristotle's craft analogy makes clear. In carpentry, the carpenter does not contribute even a miniscule amount of material to the timber when making a bed (*GA* 730b11–15). Furthermore, when Aristotle wonders what happens to the physical part of semen in *GA* II, this is because it cannot remain and become any part of the offspring, but instead evaporates (*GA* 737a12). It is also implausible to suppose that the male's unique contribution to generation is *pneuma* since the female contribution also contains it, *pneuma* being added to *all* blood at the heart (*Resp*. 479b33). A final point against this idea is that male insects do the same work as males in other species in starting off the process of generation, but they do not have any male semen; instead they achieve conception directly when the female inserts a part into them (*GA* 729b25). Thus, we must dismiss wholesale the idea that the male contributes some special or 'refined matter' in generation for Aristotle.[8]

[7] For hylozoism and soul as *pneuma* see especially Duprat (1899) 306–7, Verbeke (1945) 35 and (1978) 204, Preus (1975), Morsink (1982) 111–13, Solmsen (1957) 121, Reeve (2000), Rist (1989) and Bos (2003).

[8] See Preus (1975) for 'refined matter'.

Generally speaking, a hylozoist interpretation does not sit comfortably with Aristotle's account of living beings. He could not be the type of hylozoist who claims that all matter is alive since, for Aristotle, not all objects share the qualities required to count as genuinely alive.[9] He makes his views on thorough-going hylozoism clear in the *de Anima*, where he mentions those who 'think that soul pervades all things' and states that their view is 'absurd and irrational'.[10] It is also hard to believe that he thinks the presence of one special material is enough to make an entity alive.[11] Although he may think that life is not possible without the presence of a certain material ([1] 'it is likely that every faculty of soul has to do with a body'), no material is sufficient. Without soul as an explanatory factor which is able to establish a certain material as hypothetically necessary for it, that material could not bear the relevant relation to psychic functions.[12] Even materials within living bodies ought not to be called 'alive' in a strict sense. They only count as living under a certain description, that is, as part of a living animal's body.[13] Since hylozoism is at odds with Aristotle's outlook in biology, any interpretation of *GA* 736b30–737a17 which depends on or implies it must be rejected.

The question of what life is elicits two extreme responses – hylozoism, on the one hand, that life is a given and cannot be further analysed, and the view that life is nothing in itself on the

9 '[W]hat has soul in it differs from what has not in that the former displays life' (*de An.* 413b2). Things are said to be alive particularly in virtue of the activities of self-nutrition and generation (*de An.* 413a31–2).

10 'For to describe fire or air as a living creature is highly irrational, and yet to refuse to call them living creatures, if there is soul in them, is absurd' (*de An.* 411a15–17).

11 Those who think that Aristotle embraced hylozoism often cite sections of *GA* III in which he discusses the spontaneous generation of shellfish (e.g. Verbeke 1945, 35). There it is stated that the water and earth, which are abundant in the sea, both contain *pneuma* that can become enclosed in a shell and come to life. He concludes that, due to the presence of the *pneuma* in the aquatic matter, 'in a way, all things are full of soul' (*GA* 762a22). This comment should not be taken out of context as a definitive indication of Aristotle's hylozoism: when he says 'in a way' (*pôs*), he usually means this as a strong deterrent to taking his statement at face value. Further discussion of Aristotle's use of 'in a way' will be necessary when we face the difficulty of understanding what he means when he says that women are 'in a way' deformed men. For more on spontaneous generation see Chapter 7.4.

12 This point will be defended further in what follows in my arguments against reading Aristotle as a non-reductive materialist.

13 See Chapter 4.4 on flesh.

other hand. At the latter extreme sit other Pre-Socratic philosophers. For Democritus, for example, all that matter consists in is atoms travelling in a void. Since the atoms are not alive, it might be imagined that Democritus would have trouble explaining living things. For, if atoms on their own are not alive, how can combinations of atoms count as living? His solution is to say that life is not a mysterious quality handed to us by the gods, or present in all (or special types of) matter, but is nothing more than atoms in a certain arrangement (*de An.* 405a9–13). Aristotle found sort of account inadequate. He complains that it cannot explain living reality because it constantly relies on chance or accident. But chance and accident are not causes at all.[14]

Although Aristotle explicitly resisted what he saw as the eliminativism of some of his contemporaries, he also shared their interest in the parallels between animate and inanimate processes. For Aristotle and his contemporaries, the beginning of life is like the making of cheese, either through heating milk or adding rennet to it.[15]

> The action of the semen of the male in setting the female's secretion in the uterus is similar to that of rennet upon milk. Rennet is milk which contains vital heat, as semen does, and this integrates the homogeneous substance and makes it set. As the nature of milk and the menstrual fluid is one and the same, the action of the semen upon the substance of the menstrual fluid is the same as that of rennet upon milk (*GA* 739b21–6).[16]

Meanwhile, Aristotle's near contemporaries compare the construction of the embryo and the maintenance of the adult with baking bread and crafting pottery.[17] In one Hippocratic work,

[14] In failing to acknowledge formal and final causes, Aristotle's opponents eliminate functional natural objects from their ontology (Sauvé Meyer 1992). Scholars disagree over whether Democritus was a 'reductionist' or an 'eliminativist'. It all depends on whether he would count living things with souls as definite existents (which cannot be reduced to other more fundamental existents) or opts for them having merely an illusory standing (and thus allowing them to be eliminated entirely from strict ontology). See Warren (2002) and O'Keefe (2005).

[15] *Nat. Puer.* I.14.2; *Morb.* 4.32.1 (Hippocrates 1981, 7, 22); *Steril.* 213 (L VIII 412, II 3–6): *pêgnesthai, epagê* – coagulate or curdle. Empedocles likened the action that produced the entire living cosmos to the curdling of milk by rennet (DK 31B33). Cf. also *Job* 10:10 'hast though not poured me out as milk, and curdled me like cheese?' (KJV).

[16] See also *GA* 729a12–13.

[17] For Empedocles, construction of the cosmos was like firing pottery in a kiln (DK 31B73; Pre-Socratics [2010] 394–5). Embryonic development is likened to bread baking at *Nat.*

almost every conceivable craft is used in the hope of illuminating growth and nutrition, including metallurgy, cobbling, carpentry, house-building, cooking, statue-making and pottery (*Vict.* I.13–24, L VI 488–97). Aristotle employed very similar comparisons. When describing the construction of the embryo's body, he compares it to a procedure employed in making clay or wax statues (*GA* 743a1–4). Meanwhile, bones are likened to pottery, both being produced by the drying effect of heat, and the whole process of development compared with the techniques used by painters.

Bones (like pottery) cannot be dissolved by fire; they have been baked as it were in an oven (*GA* 743a19–21).

In the early stages the parts are all traced out in outline; later on they get their various colours and softnesses and hardnesses, for all the world as if a painter were at work on them (*dêmiorgoumena)*, the painter being nature. Painters, as we know, first of all sketch in the figure of the animal in outline, and after that go on to apply the colours (*GA* 743b20–5).

We can see why it has been attractive, then, to view Aristotle as sympathetic in such passages to materialism and eliminativism. The idea of forms as causes in his metaphysics can then be softened and sublimated in order to fit with more modern philosophical tastes. However, although he shared an interest in technological parallels to biological processes, craft examples are used differently by Aristotle. Rather than suggesting that the processes of conception, development and growth result from the interaction of inanimate elements, as he takes his contemporaries to be doing, Aristotle points to the design and agency necessary in each craft. The analogies provide compelling evidence that formal and final causes are operative in nature. Thus, his view is never eliminativist and any interpretation of *GA* 736b30–737a17 which presupposes or implies this to be the case cannot stand.

With Aristotle's rejection of both hylozoism and eliminativism in mind, we come to consider various ways of understanding the passage on the male role in the *GA*. Interpretations which result from the way in which commentators have understood the above passage fall into two broad camps, which I will call theological

Puer. 12.6, L VII 486–7. See du Bois (1988) 111–23 for an account of similar comparisons in Herodotus and Aristophanes.

and materialist. Theological interpretations, which strongly associate the male role with divine action, either argue that the male contribution to generation is material or immaterial.[18] The former can be dismissed since they are hylozoist and the latter do not survive a careful analysis of Aristotle's own theology, which I undertake in the next section. Materialist interpretations of the male role are those which focus on the material or 'physical' happenings at conception. Preoccupation with materials such as *pneuma* leads one set towards an eliminativist view of Aristotle, which cannot match his teleological commitments. Another set of materialist accounts, which explore the mechanism by which the male brings about form, are less obviously wrong. However, the idea that these mechanisms are adequate explanations on their own, without the backing of formal causes fixed in nature, fits ill with Aristotle's philosophy and so must also be rejected. Ultimately, I will argue that the male role in generation is best understood in the context of a certain type of vitalism, a vitalism which was attractive to Aristotle due to its advantage over rival viewpoints. This leaves intact a more robust role for the female in generation than many of the alternative interpretations here discussed and so fits better with his account of generation in the *GA*.

6.2 Theological vitalism

The male role as divine

One set of influential theological interpretations of Aristotle's male role focus especially on sections (2) and (3) of the above passage.[19] First, they emphasize that Aristotle compares the process of generation with celestial motion. *Pneuma*, the active physical component of male semen, is analogous to *aithêr*, the fifth element, which allows for the cyclical movement of the heavens. In the literature of the period, *aithêr* was the term used for this

[18] The male contribution of something divine and immaterial might either become the offspring's soul or act as an intermediary between the immaterial and the material, the eternal and the perishable: Verbeke (1945) 202.

[19] See, for example, Averroes in Davidson (1992) 233, Albertus Magnus (1890–99a) ch. 4, Aquinas (1952) 158 (as quoted in Allen 1985, 394) and Giles of Rome in Hewson (1975).

'bright sky', the upper region of the sky or 'heaven'. In *de Caelo*, where his theory of celestial motion is set out, Aristotle writes:

> It seems that the name of this primary body has been passed down to the present time by the ancients, who thought of it in the same way as we do, for we cannot help believing that the same ideas recur to men not once nor twice but over and over again. Thus they, believing that the primary body was something different from earth and fire and air and water, gave the name *aithêr* to the uppermost region (*Cael.* 270b17–21).

It is obvious that Aristotle was keenly aware of the religious connotations of the term '*aithêr*'. Given this, the association of this element with the *pneuma* within semen suggests to some that his account of the action of male semen is closely tied to traditional mythological accounts of generation.[20] Aristotle notes that once it has made contact, the male semen evaporates, not becoming any physical part (4). These two points are taken to support the view that, because *pneuma* is like *aithêr*, male action is similarly directed by divine forces.

Interpretations such as these make Aristotle's ideas fall in line with earlier doctrines in which it is God or the gods that make something come to life. I will argue that we must interpret the above passage in a different way. Although the male role is associated with 'the divine', what constitutes divinity here is not only different from that of many of Aristotle's predecessors, but it also shared little with the theology that was to succeed him. The commentaries and accounts of Aristotle that formed parts of the Islamic and Judeo-Christian tradition distorted his original attempts to place reproduction within *his* theological framework,

[20] '[L]orsqu'Aristote . . . trouve un lien d'analogie entre la chaleur vitale et l'éther, il puisse à une tradition bien établie qui situe dans les cieux la source de toute vie . . . ' (When Aristotle finds a point of analogy between vital heat and *aithêr*, he draws on a well-established tradition which locates the source of life in the heavens); Roussel (1980) 160. Anaxagoras was said to have thought that animals were created by the fall of 'seed' from heaven to earth (DK 59A117; Pre-Socratics [2010] 306–7). Ovid's mythological account of the creation of human beings in his *Metamorphoses* retains this idea: 'the new-born earth, freshly set off from the *aithêr*, still retained a few seeds of the heaven (*semina caeli*) from which it had come' (I. 80–81; See also *Metamorphoses* I. 363–5, 416–20). Ovid (1916). Zeus' ability to transform himself into rain and fall from heaven to earth, could account for the association of 'seeds' from heaven with fertility. See Guthrie (1957) 30. Divine breath continues to be evoked to explain the vivifying powers of semen in a seventeenth-century compendium on embryology (Anonymous 1986, 33, 36).

and instead drew from their own very different perspectives. It is necessary, first, to consider a selection of these interpretations.

In the fourth century, Augustine took *pneuma* to be incorporeal, which accounts for the Western Christian meaning of 'spirit'.[21] For him, and many subsequent Christian thinkers, because the male semen contains this incorporeal spirit, his reproductive role is like that of incorporeal intellect. Arabic commentators, who drew on such ideas, even more strongly connected the male role with divinity, only rather than the intellect belonging to the male, the divine intellect itself was taken to be active in generation. Thus male semen became a vehicle for the actions of God in the material world; divine intellect not only ensures the continuation of the motions of the heavenly spheres, but also implants souls in animal embryos.[22] Because form is something over and above the state of matter for Aristotle, it seemed plausible to these thinkers that forms had to come from the divine intellect, which 'emanated' them. It turns out that much of this interpretation rests upon an early error in translation. In section (4) of the above passage (*GA* 737a11–17), Aristotle explains that the soul-like principle is sometimes separate from matter, and mentions intellect. The only known medieval Arabic translation leaves out the 'sometimes' and so makes it seem that for Aristotle the origin of soul which is carried by the seed is always separated from matter and is, hence, divine.[23] This theory created an interpretative dilemma. Aristotle's favoured explanation of the formal cause of animal generation is that offspring become X through the agency of something that is already X (i.e. the parent).[24] Arabic interpreters had to try to explain how the general Active Intellect (i.e. God) could be the cause of offspring unlike it: how the offspring became X through the agency of something not-X. In attempting to accommodate their theory, they already had to admit this rift between Aristotle and their neo-Aristotelian innovations.

21 Verbeke (1945) 493–6. 22 See Davidson (1992) 232.

23 '. . . when Aristotle spoke of something separate from the body of the male semen and divine, which is the origin of soul, he meant an incorporeal agent that contains in itself the souls of all living beings and imparts them to sublunar matter' Ibn Bâjja, quoted in Davidson (1992) 234.

24 *Metaph.* 1032a25, 1049b30, 1034b17–18; *GA* 734a30, 734b22.

Rejecting the Arabic understanding of the male role as divine vehicle, medieval Christians still adhered to the Augustinian view mentioned earlier. One example is Giles of Rome's detailed commentary on the *GA* in the thirteenth century. For Giles, the *pneuma* contained within semen, which he dubbed 'non-organic spirit', was incorporeal. He insisted that by distinguishing and giving order to the foetal parts semen was performing an intellectual act. Semen is then divine because it acts like the divine intelligences (lesser gods) that move the heavenly spheres.[25] There are two key ways in which this misrepresents Aristotle's views. First of all, there is a misunderstanding of his statement about the relationship between what exists in the male semen and *aithêr*, the fifth element. Second, and more crucially, it fails to accord with Aristotle's own theology.

When Aristotle associates something within the semen with the fifth element, which makes up the 'more divine' upper region of the universe, he is not saying that *aithêr itself* makes semen fertile or suggesting that a small part of the most divine matter from the upper world is contained in male semen.[26] Instead, he makes an analogy between something in the semen and *aithêr* in order to aid our understanding of the process of animal generation. *Aithêr* in the upper region is the sort of stuff that naturally moves perpetually in circles, thus differing radically from the other four elements (i.e. earth, air, fire and water), which move rectilinearly (*Cael.* I.3). One interpretation of why Aristotle associates *aithêr* with something in male semen is that *pneuma* is that something and is a special earthly version of *aithêr*, an element separate from and different to the other four elements. Aristotle does not say, however, that *pneuma* is analogous to *aithêr*; rather, he says that what makes the semen fertile, or 'the hot', which is 'the nature *in*

[25] Hewson (1975) 116, 112 and 114–15. For Aristotle, the intellect (*nous*) has a special status as incorporeal and separable from material situations, but exactly how this works is not clear. Within the context of the generation of animals, references to intellect and its incorporeality are asides (*PA* 641a32-b2, Lennox 1999). Aristotle is mainly concerned with how the embryo becomes viable in all animals, the majority of which can never develop intellectual faculties.

[26] This is one of the more far-fetched interpretations of the text. See especially Preus (1975) 85, Rist (1989), Reeve (2000) 48, 59 and Nuyens (1948).

the *pneuma*', is analogous (my emphasis).[27] It seems reasonable, then, to suppose that the hotness of semen is analogous to *aithêr* not because it is an earthly version of the heavenly element, but because it is 'more divine' than those elements taken on their own.

For Aristotle something is divine if it enjoys continuous existence in the same state. Divinity comes in degrees, as is clear from his frequent use of the comparative or superlative forms of the adjective divine – *theos*.[28] The most divine thing is God, who perpetually carries out the same superior activity of thinking (*Metaph.* XII). The rest of the universe divides into two realms, the upper region consisting of the stars, planets, sun and moon and the region below the moon inhabited by perishable animals and plants and the inanimate elements (*PA* 644b25–645a2; *GA* 731b24–8). The fifth element is more divine (*theioteros*) than the other four because it facilitates the cyclical turning of the upper region, which is more divine (*theioteros*) than the lower one (*Cael.* 269a31, 288a4). The heavenly bodies are the divinest (*theioeteta*) of all visible things and so are more divine than humans (*Ph.* 196a33; *EN* 1141a36). However, humans do not fare that badly since they possess the ability to think or reason, which is 'the most divine element in us' (*EN* 1177a15). Due to their intellectual abilities, humans partake in divinity more fully than all other animals (*PA* 656a8), but other animals are not devoid of their own share. Through reproduction of the same form from one generation to the next, animals and plants share in (*metechein*) the divine (*de An.* 415a29), which is why form is more divine than matter and the male role of establishing form more divine than the female one of providing perishable materials (*GA* 732a3–9).

In the *Generation of Animals* and elsewhere, Aristotle explains that the degree of divinity natural objects are capable of (nature indicating change, rather than stability) is not the continuous existence of individuals, but the continuity of type or kind. He often refers to the ability to reproduce the same form from one

27 See also Balme in Aristotle (1972) 163.

28 *Theos* occurs in its comparative form in the above passage (*theioteros*, 'more divine' – section 1) and also at the beginning of *GA* II, where Aristotle explains why male and female are separated in higher animals, by saying that the male is more divine (*theioteros* again) than the female.

generation to the next as cyclical; types return back on themselves (*antistrephein*) in the next generation (*GC* 338b7–17).[29] Talk of cycles in the context of the perpetual reproduction of animals clarifies the analogy with *aithêr*, for both that which makes male semen fertile and the *aithêr* are materials that facilitate eternal cycles, which is how they can be associated with the divine. In medieval thought Aristotle's idea of divinity in terms of perpetuity or striving towards perpetuity was left behind and replaced by different models of divine action through male semen. These developments of his thought ought not to be associated with Aristotle himself. Not only is it inaccurate, it can skew our understanding of his views on women and the female, as I will explain in the following section.

More erroneous associations: men, intellect and action

For many Christian and Arabic medieval commentators the divinity of the male role must be contrasted sharply with the earthly and degraded position of the female role. This begins with an emphasis on the superior action of male semen and its supposed similarity to God's agency and intellect. God and intellect, in later common understanding, make things happen at a distance, not involving themselves in material processes. Recall that the male role has been likened to the agent that sets in motion a series of complex events (*GA* 734b9–11, 741b8–9). In the previous chapter it appeared that the triggering of an automatic puppet lessened the significance of the male role to some extent, when compared with his role as carpenter or craftsman. However, later commentators thought of this action as superior since it is the kind of activity God would engage in.[30]

Medieval interpretations of Aristotle not only associate the male role with intellect but also connect this in a thorough-going way

[29] See also Lennox (2001) 137.

[30] The pseudo-Aristotelian *On the Cosmos* likens the influence of God to that of the person who starts off an automatic machine: 'The most divine thing of all is to produce all kinds of results easily by means of a single motion, just like the operators of machines, who produce many varied activities by means of the machine's single release-mechanism' (*Mu.* 398b8–17; after Furley).

with an overarching gendered ontology. Albertus Magnus, the thirteen-century scholastic whose edition of and commentary on Aristotle's biological treatises were in wide circulation throughout this period was just one thinker to perpetuate such ideas. Being responsible for the growth of the embryo, intellect itself, he thought, must be transmitted by the male and be somehow contained in his semen.[31] Thus, he held that intellect belonged exclusively to men and that this was the key to understanding their role in the generation of a new animal, meaning that it must be absent from all manifestations of femaleness. This characterization of gender owes more to developments in early medieval thought than to any Aristotelian text. Albertus draws particularly from Philo of Alexandria (first century BCE to first century CE), whose complex allegories represented woman as earthly, sensual and corruptible while men were heavenly, rational and immortal.[32] Augustine of Hippo carried through the tradition of these powerful associations.[33] In this pseudo-Aristotelian scheme, a metaphysical gendered hierarchy takes hold: spirit, reason and heaven are assigned to the masculine realm and their opposites to the feminine one.

Aristotle's attitudes towards gender difference do not take the same form as these later developments. First of all, on a general level, his biological metaphysics is at odds with such gender associations. When men were exclusively associated with intellectual endeavour, early medieval writers made sure to distinguish between 'man' and 'human being'. Later, Augustine also found these distinctions crucial, using 'man' (*vir* in Latin) rather than 'human being' in his discussions of intellect. Aristotle, in contrast, did not maintain that all higher mental functions belonged only to male humans. When describing human intellectual capacities, he

[31] Huby (1990) and Allen (1985) 368; Albertus Magnus (1890–99a).

[32] Philo *On the Special Laws* I.200f.; III.178–80 (Colson in Philo of Alexandria 1937, 214–5; 1312–4); *On the Account of the World's Creation Given by Moses* (Colson and Whitaker in Philo of Alexandria 1929). See also Sly (1990) 48–53.

[33] 'Flesh therefore stands for wife, similarly spirit sometimes stands for the husband.' '[Providence] submitted the corporeal creatures to the spiritual, the irrational to the rational, the earthly to the heavenly, the female to the male.' Augustine *Tractus in Iohannis Evangelium*, Tract II.14, (1969) 71, 201–3; see also Augustine (1972) vol. 8, 277 and Allen (1985) 234–5.

uses the gender neutral term, human being (*anthropoi*). For example, in the important statement at the very beginning of his work on primary philosophy, Aristotle states that 'all humans (*anthropoi*)', rather than all male humans, 'naturally desire knowledge' (*Metaph.* 980a22). It is also the case that Aristotle's own views on intellect do not necessarily exclude women. Neither male nor female are associated by Aristotle with mind or emotion in the context of describing their biological functions.[34] It is very likely that both men and women have intellectual capacities, as both share in the form human being (*GA* 730b35). And finally, Aristotle did not think, like some moderns, that the intellectual capacities of female animals are adversely affected by their reproductive role. In fact, care of offspring is often an aid to or sign of rationality.[35]

Only humans gain intellect; only they have this element of the soul which enters from 'outside' in the process of generation and development. How, then, could all animal generation depend on it? Furthermore, intellect is not associated with the male's reproductive capacity. The male concocts semen thoroughly and thus is able to bring about the generation of another of the same form. Divinity of the male in generation consists not in its intellectual capacities, but in its ability to ensure the continuation of the kind – capacity shared by all non-human animals. There are two ways in which humans participate in divine activity: (1) reproduction along with animals and (2) intellectual activity. (2) does not link up to the discussion of (1) in any of Aristotle's writings.

The popularity of medieval adaptations of Aristotle's theory helps to explain why he has been burdened with much that does not originate in his writings.[36] What is particularly striking is the

[34] Some cite *Pol.* 1254b6–15 as evidence (e.g. Sly 1990, 52). See (P1) Chapter 1.3. The passage aims to illuminate the types of rule he is describing (despotic and constitutional; *Pol.* 1254b4–6). The male rules the female with a constitutional rule, which is a political rather than a biological distinction. Thus, although he clearly makes such associations he does not attempt to provide justification from concrete biological differences between the sexes.

[35] See especially *HA* 611a10–23 on female horses and deer, 611b33–612a2 female bears and 612b21–32 on female swallows. The human mother's feelings for her children serve as a paradigm of friendly love, which is crucial for virtue in his ethics (e.g *EN* 1159a27–32, 1166a4–6).

[36] See Coles for a convincing dismissal of medieval interpretations (1995) 65. The very notion that *pneuma* could be immaterial, for instance, is unacceptable to Aristotle, who

way that modern readings of Aristotle's theory which follow medieval interpretations distort his views on gender roles. Feminist commentators in particular are often attracted to the idea that the male, or men in general, in Aristotle's theory represent immaterial soul, divinity and intelligence.[37] The idea that the male provides and the female lacks 'active reason' or divine intellect in the process of generation is also prevalent, and depends obviously on various medieval (mis)interpretations of Aristotle's cosmology.[38] Most recently, the association of reproduction with intelligence has been extended to fill in a network of associated gender dichotomies ascribed to Aristotle himself.[39] It is thus important to keep in mind the disparity between Aristotle's works and medieval gender hierarchies in order to avoid any misapplied criticism and misunderstandings of his views on females and the female role in generation.

6.3 Materialist interpretations

In the process of generation, soul (*psuchê*) acts on the world to produce a new embodied animal. Aristotle's embryology promises, then, to explain how he thinks that soul and material body interact, an issue of great interest to all post-Cartesian philosophers of mind. Many urge that Aristotle is not a dualist.

takes it to have 'body' (*sôma*). The contrast between spirit (*pneuma*) and body (*sôma*) is first found in the New Testament (e.g. *Ep. Cor.* V.3).

37 The Aristotelian analogy of male as craftsman, for instance, is likened to God, who breathes life into inanimate materials: Tuana (1989) 150.

38 Feminist medievalists, such as Allen (1985) 104 and Sly (1990) 27, think that Albertus Magnus or Philo were merely adhering to Aristotle's doctrine when they associated the male generative role with intelligence or intellect. Grosz (1990) 155 holds that for Aristotle the female 'is unable to provide any psychical input for the regeneration of species, for she lacks the principle of active reason'. See also Hirshman (1998) 219 and Green (1992) 84–5.

39 Allen (1985) 94. Sometimes this is understood in terms of men being 'more divine' because they are a higher species who participate in reason, while women are tied to their bodily functions and are more like animals. See Wilshire (1989) 93 and Zeitlin (1996) 109. The hierarchization of gender relations in Aristotle's biology along these lines is also important to the interpretations of various other feminist philosophers, who believe that the association of earth and body with woman, heaven and spirit with male is the construct of Aristotelian reproductive metaphysics. Irigaray (1985a) 133, 141 and 162. See also Tuana (1989) who accepts the view that male is incorporeal spirit as Aristotelian and Chapter 1.1.

Indeed, his account of the soul as the actuality of an organic body potentially possessing life (*de An.* 412a20–22) indicates a move away from the soul as a 'ghost in the machine'. Of course, Aristotle did not subscribe to the idea that complex actions and events which people comprehend at the level of macroscopic experience are not 'real'.[40] However, it is considered plausible that his ideas represent a good alternative to reductionism or eliminativism, while not rejecting the truth and relevance of the physical sciences; and so Aristotle becomes a proto-functionalist.[41] Functionalism is an attractive theory of mind in that it can explain the intuitive contrast between mind and body without the need to posit (metaphysically problematic) immaterial substances. The mind, for functionalism, is something different from the purely physical only in so far as it is a description of the way in which systems operate in a functional manner; this description is not reducible to a physical description since it comprises a different level of analysis.[42] Thus, pain is C-fibres firing in terms of its physical instantiation but it is also the feeling one gets when damage occurs in the body, which then causes one to avoid such damage; it plays a functional role within a system. In order to find that Aristotle's views are in line with this modern theory, one needs to show that he was not a substance dualist and, in particular, that he did not think that immaterial form or soul are robust causes. Nussbaum, having taken up the idea of

[40] For Aristotle 'psychological or biological states and processes could not be fully explained in terms of the physical states which form their material cause. This much is agreed by all' (Charles 1988, 1).

[41] See Granger (1990) 27–8 on the vogue for characterizing Aristotle as functionalist. Putnam, a prominent proponent of modern functionalism, wrote that he 'was pleasantly surprised to find that my view was substantially the same as Aristotle's, although stated a bit more precisely with the aid of the vocabulary of contemporary scientific methodology and cybernetics' (quoted in Johansen 1998, 6 from Putnam 1975, xiv). The appropriation of authority from Aristotle for one's own theory seems to be conscious in the case of Putnam, who I once heard say, in relation to his use of Aristotle, 'the smarter I get, the smarter Aristotle gets' (Conference on Aristotle and Contemporary Science, Aristotle University of Thessaloniki, September 1997). The idea that Aristotle is a functionalist is taken very seriously by some. One example is Lewis (1988) who writes, ' . . . in regarding form or soul is a *dunamis*, Aristotle is in effect treating it as a *functional* property' (78).

[42] Modern physics assumes that all entities and causes are material, and so 'physical' and 'material' are practically synonymous. This may well not be the case for Aristotle. The study of *phusis* – nature – involves non-material as well as material entities and causes. While I acknowledge this, in an attempt to avoid confusion, I will use the term 'physical' in the modern sense as most commentators in this debate do.

Aristotle as functionalist, argues that the material causal account is a sufficient one for mental or form-related processes. On this view, the fact that Aristotle emphasizes what occurs on a macroscopic level is not thought to be part of the causal story – but, rather, a different way of presenting the same events.[43] The explanation on the macroscopic level is irreducible, but causally inert.

The idea that Aristotle's views on psycho-physical interaction fit with modern non-reductive materialist theories have inspired some scholars to argue that sufficient physical conditions of all psychological processes can be given in Aristotle's terms. Reference to soul and form will be merely heuristic.[44] Some think it attractive to find this to be the case since it avoids the antiquated metaphysics that the alternative interpretation leaves in place. Form as causally efficacious allows for mysterious entelechies (actualities) that harbour significant influence over the physical world. This also looks like a 'just so story', a simple answer to the question of why X is the way it is – because it was always going to become X![45] Despite this understandable preference for materialism, when we turn to his embryology, it is difficult to maintain that this is an accurate representation of Aristotle's view. The content of the *GA* strongly suggests that soul and form are causally efficacious and can be immaterial, and thus Aristotle turns out, in this context at least, to be a vitalist. I will explain in what follows how particular interpretations of Aristotelian mind, starting with functionalism, often have difficulty accommodating soul in Aristotle's embryology. Without modification, these interpretations threaten to undermine our understanding of male and female roles in generation in his

[43] *de An.* 403a25–403b4 on affections (τὰ πάθη) of the soul as 'enmattered accounts' (λόγοι ἔνυλοί εἰσιν) appears to provide support for this view. 'A natural philosopher would define an affection of soul differently from a dialectician; the latter would define e.g. anger as the appetite for returning pain for pain . . . while the former would define it as a boiling of the blood or warm substance surrounding the heart' (*de An.* 403a29–403b1).

[44] For teleology as merely heuristic in Aristotle see Nussbaum (1978) 69–99, Johnson (2005) 182 and also Chapter 10.

[45] This is sometimes also termed 'backwards causation' and may remind the reader of one of the most notorious criticisms of Aristotelian science adduced by Molière (1962), who has a character ask 'why does opium cause sleep?' only to receive the unhelpful scholastic reply 'because of its soporific power' (659–61).

theory. The final section will attempt to counteract objections to seeing Aristotle as a vitalist.

On the functionalist interpretation, Aristotle has two parallel ways of characterizing processes, the physical and the psychic. On the physical level, the explanation of any phenomenon is completable in purely physical terms, thus not violating the causal closure of the physical.[46] Some turn to the zoology, and particularly Aristotle's account of generation, in an attempt to substantiate this position. It might at first seem unclear how embryology could provide insight regarding a thinker's theory of mind. But although mind plays no role in the processes of animal generation, soul (*psuchê*) does. *Psuchê*, for Aristotle, indicates whatever it is that makes something alive, including a wide variety of living functions and activities. Soul is also form, *eidos*, which helps to explain how a living being generates another living being the same as itself. Generation requires the action of form/soul on the world. The male parent ensures that another member of the kind comes into being through a psychic action. As we have seen, the functionalist interpretation of Aristotle's account is that soul is a way of describing the physical functions in a living body. Thus, functionalist commentaries hold that when we consider how the new offspring comes to be due to the male agent, we must also, in addition to the formal account, find a physical mechanism that is complete or completable. Thus, the male role will be the 'motions' or efficient causes that physically form the embryo. Aristotle would thus have no need for immaterial entelechies to intervene, because the entire process by which the male parent brings about generation is a physical one.[47]

A broader disagreement underlies this perspective on Aristotle's theory of generation. Gotthelf (1976) argues that, for Aristotle, no

[46] For a general account of the non-reductive materialist stance see Crane (2001) ch. 2. Charles (1988) 20: Aristotle 'exclud[ed] reduction while also allowing for a *full* physical story of the relevant kind to be told'. Lewis (1988) 55: 'On the view I wish to defend . . . the material/efficient and final points of view allow for the possibility of different but non-competing explanations of the *same* things. Although the different explanations by themselves give only part of the full explanatory story, each point of view can be on its own terms *complete*. In particular, I argue that a chain of material/efficient causes can *fully necessitate* a given particular effect that is also explained by form/final causes.'

[47] Both Charles (1988) and Lewis (1988) follow Nussbaum in attempting to show that causation, for Aristotle, occurs only at the level of the physical.

explanation in terms of material factors, such as heat and cold or expansion and contraction, will ever be sufficient without reference to the forms in nature that direct development and growth.[48] Meanwhile, on the other side, Charles (1988) and Lewis (1988) propose that material and efficient causes can stand alone.[49] Both interpretations have their drawbacks. The reading which supports non-reductive materialism leaves Aristotelian forms and his teleological account rather limp and useless.[50] On the other hand, if forms are robust causes that must be present in order to complete physical causal chains, then Aristotle's science is unacceptable by modern standards.[51] In order to see which theory it is most plausible to accept, we need to look more closely at the debate and at the text of the *GA* and then consider how to handle any subsequent difficulties.

Charles (1988) initially argues that Aristotle's 'discussion of animal generation is at variance with what he should have said if his views had been those suggested by [Gotthelf] . . . ' (5). He notes that Aristotle is engaged in improving on, expanding and even completing the 'physical story'.

> The issue then, is this: does Aristotle's own positive account of (e.g.) animal generation seek to extend and complete his predecessor's physical account, or does it rather rule out the possibility of any full physical story? (1988, 23).

Here Charles gives us two options: either Aristotle was interested in improving the current physical account of generation or he was not, and (the implication is that) if he was not, then and only then could he have ruled out the possibility of a full physical story. This does not, however, necessarily follow. Aristotle could be attempting to extend or improve on his predecessors' physical theory

48 I take this to be a vitalist interpretation. Other more recent advocates include Matthen (1989) and Cameron (2010). Gotthelf (2012b) defends his position contra Charles (1988).

49 Berryman (2007) regards this conflict as one that the Aristotelian texts lend themselves to since there is an evident and unresolved tension between the two options therein. '[Aristotle's] use of teleology seems at some points to preclude the possibility of a sufficient material account' (352).

50 Lewis (1988) 85. One is reminded here of the complaint against the functionalist that their theory really amounts to a sort of epiphenomenalism, in which mind is inefficacious. See Crane (2001) 57.

51 For instance, for Burnyeat (1992) 16, Aristotle's science is 'incredible' in the sense of being unbelievable (by us). See also Caston (1997) 338.

while still maintaining that it cannot be completed in principle in this manner. Charles does not, however, rely entirely on such spurious reasoning. Instead he also seeks to find that there is a complete and sufficient physical causal story for the male role. First of all, he explains that '[a]t the very least, Aristotle's account [of animal generation] has several features which suggest that it is part of an attempt to sketch the outline of a full physical story of the independent physical processes involved in animal generation' (25).[52] One feature he has in mind is the fact that Aristotle refers to processes analogous to generation, such as what happens when rennet acts on milk to form cheese (*GA* 739b21–26) and the workings of automatic puppets (*GA* 734b10–17). Because 'the interactions' involved in the cheese and puppet are, argues Charles, 'purely physical' this suggests to him that the male action in generation is so also. This inference is unwarranted, however. First of all, not all features of items or events that are analogous will be identical; otherwise, the items or events would be the same and not analogous. Second, it is clearly the case that both cheese and puppet differ in a crucial respect from a foetus in that they are not alive. And so it is still possible, even probable, that soul plays a causal role in accounting for living processes for Aristotle.

Charles and his sympathizers require the assumption that in embryology (and elsewhere) talk of soul constitutes, for Aristotle, a different level of explanation from physical phenomena and that it is not involved in causation and so does not intervene in physical processes. Because it is not physical, it is not causally efficacious. Against this view, there are three key reasons to hold that form is causally efficacious even though it is not physical.

(1) Textual evidence from across the Aristotelian corpus,
(2) how efficient causation works in embryology and
(3) Aristotle's account of the female role in generation.

[52] The idea of generation being a 'physical process' seems not only to be anachronistic but also to beg the question which is supposed to be at issue here. If generation could be a 'physical process' then why would Aristotle need to refer to form in his account of how it occurs?

Let's begin with (1). At many points in his works, Aristotle speaks of the soul as causally efficacious and even as some sort of agent. For instance, the soul is the cause of the unity of individual substances such and plants and animals. The elemental powers, if left to their own devices, would result in the physical body of the animal or plant (or heavenly body for that matter) exploding or at least disintegrating rapidly (*de An.* 416a6–10, *Metaph.* 1041a32-b33).[53] Aristotle also explains that the soul uses the body.[54] Scholars of Aristotle's psychology have long recognized that such instrumentalist language occurs in many treatises. One way to continue to maintain that Aristotle was not a dualist is to relegate the instrumental language to an earlier period. Both Jaegar (1923) and Nuyens (1948) posit an earlier Platonic phase where Aristotle espouses the dualism of the instrumentalist position, which he later rejects in favour of non-dualist hylomorphism. This thesis proves problematic in light of further analysis revealing the prevalence of both sorts of characterization in the biological works and those on the soul, including the possibility that 'the soul is like a sailor of a ship' mentioned early on in *de Anima* (413a8–9).[55] When he says that the soul uses the body this is unlikely to be meant merely metaphorically.[56] As one cannot easily eradicate the dualist implications of instrumental language, it is best to find a way to make it accord with the soul and body as described in much of the *de Anima*, as form to matter.[57] In Aristotle's definition of the soul as

[53] This point is made by Granger (1990) 40–4. See also Miller (1999) 322–4, Caston (1997) 329, Sheilds (2007) 283.

[54] *de An.* 407b25–6, 415b18–20, 416b20–7, 433b19, *PA* 642a12, *GA* 789b9. '[T]he embryo's formal nature (its potential for form) is playing the central efficient-causal role, by "making use of" material-efficient agents such as heat and cold and the necessary natures of the materials from which the embryo is being constructed. According to this view, these material-level contributions, though necessary, are by no means sufficient by themselves for embryogenesis' (Gotthelf 2012b, 100).

[55] Lefèvre (1972) ch.2 and Menn (2002).

[56] Lewis's (1988) insistence that Aristotle's talk of the soul using various parts is 'strongly metaphoric' forms part of his argument that for Aristotle 'soul is not itself causally related to the processes of formation or growth' (75). Leunissen (2010) 88, n.22, in contrast, is happy to observe of *GA* 743a36-b6 that, in a literal manner, 'the formal nature of an animal uses material potentials'. This stems, no doubt, from her sympathies with Gotthelf's (1976) approach.

[57] Many Neoplatonist philosophers did not see any real difficulty with reconciling instrumental and hylomorphic characterizations of the soul in Aristotle. For a detailed exposition see Gerson (2005) ch.5.

'the first actuality of an organic natural body' (*de An.* 412b5–6), the 'organic' (*organikon*) could be translated as 'with instruments' (*organa*).[58] This would mean that instrumentalism is fundamental to his account of soul, and does not merely represent an early phase of his thinking. The texts are somewhat ambiguous and can be read to support the view that Aristotle was sympathetic to dualism in certain contexts. There is very little evidence, at the very least, of any outright rejection of dualism. Aristotle's predecessors are mostly dualist and he is never concerned in the text of *de Anima* to challenge this aspect of their theories.[59] When commentators also broaden their analysis of soul in Aristotle to texts other than *de Anima*, they find that a rejection of dualism is not so apparent.

We must now turn to (2). In order to dismiss formal and final causes as merely heuristic, one must establish that physical events are only ever truly caused by other physical events through a chain of efficient causes. Aristotle's automatic puppet analogy is taken to support the male role as a physical mechanism of this kind.

> Aristotle, it appears, is seeking to give a step-by-step account of the processes involved in the generation of animals and their parts so as to achieve a full account of the whole set of processes involved . . . Thus, for example, he compares first the action of the sperma in 'setting' the female residue with that of vinegar or rennet on milk (739b22ff), and then the formation of the organs of the body after this 'setting' with the movement of miraculous puppets (741b9–10)—whose interactions will be purely physical (Charles 1988, 26).[60]

The interactions are taken to be 'purely physical' in the modern sense of physics as matter in motion. Motions in the (male) semen, which are thought to be like Newtonian causation – bare matter undergoing local change – cause the motions, again taken to be changes in place, that fashion the embryo. Lewis corroborates this idea.

> When the form of the animal is 'written' into the semen, it is 'represented' there by the movements of the *sumphuton pneuma*. With these movements, Aristotle supplies a thoroughly materialist mechanism for transmitting form via the semen to the offspring . . . On this account, as I understand it, the form or soul of the male

[58] Menn (2002) 108–9. See also Bos (2003) ch.5.
[59] Menn (2002) 86: 'But there would be nothing surprising in Aristotle's having been at one point an interactionist dualist . . . '
[60] Nussbaum (1978) 146–52 also claims that this image is meant to show that embryonic development is the result of a 'causal mechanism'.

parent is transmitted by way of the movements in the semen that are proximately efficient causes of the offspring (Lewis 1988, 66).[61]

According to this view, changes in the embryo, which can be traced back to movements in the male semen, constitute 'a form free causal chain that fully necessitates the offspring' (*Ibid*., 64). It is very difficult to substantiate the view that these *kinêseis* are the same as a chain of local motions based on the *GA*. First of all, the male role is the efficient cause of substantial change, *hê archê kineseôs* – and clearly not a change in place. Second, for Aristotle, substantial change cannot be reduced to chains of local motion. In any case, local change is only one type of *kinêseis*.[62] Furthermore, this analysis of the male role in generation as 'a complete physical mechanism' rests on an anachronistic picture of efficient causes operating apart from form on passive matter.[63] There is no space in this perspective for the Aristotelian idea that the female contribution contains complex potentials for the actualization of soul; thus it can lead to a misrepresentation of the female role (3). Also against this position on the *GA* in particular, and on Aristotle's teleology generally, stand the arguments of Gotthelf, with his outstanding knowledge of and feel for the text. Gotthelf (2012b) supports his (1976) position contra Charles, and others who attempt to mechanize the process of animal generation. He focuses on *GA* II.6, where successive stages of embryonic development are detailed. Clearly the final result, the production of a new animal the same in form as its parents, is what guides the account. There is no attempt to give a material-efficient causal story.

61 'I have argued that form has in Aristotle's theory a totally materialist mechanism by which it is transmitted from parent to offspring', Lewis (1988) 68.

62 I have already discussed various difficulties with translating *kinêsis* or *kinêseis* as 'movements' in Chapter 5.1. What Lewis refers to as 'movements' must be the *kinêseis* Aristotle details in *GA* IV.3, relating to the development of parts resembling parents and ancestors.

63 Matthen (1989) complains along similar lines, with reference to Charles (1988) and Lewis (1988) that ' . . . some recent commentators come dangerously close to making the suggestion that Aristotelian efficient causes are specified in a Newtonian way' (162, n.2). Furthermore, he notes that the 'discovery of DNA meant that genes and genetics can now be shown to be firmly rooted in the Newtonian scheme of local causation' whereas 'Aristotle could never have accomplished a grounding of his powers in a scheme of local causation, because there is no non-teleological base to which he is able to retreat' (162–3).

The drive to find a mechanistic account comes from a preoccupation with non-reductive materialist theories. However, this motivation has become less credible given the increasing recognition that characterizing Aristotle as a proto-functionalist is insupportable.[64] One early anti-functionalist critique focuses on what happens during perception and argues that psychic processes simply cannot be explained in physical terms for Aristotle (Burnyeat 1992). Since perception is the second actualization of soul, whereby the sense organ is actualized in such a way as to experience a perception of a certain sort, this is not a process that has a physical base.[65] Part of Burnyeat's point is that soul operates for Aristotle not through a physical mechanism but through powers and potentials. These are potentials for form, the realization of complex activities of soul.[66] Burnyeat explains that 'in a certain sense an animal's perceptual capacities do not require explanation. For Aristotle such capacities are part of an animal's life and in Aristotle's world the emergence of life does not require explanation. For Aristotle, it is the existence of life which explains

[64] Robinson (1983) is an early negative response to the idea of Aristotle as functionalist. See also Granger (1990), Olshewsky (1992), Burnyeat (1992), Caston (1997) and Miller (1999). When scholars reject the idea that Aristotle was a vitalist, they sometimes do so in order to include his point of view in modern debates. For instance, Charles urges that his line of interpretation 'represents Aristotle as giving reasons for rejecting reduction which are entirely unaffected by . . . subsequent scientific developments', such that if this interpretation is correct 'Aristotle's views would remain of direct relevance for contemporary twentieth-century discussions of non-reductionist versions of materialism' (4). Because functionalist theories of mind and the metaphysics behind them are now going out of favour, assigning such views to Aristotle does not necessarily help the modern philosopher to accept him. In any case, why try to fit Aristotle into a modern 'pigeon hole'. See Sorabji (1974) 84.

[65] The eye jelly does not have to turn red in order for there to be a perceptual experience of redness, contra Sorabji (1974). Burnyeat remarks that, '[o]ne might say that the material of animal bodies in Aristotle's world is already pregnant with consciousness, needing only to be awakened to red or warmth'(1992, 19). Granger (1990) argues that Burnyeat's interpretation ought to be rejected because it turns Aristotle, unacceptably, into a hylozoist (46). However, this charge is unfounded. For Burnyeat, the matter that is 'pregnant with consciousness' refers to the liquid of the eye which is part of a living body. Caston (1997) 338: 'Matter, for Aristotle, is not "pregnant with consciousness" (*pace* Burnyeat 1992, 19). If, on his theory, seeing takes place in transparent material, it does not occur in just any sample of transparent material – it occurs only when this transparent material is part of a certain complex and functioning being'. I do not think that Burnyeat would disagree with the interpretation expressed in that second sentence.

[66] These are arguably similar to Gotthelf's 'irreducible potential for form'. As Johansen (1998) notes, 'a potentiality can be realised without a material change' (11).

why animals have the physical constitutions they do, not the other way around' (22).[67]

At the heart of Burnyeat's 'spiritualist' position is the charge of anachronism, an apt criticism in light of the assumptions concerning efficient causation just detailed. However, when it comes to the process of perception, there is a key difficulty for the 'spiritualist' position. It leads to the conclusion that perceiving involves no material change at all, which is at odds with many Aristotelian texts on psychology.[68] Many therefore understandably reject the idea that Aristotle was a 'spiritualist', particularly with respect to perceptions, feelings and other mental states. One can avoid the difficulties of fitting his view to functionalism while not resorting to 'spiritualism' by retaining that, for Aristotle, materials necessarily underlie mental states. Two interpretations of this sort demand our serious consideration. The first fits in more modern philosophical categories, while the second suggests that Aristotle's theory can provide an alternative to the modern choice-set.

(1) Caston (1997, 2005, 2007, 2009): soul as emergent causal power[69]
(2) Charles (2009a, 2009b): soul as inextricably psycho-physical; mind/soul and body cannot be separated in being or in thought[70]

Both (1) and (2) continue to rely on the necessity for a material underpinning of the psychic in Aristotle, as they wish to answer the difficulty of psychic-bodily interaction without rejecting the causal closure of the physical. I will mainly focus on (1) in what follows, touching only briefly on (2), and finally presenting

[67] In so far as he is talking here about potentials or powers and these are not themselves material, Burnyeat's interpretation is not 'unabashedly physicalistic' as Granger (1990) alleges (46).

[68] Burnyeat (1995) 429–30 suggests that perception ought to be described in purely formal or phenomenal terms. For two clear attacks on this viewpoint see Caston (2005) 260 and Charles (2009a) 27–8.

[69] Heinaman (1990) is a possible precursor to this view.

[70] 'In Aristotle's account, "passions of the soul" ... are, I argue, inextricably psycho-physical, non-composable into two separate types of activity, one purely psychological, the other purely physical. More specifically, the apparently purely "psychological" activity type cannot be defined as the one it is without reference to some specific type of physical process ... His account, correctly understood, represents a distinctive and interesting alternative to the standard options of post-Cartesian philosophy (materialism, functionalism, dualism and spiritualism)' (Charles 2009a, 1). This position is dubbed 'extreme hylomorphism' by Caston (2009).

reasons to reject both with reference to the nutritive/generative soul as evidenced in the *GA*.

Modern emergentism comes from a broad attempt to secure the autonomy of the special sciences, that is, as a form of anti-reductionism – which has it that chemistry cannot be reduced to physics or biology to chemistry due to the ontological relevance of properties that emerge at each level.[71] So, for instance, it is thought that there are chemical powers which physical particles do not have until the relevant complexity results in their emergence.[72] Furthermore, emergent properties are unpredictable and must be accepted on faith.[73] In the ancient context, emergentism is evident when drugs are taken to have powers which their constituent ingredients do not contain.[74] These emergent properties are also causally efficacious.[75] Caston's (1997) account of Aristotle's theory as emergentist (1) begins in antiquity, with the 'harmonia' theory found in Plato's *Phaedo* – the idea that the soul is like the attunement of a musical instrument. Aristotle's theory of soul shares certain features with this account, in particular that the soul cannot survive the death of the body and is dependent in some sense on the existence of the body. But it also differs in several ways, most importantly because it gives soul more powers than the bare epiphenomalism of the harmonia stance. Caston argues that emergentism best explains Aristotle's disagreement with harmonia theory.[76] For him, the key problem for non-reductive materialist accounts is that they do not allow

[71] See Fodor (1974). [72] See Broad (1925) and O'Connor and Wong (2012).

[73] Alexander (1920) explains that emergentism must accept emergent properties with 'natural piety' (a phrase borrowed from to the poet William Wordsworth) 47. See also Caston (1997) 318, n.20.

[74] See Caston (1997) on Galen's 'four-fold drug' (350).

[75] For the emergentist, 'the mental supervenes on the physical, while having distinctive causal powers of its own' (Caston 1997, 337). Caston also describes emergentism in a more technical way. It requires both of the following: 1. that the mental supervenes on the physical and 2. that some events can have some effects in virtue of falling under a mental type; and some of these effects lack a complete physical cause (317). Emergentism is still viable in the philosophy of mind. See Crane (2001) and his account of how mental causation is not over-determined for the theory (62–3).

[76] 'Aristotle thinks that vital and conscious behaviour cannot be completely explained in terms of the tempering of the body, even if it always accompanies such behaviour and cannot be found apart from it' (Caston 1997, 346).

for the way in which soul can be a cause on Aristotle's view.[77] Thus it seems to make sense to understand soul for Aristotle as emerging from the complex materials of the living body, and then supervening on that body and being causally efficacious with respect to it.[78]

Aristotle's embryology, and specifically the actions of soul in development, pose difficulties for the emergentist position. Although Caston rejects a functionalist analysis, he agrees with it in so far as he thinks that for Aristotle there are material processes that underlie all psychological states.[79] In order for this to be so, Aristotle would have to hold that soul only 'emerges' once the material preconditions for its existence are there. Caston (1997) argues for this being the case:

> More generally, Aristotle believes that bodily states provide non-trivial sufficient conditions for the generation, persistence, and demise of individual souls. ... [O]n Aristotle's view, the living animal cannot exist without a soul and the soul cannot exist in actuality prior to the living animal (*On the Generation of Animals* 2.1, 735a6–7). The specific alterations he mentions, moreover, evade familiar homonymy problems, since the matter that is heated or rarified in the production of the living animal is not the animal's living body, but matter that pre-exists it. Independently identifiable bodily changes thus constitute a sufficient condition for the generation of the soul (335–6).

The evidence from the *GA* is much less straightforward than this suggests.[80] For Aristotle to be an emergentist, *psyche* (mind or soul) would have to emerge once the material body has become sufficiently complex. But when a new animal comes to be, according to *GA* II, it has a soul to begin with, before its body has developed.

> [N]othing generates itself, though as soon as it has been formed a thing makes itself grow. That is why one part is formed first, not all the parts simultaneously.

77 Caston (2007) 325: 'Aristotle regards psychological states and the soul itself as efficient causes. Functionalists, in contrast, locate the real causal power in the underlying material states.'

78 For the emergentist 'soul is a causal power that comes to be in a body when it reaches a certain level of organisation' (Miller 1999, 327).

79 Caston (1997) 338: 'Aristotle is no vitalist. Rather he sees form as inseparable from the disposition of matter; something that arises whenever one finds matter in the appropriate arrangements' and (1997) 337: 'On this view, which properties a living thing has, including its causal powers, will be fully determined by its elemental constitution.'

80 Miller (1999) also doubts that emergentism can make any sense in the context of Aristotelian embryology (332).

And the part which must of necessity be formed first is the one which possesses the principle of growth: be they plants or animals, this, the nutritive, faculty is present in all of them alike (this also is the faculty of generating another creature like itself, since this is a function which belongs to every animal and plant that is perfect in its nature). The reason why this must of necessity be so is that once a thing has been formed, it must of necessity grow. And though it was generated by another thing bearing the same name (e.g. a human is generated by a human), it grows by means of itself. So then, since it makes itself grow, it is something: and if indeed it is some one thing, and if it is this first of all, then this must of necessity be formed first. Thus, if the heart is formed first in certain animals . . . we may suppose that it is the heart (or its analogue) which supplies this principle (*GA* 735a14–26).

The new soul in an embyro's heart directs the development of its body when it is as yet unformed. An emergentist interpretation can try to get around this by pointing out that the new animal's soul is not existent actually but only there potentially, until the correct organs are in place. But what, then, will be causing the development of the organs? As earlier explained, once a new animal's nutritive soul is actualized, the whole soul (nutritive and sensitive) ensures the development of the right kind of blood and tissues to best facilitate the operations of the sense organs which will later develop.[81] If the sentient soul is not there from the start (i.e. it hasn't yet emerged) then it is not clear how the sense organs develop.[82] The emergentist interpretation might then propose that it is not the new animal's soul that ensures the development of the correct parts, but that of its parent.[83] Also, for the emergentist, the male's soul does not directly cause the soul of the offspring to come to be in matter. Instead the male animal acts as an agent and, using his soul, he orders matter, which, when it becomes increasingly complex, will acquire the emergent properties of a living body with a soul.[84] This view still does not mesh with

81 Chapter 4.5.

82 As detailed in Chapter 6.3, specific soul faculties cannot be actualized until the organs are present. So, for example, the actualization of the new animal's sentient soul requires the presence of the sense organs (*GA* 741a10–12) and its specific locomotive capacity for walking will require feet (736b21–2).

83 Soul is put in place by the parent, or parents (*GA* 736a32–36b2).

84 From correspondence with Victor Caston, September 2012. Emergentism being a type of property dualism has it that psychological properties are not free standing, but belong to the whole person; thus, the person is the cause. On an emergentist interpretation of Aristotle, individual changes are not referred back to soul; soul is rather a 'standing cause' Caston (1997).

Aristotle's embryology, where form or soul must be present in the new animal itself in order to ensure the development of the nutritive and sentient parts of soul.[85]

Emergentist (and other anti-substance dualistic) interpretations of Aristotle could challenge this picture by the thought that the embryo's soul itself is not an agent in its development but rather that the embryo is this, acting with its soul. *De Anima* I explains that the soul is not an agent; the soul doesn't pity, get angry or think but rather the person does these things with her soul.

> More legitimate doubts might remain as to its [the soul's] movement in view of the following facts. We speak of the soul as being pained or pleased, being bold or fearful, being angry, perceiving, thinking. All these are regarded as modes of change, and hence it might be inferred that the soul is changed/moved. This, however, does not necessarily follow. We may admit to the full that being pained or pleased, or thinking, are changes (and of them a being changed), and that the change is originated by the soul. For example we may regard anger or fear as such and such changes of the heart, and thinking as such and such another change of that organ, or of some other; these modifications may arise either from change of place in certain parts or from qualitative alterations . . . Yet to say that it is the soul which is angry is as if we were to say that it is the soul that weaves or builds houses. It is doubtless better to avoid saying that the soul pities or learns or thinks, and rather to say that it is the man who does this with his soul. What we mean is not that the change is in the soul, but that sometimes it terminates in the soul and sometimes starts from it, sensation e.g. coming from without, and reminiscence starting from the soul and terminating with the changes or states of rest in the sense organs. . . . That the soul cannot be changed/moved is therefore clear from what we have said . . . (*de An.* 408a32-b30).

For many this passage indicates that Aristotle could not be a substance dualist, because the soul is not an agent, the person, as a combination of body and soul, is such. The 'extreme hylomorphist' view (2), for example, must reject any possibility of the soul acting on its own.[86] Thus Charles (1997, 18) presses on the weaver image. A person cannot weave without her body and thus he understands that, for Aristotle, no psychic power or activity can be characterized without reference to the materials involved.[87] It is

[85] '[T]he natural *kinêsis* exists in the product [i.e. the embryo] itself' (735a4).

[86] See also Frede's anti-dualistic interpretation (1992) 103–4.

[87] 'Weaving is a clear example of a process or activity, which essentially involves moving one's body. One cannot define it in a purely psychological way without reference to the

crucial in assessing these positions to consider the context of the above passage and thus what its intended message might be.

The passage appears as part of a dialectical exercise of sorting through the views of others and refuting many of these. At this point, Aristotle is concerned to overturn the Platonic position that the soul is in constant change or motion.[88] The first sentence indicates his focus: he says there that activities such as perceiving and being angry suggest that the soul is 'changed' or 'moved', presumably because they are affections (*pathei*), things which affect something else. In this context, he argues that this is not the case; the soul is not itself moved or affected. Emotions and sensations move (or impress themselves on) the entire person. He is not arguing that this is the central way to set out what the soul is like – that all soul activities are always like those of the weaver or builder.[89]

The passage can, in fact, be found to be consistent with instrumental language. If the soul doesn't weave, but the person weaves with her soul, then her soul must direct her actions towards an end (i.e. a finished garment). One might even say that 'the soul-body composite' weaves 'through one part of itself, namely the soul' (Menn 2002, 101). One can read the passage to say that the soul of the agent is using the body, for example, the hands, in order to achieve its ends and thus that this soul is a separate agent. Furthermore, what is expressed in this passage may apply to weaving and getting angry but it does not follow that it is the case for other psychical activities. In the case of weaving, one does

hands, fingers and loom of the weaver . . . Weaving, like being angry, is an inextricably psycho-physical process' (Charles 2009a, 19).

88 This constitutes part of a broader argument that the soul is a primary source of change or rest, and, as such, is itself unchanged. It could only be changed or moved incidentally, as when a person moves her body (*de An.* 406a3–12, Cf. *Ph.* 256a4-b3 and Menn [2012] 435–6). Aristotle's concern to establish the soul as a sort of unchanged changer, accords with the soul as described in the *PA* and the *GA*; Cf. *de An.* 415a26-b2, 416b23–5. Menn (2002): '[T]he primary efficient cause of nutrition is distinguished from the thing which is increased . . . in nutrition; the primary efficient cause of nutrition is thus not the body but an unmoved mover. That such a cause is necessary follows from the definition of the nutritive power as "a power of preserving the thing that possesses it, *qua* thing that possesses it" (416b17–19): this is therefore an efficient cause which . . . constantly produces the same effect, which for Aristotle shows that the cause must itself remain in the same persisting state, i.e. that it is an unmoved mover' (119).

89 Indeed, contemplation will not fit this model. 'While Aristotle denies that the soul alone is the subject of the complex *pathos* of sensation or anger . . . he never denies that the soul alone is the subject of the non-kinetic activity of *noein*' (Menn 2002, 101).

need to be embodied and be a psycho-physical living entity in order to weave. However, this is not like the action of soul in natural generation. While it makes some sense to say something like 'the soul doesn't cause a new animal to come to be [strictly speaking], the person [i.e. father] does with his soul', it is a different situation when the male semen has left the scene and evaporated. At this point, the embryo's own soul is the cause of the changes that take place to develop its body (*GA* 735a4, 14–15), and thus it is not the case that 'the embryo's nutritive soul doesn't cause the development of the parts of its body, the composite animal does' – the composite animal simply cannot do so, as it does not yet properly exist. All this points to the presence of something different from the materials, a potential for form or a vitalist principle, which is already present and can bring about these effects. The same points Gotthelf (2012b) so aptly makes against Charles (1988) apply also against the anti-substance dualist more generally. That is, the account of generation in Aristotle's *GA* is not completable without reference to soul because the potentials involved are not material potentials but formal ones – the potential to become a new animal of the same type as the parents. Despite Aristotle's obvious interest in the materials and the actions and reactions of materials at the lowest levels, descriptions of physical processes in embryology are always incomplete explanations of substantial change without form or soul.

Earlier on I gave a detailed account of the correspondence between bodily composition and psychological activities and capacities. This would appear to be close to a thesis espoused by emergentist and other anti-substance dualistic readings of Aristotle: matter must always underlie psychological states and processes. As emphasized in Chapter 4, the blend of an animal's body and blood look to be mostly necessary (although of course not sufficient) for psychological activities and capacities, such as the emotions of fear and anger (*Resp.* 479b23, *PA* 650b27, 650b36). Fear corresponds to a cold blend, while anger occurs mostly in instances where the blood is hot and fibrous.[90] It is these

[90] See Chapter 4.5.

sorts of descriptions both in Aristotle's biology and psychology that lead many towards the idea that he would have been amenable to a non-reductive materialist stance. While I do not deny that often matter in and of the physical body crucially corresponds in this manner, in order for Aristotle himself to have advocated a thorough-going non-reductive materialist or anti-substance dualist position, he would have had to hold that material situations are *always* necessary for psychological states and processes. Given that the psychological is so broad for Aristotle, one cannot find this to be the case. Particularly, when we look beyond 'the mental' to nutritive soul, in the case of generation, underlying material interactions are not what Aristotle is giving us. It is not the arrangement of menstrual blood that makes it generative but rather the fact that it is blood imbued with specific potential for form by the nutritive soul of the animal.

Here it is instructive to consider the difference between art/craft and nature. The craft of weaving requires the go-between of the artisan (weaver) and her body, and its parts, such as her hands, in order to bring about its goals. In contrast, nature can work directly without the need for an artisan or her body. It, therefore, does not require a chain of physical reactions to bring about its goals. In a passage discussed earlier, Aristotle notes that in some animals the male has no semen; in these cases the process of generation is more like the art of pottery when the artisan shapes the matter with her hands. Only, there are no hands and no semen; nature works directly.[91]

The idea that, for Aristotle, complex materials underlie everything psychological works best when it comes to understanding certain intentional states. Let's take the experience of seeing something to be red. Here, we have the second actualization of the sentient soul, that is, the exercise of the sense organs in response to external stimuli. This experience is qualitatively different from the

[91] See *GA* 730b10–31. '[I]n Aristotle's paradigm case in the *Generation of Animals* (the male seed in those animals whose males emit seed), there is nothing analogous to a hand guiding the instrument. Once the seed has been emitted it works on its own without further contact with the father's body; the motions of the art are simply present directly in the instrument, and Aristotle seems not to feel a need for any further explanation of how this can happen' (Menn 2002, 125). For a fuller exposition of how nature doesn't need anything equivalent to an artisan see *ibid.* 135–9.

nutritive capacity and the first actualization of the senses. It may well be dependent, therefore, on materials underlying it both in the body and in the external environment. Since materialist readings do not accord with all the actions he grants to the soul in his philosophy, it is worth separating out different sorts of psychic function and activity. In the *de Anima* and in many zoological works the soul is divided into (a) nutritive/generative (b) sentient/desiring and (c) intellectual (*de An.* II.3, *GA* 736b14–16). (c) is problematic for any materialist reading of Aristotle since he mentions at several points that it has nothing to do with the body and is immortal and divine.[92] This explains why many discussions of Aristotle's theory of mind mostly focus on (b), which is closest to the modern characterization of 'the mental'. (a) is brought into discussions of Aristotle on mind in an attempt to support different positions. However, what is the case for (b) may well not be the case for (a) and vice versa. So, for instance, Charles (2009a) argues that all processes in living bodies must refer to goals which are related to the soul, and so are inextricably psycho-physical.[93] However, this does not show the flip-side: that all psychic processes are necessarily physical. In fact, it is easy to see now that (a) suggests that there are ways in which the lowest portion of soul is not always best described in Aristotelian terms as physical. In its important role as unchanged changer, of maintaining the living reality of the animal through nutrition, and then reproducing it through generation, this part of soul uses the body and ought to be conceptualized as a separable reality with a causal power.[94] It is an immaterial causal agent in the processes of nutrition and the development of a new living being. Although there will be processes in living things that involve both the body and the soul (i.e. those he notes which are 'common to body and soul' *de An.* 403a4), there must also be some that are not so, but happen to one without the other; otherwise, why would he need to

[92] In *de An.* the active intellect is separable from matter (III.5). The very section of the *GA* under discussion in this chapter, *GA* II.3, posits that intellect (*nous*) is an aspect of soul which enters the embryo from outside and has no material underlying it ('no bodily activity has any connection with the activity of *nous*' 736b24–29, Cf. 737a9–10). See Miller (2012).

[93] Charles (2009b, 302–4) is now in agreement on this with Gotthelf (1976, 2012). See also Johansen (2012) 150–5.

[94] See Menn's (2002) interpretation of the *GA* (125).

specify the subset of what is common to body and soul? Something that happens to the body without the soul might be decay of certain kinds, for example, the hair going grey (see *GA* V). As for what the soul can do without the body, Aristotle notes there are affections which are 'peculiar to soul' (*de An.* 403a5–8). Scholars are generally quite uncomfortable with Aristotle being a 'spiritualist' or a 'vitalist'. But when it comes to the intellect (*nous*) and the lowest portion of the soul (*threptikon*), responsible for the creation of new life, Aristotle's views must correspond to these terms. It remains an open question with respect to (b) whether emergentism or even 'extreme hylomorphism' better captures Aristotle's position than 'spiritualism'.[95] However, when it comes to (a), the concern of this book, one must accept an Aristotelian 'vitalism'.

A preoccupation with the mechanics of the male role, stemming from a sympathy with non-dualistic interpretations of Aristotle, particularly the functionalism of Charles (1988) and Lewis (1988), is often tacitly accepted in commentary on the *GA*. When readers imagine that the physical happenings in generation can be characterized in terms of a closed system of efficient causation, this leads to an exaggeration of the mechanistic predominance of the male role.[96] For if the male forms and shapes the offspring through 'physical' means, then there is nothing left for the female matter to do. It will be like Newtonian, empty and entirely passive, matter, not Aristotelian matter, full of powers and potentialities. It is important, then, to establish that 'physical' or mechanistic characterizations of Aristotle's male role are inaccurate so as not to end up misrepresenting the female role. This is what I undertake in the next three sub-sections before returning to the implications of a vitalistic reading of his embryology.

[95] The way in which the soul moves the body as a result of its response to desire or decision (*de An* III.9–10, Nussbaum 1978) may blur the kind of distinction I am trying to make. The source of change may exist in (b) as well as (c). However, as perception fits best with a hylomorphic or emergentist theory of soul, being a passive capacity, it does not appear to involve anything non-material, as suggested by Caston and Johansen, who focus on this faculty. The introduction of desire by Charles (2009b) may run into difficulties, however.

[96] Charles' newer interpretation (2009a, 2009b) could account for the complexity of the female contribution to generation, although it may have difficulty distinguishing the male from the female one since both will have to be inextricably psycho-physical.

Pneuma *theory*

Various early thinkers privileged air (*aêr*) or *pneuma* in their explanations of natural phenomena. The author of the Hippocratic *On the Sacred Disease*, for instance, attributes insensibility to the blocking of *pneuma* in the body.[97] Meanwhile, Diogenes of Apollonia described semen as the 'foam' (*aphros*) of blood, foam being a close analogue of *pneuma* – a liquid infused with air (DK 64B6).[98] Theories about the role of *pneuma* in the human body were to become very important in a tradition of medical writing after Aristotle which culminated in a comprehensive theory of different varieties of *pneuma* found in Galen. Galen's division of *pneuma* into natural, vital and animal was to remain influential well into the early-modern period at which time these 'spirits' were widely believed to control the body. Because he stands between these two traditions, many have thought that Aristotle's theory of the material nature of the body builds on that of his predecessors and leads to the views of his intellectual descendants.[99]

This idea of a scientific progression of *pneuma* theory is brought in to support certain materialist interpretation of the male role. If it can be found that Aristotle's ideas extend and complete his predecessors' theories of *pneuma*, then this supports the view that he intended to give an independent physical account of the male role in generation (Charles 1988, 23). Because *pneuma* is a material stuff, Aristotle is thought 'to characterise what is distinctive about *pneuma* in independent physical terms' without reference to the soul of the animal (Charles 1988, 30). As tacit support for Aristotle as functionalist, a focus on *pneuma* supposedly reveals a scientist immersed in the latest research on

97 Hippocrates *Sacr.* Similar evidence from the physician Diocles is cited by Verbeke (1978) 204. Empedocles emphasizes the role of air and heat in the development of the *cosmos*. See Reiche (1960).

98 As Aristotle himself points out, the name Aphrodite looks as if it comes from *aphros*, foam (*GA* 736a19–22). This association first appears in the literature of Hesiod (*Th.* 188–97).

99 Jaeger (1913) thinks that Aristotle received the *pneuma* 'doctrine' from his medical predecessors, as do Verbeke (1978) 204, Freudenthal (1995), Bos (2003), Coles (1995) and arguing against this point of view Tress (1999). For an overview, see Balme (1972) 160–3.

material mechanisms and thus one that does not think that forms or souls are necessary in these explanations.

Many of these assumptions are made when focusing on *pneuma* in Aristotle's embryology.[100] Aristotle is considered to be a slightly more sophisticated scientist than his predecessors because he realized that another inanimate physical element was needed in order to map out the correct mechanisms. This 'new' element is the innate (or *sumphuton*) *pneuma*, a special sort of material (which veers rather close to hylozoism). In association with a brief discussion of the part that *sumphuton pneuma* plays in other biological contexts, a comprehensive theory of the material basis of the operations of the soul in Aristotle is thus envisaged.[101] It is very difficult, however, to find any such systematic theory of *pneuma* in Aristotle's works.[102] *Pneuma* can mean many different things in different contexts: it is used to refer to atmospheric wind (*MA* 699a1; *Mete.* 349a13–25, 365b36); breath, which is inhaled and held (*GA* 775b1–3; *Pol.* 1336a34–9); or simply to hot and wet air, which results necessarily when fluids are heated (*GA* 742a15; *Mete.* 387a25). *Sumphuton pneuma* is also described as that which is used to produce movement of the limbs (*MA* 10) and as what some of the sense organs are made of (e.g. *de An.* 420a9–12; *GA* 744a2).[103] Because there is no system present in Aristotle, in order to portray him as the link between the Pre-Socratics and Hellenistic philosophers and physicians, an interpreter has to 'reconstruct' what they believe he must have had in mind.[104] But nothing that we read in our Aristotelian texts can show us that he planned a system like that of his successors. Even if later thinkers filled in what they took to be gaps, this does not show that Aristotle

100 I will not discuss the problems with viewing science (and especially ancient science) as 'progressive' here, but see Connell (2000) 405–10.

101 Beare (1906) 334–6, Jaeger (1913), Peck in Aristotle (1942), especially 576–93, Verbeke (1978), Nussbaum (1978) Essay 3, Rist (1989) xvii, 33, 131–4, Preus (1990) 484, Freudenthal (1995) and Bos (2003).

102 Düring (1966) 343. See also Verbeke (1978) 191 and Nussbaum (1978) 160 who seek a systematic theory of *pneuma* but are disappointed with the lack of information in the texts of Aristotle. Bos (2003) endeavours to claim the pseudo-Aristotelian *Spir.* as authentic.

103 For comprehensive references to other works see Aristotle (1942) Appendix B.

104 The idea that Aristotle was going to develop a theory, but had not got around to it when he died, has also been suggested. Nussbaum (1978) 143: 'The theory is internally obscure, one of a series of cryptic pointers towards a fuller account of this *pneuma* that Aristotle may have planned . . . '; Rist (1989) xvii.

felt that anything was missing or that he was leaving an almost completed system of pneumatics for his successors.[105] It is difficult when using such methodology not to make too much of what he says with reference to *pneuma*, and 'reconstruct' what was never there.[106]

The special qualities of *pneuma* in the body are often said to come from the fact that it is innate (*sumphuton*). But Aristotle does not attribute any special qualities to inborn air, which he says is present due to necessity – embryonic material is fluid and has to be heated up in the process of generation (*GA* 742a15).[107] The presence of *pneuma* in the male semen is not *sumphuton* in any case, but is a necessary result of fluid being heated by the male animal's soul after puberty. In general, Aristotle's idea of *pneuma* is gleaned from the observation of rotting organic matter on the one hand, which tends to produce steam (*Mete.* 387a25) and from the observation that people hold their breath when they exert themselves (and when they are about to 'eject' something from their bodies), on the other hand.[108] Aristotle could (and I assume did) decide where *pneuma* comes from and what it can do in bodies, without having to agree with a comprehensive theory of *pneuma*. In Aristotle's biology, *pneuma* is necessary for strength (*GA* 775b1–3; Cf. *Pol.* 1335a34–9), playing a role in pushing out or excreting things from the body. The holding of breath (*pneuma*)

[105] The interpreter often presupposes knowledge of later *pneuma* theories, which *were* comprehensive and systematic when 'reconstructing' Aristotle's supposed theory. Freudenthal (1995) sets out to 'reconstruct' Aristotle's *pneuma* 'research programme' (112), assuming that this can be done without anachronism. See also *ibid.* 120: 'Aristotle ... wished to frame an integrated physiological account ... ' and (135): 'global project' (182). Freudenthal criticizes other commentators for not succeeding in 'making sense of Aristotle's pronouncements on pneuma as a global and comprehensive project' (111). Freudenthal can be criticized, in turn, however, for assuming that *pneuma* was indeed 'a global and comprehensive project' for Aristotle, on such scant evidence. See King (2001) 7.

[106] Sometimes these conjectures are made on the basis of the probable content of Aristotle's 'lost' work *On Generation and Nutrition*. See Coles (1995) 76.

[107] The fact that it is inborn need not lend it special status, since there are many other matter of fact properties of animals that Aristotle considers to be *sumphuton. Metaph.* 993a1: the sentient faculty of the soul; *Mete.* 382b10–12: water, cf. *GA* 750a9; *EE* 1230b17: the propensity to be overcome by pleasure; *Poet.* 1448b5: imitation.

[108] I am supported in this by Balme: 'Aristotle [did not] credit *pneuma* with the special powers that his medical contemporaries postulated. He defines it simply as "hot air" (*GA* II. 736a1), and confines its role to actions that can be explained from the natural properties of heat and air' (Balme 1987a, 279).

is required to release semen and other residues (*HA* 586a18, *GA* 737b28–738a1), as it is in parturition (*GA* 775b1–2, *HA* 587a4–5). He does not refer to the inborn variety of *pneuma* in these instances but rather to the force that inhaled air can provide in order to move items within the body. He also uses the term for wind, a material in the external environment which possesses the most motive force (*Mete.* 365b36). What is actually written in the embryological context is that semen contains the 'so-called hot', which is 'enclosed in the nature in the *pneuma*' (*GA* 736b35), which would be an extremely convoluted way to introduce the discovery that *pneuma* lies at the heart of a generative mechanism.

One reason to resist attributing to Aristotle any desire to develop a fully worked out system of pneumatics is that his philosophical outlook differs fundamentally from that of those who earlier and later placed such emphasis on this material. Many *pneuma* theorists are hylozoist. The Hippocratic *On Fleshes*, for instance, posits that *pneuma* is a cosmic principle that pervades all things.[109] For later Stoic thinkers also, *pneuma* is the material substance of soul, which exists throughout the universe. The *pneuma* contained in male semen is a parcel of this divine matter.[110] In the pseudo-Aristotelian *On the Cosmos*, which is probably part of the Stoic tradition of thought, the author indicates that *pneuma* is 'the vital and generative substance which is found in plants and living creatures, and permeates all things' (*Mu.* 394b10). Vitalism differs from hylozoism. For the hylozoist reference to a certain type of matter is properly explanatory of life and its manifestations.[111] The vitalist, in contrast, posits a fundamental ontological distinction between matter and life. For Aristotle, for example, matter is not able to be alive without the presence of soul, which is the source of life and is irreducible to matter.

[109] Hippocrates *Carn.* See also *Nat. Hom.* 12; *Cord.* 6; Solmsen (1957) 119.

[110] See Sedley (1999) 384, 389 and 482, Verbeke (1945) 35 and *idem* (1978) 204. Galen's historiography in this instance appears to be more accurate. Galen understands that, for Aristotle, the male does not contribute any matter. He maintains that Aristotle should have let the male contribute something material (i.e. *pneuma*), which would have placed him in agreement with Galen's own stance (*Sem.* I.5, 18–19).

[111] Burnyeat (1992) was accused by Granger (1990) of being a hylozoist, which is inaccurate. See Chapter 6.3.

Other thinkers who employed a system of pneumatics are closer to reductive or non-reductive materialists. As the previous sub-section emphasized, those who wish to find that Aristotle falls into the latter camp, see form and soul as causally inert, leaving *pneuma* do all the real causing in generation. That view is untenable, given that we find Aristotle explaining, at the very end of the *GA*, that *pneuma* is merely a tool.

And so there is nothing to prevent the teeth being formed and being shed in the way [Democritus] says; but it is not on that account that it happens, but on account of the final cause, the end; those other factors are causes *qua* causing change, *qua* instruments, and *qua* material, since in fact it is probable that nature makes the majority of her productions by means of *pneuma*, used as an instrument. *Pneuma* serves many uses in the things constructed by nature, just as certain objects do in the arts and crafts, e.g., the hammer and anvil of the smith (*GA* 789b5–11).

The material-efficient causal explanation is dependent on there being a form and an end. *Pneuma* is merely an instrument that nature employs to achieve such an end.

Those who think *pneuma* has special significance propose that Aristotle resorts to it as a way to 'fill gaps' in his physical theory, which normally only allows for four sublunary elements (earth, air, fire and water).[112] The assumption is that Aristotle was faced with grave *physical* problems when encountering life processes and so reached for a solution in terms of another physical element.

The sperma is '[only] a residue of nourishment'. This is surely not a suitable vehicle for the soul functions. An agonising predicament (Solmsen 1957, 121).

The predicament Solmsen describes is one felt by modern commentators, aware of advances in genetic science and neurology, who feel that Aristotle is not addressing the issues that we would like him to. He is not providing anything like an adequate account of the material basis for the functions of the soul or mind. But there is nothing odd about this because Aristotle does not give us, and

[112] Solmsen (1957) decided that in trying to explain how soul gets from father to offspring 'Aristotle is able to announce a new discovery' (i.e. of *pneuma*, 119). For Nussbaum (1978), *pneuma* solves another very difficult problem: the existence of voluntary movement in animals (143–4 and 163). See also Charles (1988) 31, Verbeke (1978) and Peck in Aristotle (1942).

does not try to give us, a completable physical picture of the actions of soul in generation and so did not think he needed any physical vehicle for soul functions.

The same inclination that inspires Solmsen and other *pneuma* theory supporters fuels attempts to fill in or elaborate the physical make-up and qualities of pneumatized semen in Aristotle's theory, by comparing it to boiled milk, for instance.[113] It is telling that Aristotle himself was not interested in the exact workings of *pneuma* in the process under analysis. He also does not focus on exactly how male semen comes to have tiny bubbles, whether these bubbles rise or dissolve and what precisely they do to the menses. With his teleological outlook and his vision of a world peopled with forms, continuously reproducing themselves, he is happy to give only a sketchy picture of the material basis of these processes. He was not waiting for a new scientific discovery; and so his approach to explanations of living phenomena, even the start of life, is different from any post-Enlightenment stance.[114] For Aristotle, *pneuma* is not a mysterious life-giving stuff; it is just another material. From the text of the *GA*, the most we can deduce about *pneuma* is that it is a necessary but not a sufficient condition for generation. It is not the unique qualities of *pneuma* that make the male principle able to start life; this is due rather to nature and soul.[115]

113 Freudenthal (1995) 120–5: 'Aristotle's view seems to be that in the living body, the action of vital heat on the blood transforms part of it into *pneuma*, which, contrary to what happens when ordinary heat acts on a liquid containing earth and contrary to what is implied by Aristotle's physics, does not separate off the remaining blood; rather, the *pneuma* remains suffused *in* the blood, as it can be observed to do in milk' (125). See also Coles's attempt to guess at how pneumatized semen could physically determine the embryonic body (1995), especially 63–4.

114 'Perhaps some contemporary philosophers with materialist sympathies find it easier to regard Aristotle as a respectable ancestor if he can be seen as conforming to the pattern of non-reductionistic materialism. However, with respect to the evolution and changes of ideas it is often instructive to see differences ... we do not think Aristotle's respectability has much to do with whether he did or did not happen to say some things in the course of his theorizing that coincide with what some scientists and philosophers happen to be saying in our own time' Code and Moravscik (1992) 144–5. Matthen (1989) also encourages modern readers to consider the very different background assumptions Aristotle would have been making when constructing his theory of animal generation (159).

115 Where there is life there is bound to be *pneuma*. It is part of the composition of male semen by necessity because it is always present when fluids are heated (*GA* 735b14–16). It is also in the blood of the mother, which becomes the body of the

Vital heat

Another set of interpretations seek to find Aristotle providing a complete physical picture of generation and living processes, through his use of 'vital' heat. Section (2) says that male semen is fertile due to 'the so-called hot'. It continues:

> This is not fire or any other such power, but that which is enclosed in the semen and in the foamy *pneuma*, i.e. the nature in the *pneuma*, analogous to the element of the stars. (3) Because of this, fire generates no animal, and we do not see anything taking shape either in wet or dry fire.[116] But the heat of the sun and that of animals (does generate), not only through semen, but also through any chance residue different in its nature; this also has the principle of life (*zôtikên archên*). From these things it is clear that the principle of heat in animals is neither fire nor does it come from fire (*GA* 736b35–737a8).

Fire alone cannot explain fertility. If we were to place menstrual blood in a very hot oven or on a fire, an embryo would not result.[117] Perhaps, then, Aristotle thinks there is a special kind of heat found only in semen and living things, which explains the formation and maintenance of the bodily parts.

If 'vital' heat is what explains development and animation, then it should be found to be a necessary and sufficient condition for these living processes. It cannot, however, be a sufficient one since 'vital' heat is used by Aristotle not only to explain animate but also to explain inanimate processes. So, for example, when milk is curdled in the cheese-making process (*GA* 739b24), 'vital' heat present in the fig juice which sets milk into cheese. If this interpretation accepts that 'vital' heat is not a sufficient, but only a necessary condition, and continues to reject the causal efficacy of soul, then it cannot account for what makes one entity or process living and another not. Why doesn't the cheese count as a living being? As with any other necessary condition, on Aristotle's theory, it is merely a basis for psychic functions. The soul needs an organic

offspring, and as part of the initial *conceptus*, it is used by soul to articulate that body (*GA* 741b37).

[116] Although Aristotle does appear to allow that animals can be spontaneously generated from fire (*HA* 552b11–12).

[117] Balme in Aristotle (1972) 71: 'the chief factor he invokes to explain biological phenomena is "vital heat", which he does not distinguish from the action of fire (one of the ordinary four elements) except when it is associated with pneuma in effecting reproduction'. Freudenthal (1995) 26. See also Bolton (1987) 163.

body capable of living. But the soul must be there in order to render any such material truly 'vital'. For Aristotle, the state of affairs in the cheese is not due to physical inadequacies but to the lack of soul. 'Vital' heat in the fig juice cannot make a heart of a new animal, because the juice has no soul. The vitality of the heat has been unhooked from its mooring and so has no power to maintain living functions in this case. It merely thickens the milk.[118]

Part of Aristotle's point in using comparisons with cheese-making and other food manufacture is actually to oppose the type of view that is here being ascribed to him (see Chapter 4). As he emphasizes again and again, the products of crafts do not stand on their own; they all require human goals and agency. The statue-makers and potters may put some bits of clay together and take other bits away; they moisten the dry and dry the moist; in doing so they are controlling nature in order to produce an ordered and well-crafted product. They are not at the mercy of elemental forces, but instead put these powers to use in the appropriate manner, moulding and influencing them in specific ways. Thus, when Aristotle compares generation to cooking, he wants us to think about the chef rather than the vitality of the heat – for the presence of the chef is what makes the heat have the effect it does.

> This [vital] heat resides in the spermatic residue, and the *kinêsis* and actuality which it possesses are in amount and character correctly proportioned to each part. If they are at all deficient or excessive, to that extent they cause the forming product to be inferior or deformed. The same is true of things that are 'set' by heat elsewhere than in the uterus; e.g. things which we boil to make them pleasant for food. The only difference is that in this case the correct proportion of heat to suit the *kinêsis* is supplied by us, whereas in the other it is supplied by the nature of the parent (*GA* 743a27–34).

When a chef gets to work, she doesn't expect the heat of the oven alone to create a healthy and delicious meal. Instead she will have to mix ingredients carefully, heating them, perhaps gradually or at

118 Another crucial difference between the rennet and milk example and the mixture of male and female seed emerges later on in Aristotle's discussion of multiparity in *GA* IV. In animal generation a greater amount of menses and male semen can (if the circumstances are appropriate) result in the development of many separate individual animal embryos. In contrast a greater amount of rennet and milk will only ever make a bigger block of cheese. See the discussion at *GA* 771b19–772a25, which ends with the insight that 'the parallel instance of milk, which was cited, is not comparable'.

differing temperatures at various different times; she must impose order and limits, operations that heat alone cannot bring about, except perhaps by chance.[119] Nature in generation is like that. Vital heat does not impose any sort of limitation, the animal's soul (its 'nature') must do so.

As Aristotle explains elsewhere, in the strictest sense the fire which does the work of digestion in animals does not differ in kind from external heat (*Resp.* 474a28). The seat of the soul, the heart, is 'set aglow with fire', the same fire as found elsewhere (*Juv.* 469b16).[120] The reason the former heat is 'vital' is that it is contained within the living system of a particular animal and used as an instrument.[121]

> And so in speaking of an axe or any other instrument, we should not say that it was made solely by fire, so we should not say this about a foot or a hand . . . As for hardness, softness, toughness, brittleness and the rest of such qualities which belong to the parts that have soul in them – heat and cold may well produce these, but they certainly do not produce the *logos* in direct consequence of which one thing is flesh and another bone; this is done by the *kinêsis* which derives from the generating parent, who is in actuality what the material out of which the offspring is formed is potentially. Exactly the same happens with things formed by the processes of the crafts (*technai*). Heat and cold soften and harden the iron, but they do not produce the sword; this is done by the movement of the instruments employed, which contains the *logos* of the craft . . . (*GA* 734b27–735a2).[122]

According to Aristotle, 'physical' powers outside living systems do not produce functional products without the guidance of soul;

119 Matthen (1989) explains of embryological development in Aristotle. 'The series as a whole and derivatively each event in it cannot be understood merely as a heating or cooling, but has to be understood in terms of the end' (171).

120 ἐμπεπυρευμένης. See also *Resp.* 478a29: 'Generally, the nature of animals needs cooling because the soul in the heart is emblazened (καταψύξεως μὲν οὖν ὅλως ἡ τῶν ζῴων δεῖται φύσις διὰ τὴν ἐν τῇ καρδίᾳ τῆς ψυχῆς ἐμπύρευσιν)'.

121 Alexander of Aphrodisias, an early interpreter of Aristotle, recognized the difference between vital heat and external heat in this manner in his commentary on the *de Anima*. He writes: 'For the hot is that which produces change, but the change which the hot initiates is neither limited nor regular. That which feeds itself, on the other hand, orders, provides a guided way and has a certain limit, which is something that is proper to soul and not to fire' (κινητικὸν μὲν γὰρ καὶ τὸ θερμόν, ἀλλ' οὔτε ὡρισμένη οὔτε τεταγμένη κίνησις, ἧς προηγουμένως ἀρχὴ τὸ θερμόν. τὸ δὲ τρεφόμενον τάξει τινὶ καὶ ὁδῷ πρόεισιν καὶ ὅρον ἔχει τινά, ἃ οἰκεῖα ψυχῆς, ἀλλ' οὐ πυρός; 1887, 35.5–8). See also Matthen (1989) 171 n.6: 'There is no evidence . . . that when he uses such appellations as "soul-heat" Aristotle is trying to mark off a particular variety of heat from another'.

122 See also nature as a painter at *GA* 743b20–5.

just as heat, or rather 'fire', does not generate an axe – this is the province of the craftsman, who uses heat as a tool (*PA* 652b10–15). In speaking of an axe, we should not say that it was made solely by fire, so we should not say this about a foot, a hand, or flesh. Thus, we could never give a complete account of generation in terms of 'vital heat' since its vitality is due to the organizing actions of soul or form.[123]

Puppets and genetics

As we have seen, positing *pneuma* or 'vital heat' as special materials which Aristotle leans on to explain generation is not a plausible way to characterize his account. The materialist interpretation does have another way in which it can seek to avoid giving soul any robust causal role in Aristotle's theory of generation. It can take the automatic puppet analogy to imply that generation is a 'physical' process. In order to assess this strategy, we must consider more closely what message or messages the analogy is intended to convey. The analogy is brought in as a sort of reply to a difficulty for Aristotle's theory: how can it be that the part that fashions is external to offspring, when nature normally requires an internal source of change (*GA* 734a33-b4). He explains:

> Well, we must endeavour to solve this difficulty. Maybe what has been said was stated without qualification, which ought to be qualified: e.g., if we ask, in what sense exactly is it impossible for the parts to be formed by something external? We see that in one sense it is possible, though in another it is not. Now it makes no difference whether we say 'the semen' or 'that from which the semen comes' in so far as the semen has within itself the *kinêsis* which came from the generator. And it is possible that A should move B, and B move C, and that the process should be like that of the 'miraculous' automatic puppets: the parts of these automatons, even while at rest, have in them somehow or other a potentiality, and when some external agency sets the first part in movement, then immediately the adjacent part comes to be in actuality. The cases then are parallel: just as with the automaton in one way it is the external agency which is causing the thing's movement—viz, not by being in contact with it anywhere now, but by having at

[123] 'Aristotle did not regard the automatic physical interactions as capable of producing animal tissues and organs; for when the elements act without being used by nature or soul, they do not impose limit and definition upon themselves' (Balme 1987a, 281).

one time been in contact with it, so too that from which the semen originally came, or that which fashioned the semen. Not by being in contact with it still, but by having once been in contact with it at some point. In another way, it is the *kinêsis* resident within, just as the activity of building causes the house to get built (*GA* 734b5–17).

There are two main ways ancient automata could have operated. In the first, the puppet is like a wind-up toy where one part will move another, which will then move another part and so on. The second type of automatic puppet requires only one movement to create instantaneous action in the parts like the wooden dolls whose arms and legs move up when a string at the bottom is pulled.[124]

If Aristotle intended the first sequential model, then the male semen makes only one part, the heart, and the heart makes a subsequent part (e.g. the liver) and then the liver makes the next part and so on.[125] Although this reading gains support from the fact that Aristotle speaks explicitly of one part being formed after another starting with the heart (*GA* 735a15), it seriously limits the male's role. His action would be one of simply switching on a complex machine, an 'accidental' cause.[126] It also does not capture the idea that the male achieves a single potentiality for substantial change. Thus, the second model is more plausible. If this is the sort of device Aristotle intended us to think of then the male action in generation would simultaneously trigger every development in the embryo.[127]

Many assume that the second model indicates that the role of fashioning the offspring is exclusive to the male. This reading often likens Aristotle's view to modern reductive accounts, such as genetics, and employs modern metaphors, particularly to computer programmes.[128] The idea is that Aristotle thought along the same lines as a modern developmental biologist but lacked satisfactory

124 This distinction and the examples are taken from Henry (2005).
125 Katayama (1999) 83. Alexander of Aphrodisias as detailed in Henry (2005) 35.
126 *Ph.* 255b24–7 and Henry (2005) 32–4.
127 '[T]he nature of the embryo must be a single potential for the formation of the organism as a whole and not a succession of distinct potentials for the formation of individual parts . . . ' Henry (2005) 36–7.
128 Henry (2005) 40: 'a pre-programmed automaton in the modern sense (one that owes it movements to the execution of a computer programme) could supply Aristotle with the kind of device he needs'. See also Delbruck (1971) and Freudenthal (1995).

materially reductive explanations and complex enough machines with which to compare these. One commentary explains thus:

> The gist of the idea is well captured by the following, anachronistic yet not misleading, metaphors: the movements inhering in blood, semen, etc. embed the *program* for all the distinctive parts of the animal in question; specifically, by virtue of these movements the semen has an *informational power* allowing it to transmit to the offspring the 'program' inscribed in the sire's blood.[129]

Genes are taken to contain information transmitted to offspring, about the relevant physiology and anatomy of the organism.[130] And this information is like a computer programme, immaterial and comprehensive. If passages on the male role in Aristotle's account are taken in this way, then only the male could have any 'genetic' input. It is thus imagined that, for him, the generation of animals is 'a process in which the form as stored in the semen is read into the matter as catamenia by way of a pre-programmed sequence of mechanical and chemical operations' and a copying mechanism.[131]

Elaborating Aristotle's automatic puppet image in this manner creates a serious impediment to our understanding of male and female roles in his theory. The idea that the male role is a genetic input and like a computer programme, renders the male completely responsible for all specifications of the offspring, making the female role entirely empty and indeterminate. This contrasts with what we know of his theory: specifically, that the female role is not like this.[132] This does not mean that we have to reject the automatic puppet image as inaccurate or counterproductive for

[129] Freudenthal (1995) 28. See also Nielsen (2008): 'Socrates' seed, as efficient cause of conception, carries both what we would call the genetic offprint of the human species and the genetic offprint of Socrates himself' (401).

[130] Describing Aristotle's male role in terms of 'transfer of information' is common. See, for instance, Dean-Jones (2012) 187 or Reeve (2012) 6: 'Despite its importance, no focused discussion of *pneuma* occurs in Aristotle's extant works. While this makes it difficult to determine his views with confidence, by piecing together what he does say, a reasonably clear picture emerges. From its role in embryology alone, for example, we can see that *pneuma* transmits movement by being itself in movement . . . because the movements it imparts are formative, they must be complex and various – able, as geneticists now put it, to *code for* all of an animal's parts.'

[131] Furth (1988) 119 n.26 and 139–41; See also Delbruck (1971), Sedley (2007) 179–81 and Mouracade (2008b).

[132] In fact, as I will detail in Chapter 9, both parents 'transmit information', or rather potentialities (*dunameis*), to offspring through their contributions.

Aristotle. It just needs to be understood differently in this context. Even if the second model is what he intended us to think of, it is not necessarily the case that the male is responsible for all subsequent changes in the embryo after conception.[133] Explicitly, it is employed to aid our understanding of his measured response to the particular difficulty. To the question of whether generation is externally or internally caused, Aristotle answers that it is, in a way, both. Conception and subsequent embryonic development come from the joint interaction of an external agent (the male) and the type of material that is poised to become the new animal (the female contribution which will become its living body). An automatic puppet is ready to change in complex ways.[134] The image aims to convey the idea that the material in generation is similarly ready to change in complicated ways.

Likening Aristotle's automatic puppet image to the role of genetic information in modern theories is not that helpful for various reasons. First of all, the context is completely different. For modern science matter is inert. Energy, action and reaction come about due to the motion of tiny particles of matter. This may be familiar and comfortable to us but it is not Aristotelian. Next, the idea that genetic input can be compared to or spelled out in terms of information is quite erroneous. In modern science, many feel these metaphors to be incorrect characterizations of the role of genes in development, particularly as they obscure the important part played by the structure of the rest of the cell[135] and the animal's whole body and environment.[136] Thus, it is inaccurate to imagine that a cell's DNA contains information or a programme. These criticisms show that modern genetic science has something in common with the Aristotelian perspective. For Aristotle, matter is not an inert lump awaiting specific instructions in every regard. The more accurate picture of an interaction between the different

133 As with *pneuma* and vital heat, it is clear that Aristotle does not encourage the reader to find any exact mechanism in the above passage (*GA* 734b5–12).

134 Gotthelf (1976) 241, n.30.

135 The DNA is only contained in part of the nucleus of a cell (and in the mitochondria).

136 Even for modern genetics 'having the right kind of DNA is surely part of the answer, but so too is it necessary for that DNA to be housed in the right kind of cell, in the right kind of animal, and with the right sorts of material and environmental inputs . . . ' Vinci and Scott Roberts (2005) 205.

parts that are required in the process, and in particular the background material conditions, concurs that the role of the female (or ovum) is not at all inert.[137]

There are still, however, many ways in which modern biology and Aristotle's approach to life differ significantly. In modern biology, images which compare the effects of DNA to computer programmes have some purchase since it is taken that biological phenomena, including the development of the foetus, can be explained in terms of chemical mechanisms. In short, modern biology is materialist and non-vitalistic. Although we cannot at present create an animal embryo by mixing inanimate chemicals, this is taken to be possible in principle.[138] For Aristotle, in contrast, living things must come from the living (*Ph.* 193b12). Parent animals have to use their souls, which they got from their parents, to produce potentially living materials: soul is always there to begin with. The correct materials are present only because they play a role in the bodily system that is *already in place*.[139] It is this commitment rather than any interest in *pneuma*, vital heat or automatic mechanisms, which best encapsulates Aristotle's male role.[140] Soul comes to be because soul already exists.[141]

[137] Rather than viewing genes as blueprints, advances in scientific understanding of embryogenesis and recent feminist approaches to biology emphasize that '[t]he embryo actively makes over its environment, engaging with its own development; it is thus a self-organising entity, rather than a passive victim of genetic inheritance' (Birke 1998, 200). The new animal, and possibly also the maternal matter it develops out of, ought to be the real centre of our attention in embryology: this is where the action takes place. See also Fox Keller (2000) ch.3 and Vinci and Scott Roberts (2005) 203, n.9.

[138] Many biologists regard life as consisting in irreducible emergent properties, produced by the phenotypic body in contact with the environment. However, in popular parlance, life is just DNA, the 'code' which numerous current projects promise to crack.

[139] 'Whatever is formed either by nature or by human craft, say X, is formed by something which is X in actuality out of something which is X potentially' (*GA* 734b22–3). The possible exception of spontaneously generated animals will be discussed in Chapter 7.4.

[140] Focusing on the physical explanations he provides deflects attention from what may seem unpalatable about Aristotle to the modern mind: his firm commitment to the ontological relevance of form and its goal-directed behaviour in the world. For example, Cameron (2010) takes it that teleological directedness is an intrinsic aspect of biological nature for Aristotle (1098). See also Lloyd (1983): 'The chief benefit for the natural philosopher in the study of animals is, in Aristotle's view—the discovery of form and finality in nature . . . ' (215).

[141] Balme (1987a) 283: 'Aristotle, like Plato and probably all the ancients, sees the primary actions of the elements as quantitatively indeterminate until limit is imposed upon them. If limit is naturally imposed, not at random but regularly and usually, it must

Aristotle did not need genes because he had forms. For the modern biologist, types can and do change and an account is needed to explain the processes that lead to stability – why cats can, through their unique genome and the mechanisms brought about by deoxyribonucleic acids, produce other cats. Aristotle, in contrast, assumes that fully grown tokens of all types of animal have always existed and that animals are just the sort of entity that naturally attempt to emulate divinity through repetition of type.[142] Since he thinks that change in type is impossible, he has no pressing need to provide a mechanistic account of how animals of one type remain relevantly similar from one generation to the next.[143] Aristotle did not seek to justify his assumption that a certain type of animal will produce another of its kind in the next generation; and so he did not require a mechanism to explain this.[144]

6.4 Aristotle and vitalism

Vitalism is a pejorative term, used primarily of ideas that people from the early-modern period onward have felt a pressing need to reject. Forms as causes in nature have been superseded by highly successful modern scientific techniques, based on mechanistic modelling. From the modern perspective, we cannot help but think that bodies are made of complex materials, about which our science has discovered a great deal – and which serve to explain all that we wish to understand in biology. When observing that animals strive to live, prosper and reproduce, it is our sincere concern to find that these teleological tendencies in the natural world accord fully with the chemistry and physics that we

come from a pre-existing source; this he interprets as an enclosed system of movement of air earth fire and water; and the only place where it can be is in the parent.'

142 For Aristotle 'it is an irreducible fact about the natural world, not further to be explained, that it so governs itself as to preserve in existence the species of well-adapted living things that it actually contains' (Cooper 1987, 249). Balme in Aristotle (1972) 97 and Balme (1987b) 298 thinks that Aristotle failed to show the impossibility of degradation, evolution and transformation of type. See also Granger (1987) and Lennox (2001) 89. This is questioned by Henry (2006b).

143 Because Aristotle believed in the eternity of species or form-kinds in nature, there is no difficulty for him about a first cause or a beginning. The causal chain of souls goes back *ad infinitum*.

144 Code (1997) 132–4.

accept.[145] From this perspective, there are no causally efficacious non-material forms present before a process begins that have a reality of their own. But we must place ourselves back into Aristotle's world, and ask from *that* perspective whether vitalism would have made sense. The advent of a material account of the world as a complete system in modern science made the existence of a non-material mind or soul problematic, particularly as it left no room for any account for how a non-physical entity could affect the body or cause a physical reaction. Aristotle, however, does not imagine that what we call material causes comprise a closed system; matter is not uniform and depends on its relation to the relevant form, which gives it its character. Thus, the post-Cartesian mind/body problem would have been foreign to Aristotle and he could not share our repugnance for vitalism.[146]

Even on our own terms, Aristotelian vitalism may not be as bad as feared. For us, vitalism is distasteful because it is intellectually suspect and because it can lead to a disinterest in scientific investigation. It presupposes rather dubious backward causation; for any event, all one need to say in terms of explanation is that it was designed to end up as it did – because the form was present all along.[147] An Aristotelian commitment to an explanatorily basic goal-directed element in nature can be brought in too soon when attempting to work out how something occurs, thus halting any advancement in the physical and biological sciences.[148] On this point, there are two things to say. First of all, Aristotle is not presently impeding scientific progress. There may be a salutary parallel here with his sexism. As we saw earlier on in this book, the sexism we find in Aristotle need not stop us from studying his works if it has no current negative effects. Why not treat his vitalism similarly? There is no good reason to worry that if we

[145] Our materialist and reductionistic tendencies are also highlighted by Cameron (2010) 1099.

[146] See also Morsink (1982) 164: '[I]n the absence of knowledge of these microscopic processes of molecular biology, vitalism is a viable theory because it seeks to account for the obvious telenomic character of living beings and their parts. And this is precisely the position of Aristotle.'

[147] Intriguingly, advances in quantum theory have led some philosophers to rehabilitate the notion of backward or retro-causation. See Price (1996) 170–1.

[148] See Berryman (2007) 363, who cites the Stoics.

study this aspect of Aristotle, it will negatively impact current scientific practice. Admitting such views are outmoded does not make it incorrect to assign them to him. Second, his commitment to robust causally efficacious forms in nature did not prevent Aristotle from being very interested in material happenings in biology. In fact, seeking an ordered whole inspired him to study the intricacies of animal bodies: a complete explanation requires taking into account all four causes, including material and efficient ones. The powers and potentials come from soul/form and are for future form but they are also manifested in the bodily parts, blood and residues, such as the potential for change which the semen carries within its material body.

One can find developments in early 'chemistry' in Aristotle's works, when he discusses what happens when formal and final causes are obscure or even non-existent. Although he had a place for explanations in terms only of material and efficient causation, we must be cautious not to misrepresent these. He distinguishes between intrinsic efficient cause (or efficient cause proper) and accidental efficient cause (which is, in a sense, not really an efficient cause at all). There are two ways to characterize this difference. First, the agent bringing about an effect can be viewed in two ways. In the second, the end achieved is a good one but there is, strictly speaking, no efficient cause for it, in so far as the source of the change is not directed towards the end (this is a classic Aristotelian instance of chance). An example of the first occurs at *Ph.* 196b24–30, where Aristotle explains that the individual features of the builder, such as the fact that he is pale or musical, are accidental and so are not properly speaking the efficient causes of a house. An example of the second type is that of a confectioner who aims to make tasty food but also happens to produce a healthy meal. In this instance the source of the change/production is the confectioner; so in a sense the confectioner is the efficient cause of the healthy food. However, in a sense she is not, since the healthy food did not come about due to any cause directed towards it as an end.[149] For Aristotle, efficient causation

[149] *Metaph.* 1027a2–7. This is presumably why he says at *Ph.* 255b23–5 that accidental efficient cause is a cause in a sense and in a sense is not a cause. Sauvé Meyer (1992) and Scharle (2008) 35–9.

cannot, strictly speaking, stand on its own without formal and final causes and so there is almost nothing in his works that could count as material and efficient causes apart from teleology.[150]

In order to understand why Aristotle opted to be vitalistic, we need to exercise some charity and consider more carefully what alternatives he had. Looking around, Aristotle and his contemporaries could hardly have been aware of subtle evolutionary changes in animals. Instead there is an evident persistence and perpetuation of types of living being. Even though some of his predecessors referred to physical causes to account for the development of complex functional wholes, and may have done so with some success (e.g. Empedocles, Democritus), they could not account for the continuation of kinds. The fact that one type of living being will consistently produce seed which results in a living being of the same type suggested strongly to Aristotle that animal generation requires the continual action of souls on the material world.[151]

Democritus provides the type of causal story with which the modern mindset is more comfortable, since he seems to propose material and efficient causes on their own. However, given that on Aristotelian terms, efficient causes are intrinsic to nature and necessarily directed towards formal and final causes, he fails.[152] Furthermore, despite our modern preference for Democritean approaches, if we place ourselves back with Democritus and Aristotle, Aristotelian explanations would have been more persuasive and satisfying. Democritean material explanations attempt to account for macroscopic objects, such as animals, in terms of minute particles. The problem with this is that it eliminates natural substances. Democritus thus lacks the apparatus to acknowledge

[150] Thus it makes little sense to posit any chain of efficient causation for Aristotle, since each requires underpinning by formal and final causes. See *Ph.* II.9 and Sauvé Meyer (1992) 818.

[151] The frequent refrain of *anthrôpos anthrôpon genna* runs throughout Aristotle's metaphysical writings. See, for example, Metaph. 1032a26 and references in Connell (2001) 316, n.78.

[152] Sauvé Meyer (1992). One of Aristotle's key opponents, Empedocles, only offers accidental efficient causes – and so he does not really give any cause at all. Aristotle seems to have misrepresented Empedocles' point of view on this matter, however. After all, he also has unanalysable teleological principles as part of his cosmos, namely Love and Strife (Berryman 2007, 354). On Empedocles as 'creationist' see Sedley (2007).

the fundamental reality of living beings, and so resorts only to very limited physical explanations of the phenomena of life. How is it that an undifferentiated and singular egg develops into 'an organism with a regular and differentiated structure'?[153] Certainly one can see why a vitalistic answer – that is, that a non-material vital principle directs development – is attractive.[154] At the time Aristotle wrote, the answers given by his non-vitalist contemporaries would have seemed woefully inadequate in comparison. If he did not espouse the type of vitalism that he did, that is, the idea that forms exist before embryos develop and direct that development, he did not have much else at hand to satisfactorily explain animal generation.

[153] See Bechtel and Richardson (1998).

[154] Vitalism is a metaphysical rather than a scientific hypothesis since it is unfalsifiable. Although out of fashion, vitalism is not an entirely implausible position to maintain, particularly in developmental embryology. During the last half of the nineteenth and early part of the twentieth centuries, Hans Driesch (1867–1941), an eminent embryologist who discovered what we now term 'stem cells', espoused the view that embryos contain entelechies that make them capable of self-determination (see Driesch 1914b). Charles (1988) does not want Aristotle to be a vitalist because then his theory 'will prove outmoded when science develops to a point where physical sufficient conditions can be understood with reference to organisms as a whole' (3). In light of this discussion, we have to admit that Aristotle's view would become outmoded in this circumstance. But the question remains, will our science ever reach that point? See also Morsink (1982) 165: 'if vitalism thrives on ignorance, there is still plenty of food left even today'.

PART IV

GENERATION IN LOWER ANIMALS AND PARTICULAR INSTANCES

Introduction

The following maps out the last three Books of the *GA*:

GA III: Generation and embryonic development in egg-laying kinds, i.e. birds (chs. 1, 2, 6), selachians (which are internally egg-laying, externally live-bearing) (chs. 2–3), other fish (chs. 3, 4, 5, 7), quadrupeds (e.g. lizards) (ch. 2) and snakes.

Covering all oviparous kinds with differences explained: general structure of an egg (chs. 1–2), growth of egg (ch. 1), incubation of egg (ch. 2), and the causes of wind-eggs (ch. 2). Argument for fish having both sexes and for the necessity of copulation in fish; the role of milt sprinkling (ch. 5).

Various difficulties and exceptions explained, e.g. an extended explanation of why fish sprinkle milt and birds do not, with reference to the relative perfection of the eggs (chs. 6, 7).

Cephalopods and crustaceans (ch. 8).

Insects and larvae (ch. 9).

Bees, hornets and wasps (ch. 11); Testacea (ch. 11).

GA IV: Sexual Differentiation (chs. 1–2) (applicable only to the most perfect animals, i.e. not to insects and testacea).

Resemblance and heredity (ch. 3).

Deformity/monstrosity (chs. 3–4).

Superfoetation (ch. 5), uterine tumours (ch. 7).

Lactation (ch. 8), parturition (ch. 9), length of gestation, seasonal and dietary factors (ch. 10).

GA V: Variations in animals: sleep, eye colour, keenness of sight (ch. 1).

> Keenness of sight and hearing (ch. 2). Variations in hair, including balding and graying (chs. 3, 4, 5); changes in hair colour due to seasonal and dietary factors (ch. 6); voice (ch. 7); teeth (ch. 8). Teleological explanation contrasted with Democritus' views on the formation of teeth (ch. 8).

In these Books Aristotle's descriptions and explanations of generation become more specific in two different ways.[1] First of all, he concerns himself with a wider variety of animals with differing generative methods. Second, he says more about what occurs in specific instances of generation, rather than giving only a general outline. These two techniques roughly divide the topics under discussion in *GA* III and *GA* IV–V.

GA IV provides Aristotle's most detailed account of sexual differentiation and his most comprehensive discussion of hereditary resemblance. It also discusses deformity, superfoetation, uterine tumours (i.e. false pregnancies), lactation and parturition. Here Aristotle uses the same explanations to cover all animals to which they apply. All live-bearers and egg-layers have separate sexes, as well as many larvae-producers. Meanwhile his theory of heredity applies to every generated animal. When it comes to deformity, this is something that could occur in any animal, although it is more common in some, and finally he ends on the peculiar conditions of superfoetation (which occurs in only two species: human and horse) and uterine tumours (which occur only in one: humans). Although reference to the standard account remains strongly present, and the teleological focus continues under the surface, more and more appeal is made to material necessity. One can see why Aristotle has placed these topics here as the whole account benefits from what has come before, particularly *GA* III, which discusses male and female roles in lower animals.

Chapter 7 will concentrate on *GA* III and so generation in 'lower' animals. Here, at the heart of Aristotle's project to explain generation in all animals and in all situations, male and female roles are further elucidated. Chapter 8 will detail Aristotle's

[1] See Chapter 2.4–5.

account of the differentiation of the embryo into a male or female offspring. In this context, I will discuss how he characterizes the females in his biological works. Chapter 9 will then explain Aristotle's views on hereditary resemblance and argue that these do not clash with the account of generation so far presented. Indeed, the earlier characterization of the female contribution to generation as specific material, containing specialized potentials, as well as further details about the female-only egg in *GA* III, are enhanced and rounded out by what we find in *GA* IV. Sexual differentiation, heredity and deformity are, according to *GA* IV, due to the same set of causes (769b3–9),[2] which occur when male and female generative residues meet in particular instances of generation. The material situation at conception requires a certain balance (*summetria*) but may also be subject to different types or qualities of balance (not all *summetrias* are equal). When it comes to heredity, acknowledging that Aristotle maintains a complementary model, and that the female contribution has powers which are able to shape offspring, is essential to our understanding of his ideas. Rather than dismiss or suppress the positive aspects of the female role in the latter Books of the *GA* it increases our understanding of Aristotle's biology to acknowledge them.

[2] 'Still it is not easy, by starting with a single mode of cause, to explain the cause of everything–(1) why male and female are formed, (2) why female offspring often resemble the father and male offspring the mother, and again (3) the resemblance borne to ancestors, and further (4) what is the cause why sometimes the offspring is a human being yet bears no resemblance to any ancestor, (5) sometimes it has reached such a point that in the end it no longer has the appearance of a human being at all, but that of an animal only–it belongs to the class of so-called monsters.'

7

GENERATION IN LOWER ANIMALS

7.1 Aristotle's teleology and ordering of animal kinds

GA I and II give a general account of animal generation, partly by concentrating on the paradigm instance in live-bearing animals. *GA* III then gives the reader a more detailed picture of generation in non-live-bearers, moving from perfect egg-layers (birds) to non-perfect egg-layers (mainly fish) to larvae-producers, and then detailing bees as exceptional within that class. The Book ends with an attempt to explain how it is that generation occurs in an entirely anomalous fashion in testacea. The same pattern is apparent in certain portions of *GA* I and II, where Aristotle wishes to elaborate differences and exceptions by moving away from an account covering most animals. As before, here also, many exceptions require explanations in terms of material necessity – including external influences, such as the weather and food consumed. It is also important to realize that the animals he describes here are, in some sense, incomplete – they fail to reach an ultimate end or *telos*. In order to grasp how it is that animals can be incomplete, we must understand what it means for an animal to be complete and to reach its proper end. And how we grasp that idea depends on how we read Aristotle's teleology.

The nature and extent of this teleology has been much disputed. At one extreme, everything within Aristotle's world, including the elemental powers (the *dunameis* of hot, cold, wet and dry which compose the elements: earth, air, fire and water) are directed towards one end – striving to be as much like God as possible. On this reading, human beings are the focus of this teleology as beneficiaries. This view is supported by only a couple of passages in the corpus which do not occur in the zoological works.[1] Detailed descriptions found within the zoology provide little evidence of this sort of overarching *telos*.

[1] It is based on specific interpretations of *Metaph.* XII and *Pol.* I.12. See Furley (1985), Sedley (1991) and (2007) ch.6.

Natural philosophy, after all, must explain 'how it is better so, not absolutely, but in relation to each thing's being' (*Ph.* 198b9). 'Nature makes nothing in vain, but always regards what is the best possible for each thing, preserving the individual substance and that which it is in itself' (*IA* 708a9–12). In light of such statements, it is safe to assume that, in the zoological context, Aristotelian teleology is focused around individual animals and, connected to this, at the perpetuation of particular kinds.[2] Each individual animal, then, will have a set of three types of goal, unified in its essence, as set out in Table 7.1.

Table 7.1 *Goals of living beings*

(i)	survival (i.e. nourishment, self-defence); beneficiary = the individual animal[3]
(ii)	flourishing (i.e. sentience, intelligence); beneficiary = the individual animal[4]
(iii)	reproduction;[5] beneficiary = future individuals and possibly also individual reproducers.[6]

[2] Aristotle's 'teleological explanation . . . must be applied not absolutely but with respect either to an individual entity or a specific kind' (Johnson 2005, 93). See also *IA* 704b15–18 and *PA* 665b2–5: 'each creature is provided with such (parts) as are suited to its special mode of life and motion'. *Metaph.* V.16: 'For each thing is complete and every substance is complete when in respect of the form of its own proper excellence no portion of its natural magnitude is deficient' (1021b20–2) and discussion in Johnson (2005) 84. 'Perhaps even in inferior living beings there is some natural good stronger than themselves which aims at their peculiar good (τοῦ οἰκείου ἀγαθοῦ)' (*EN* 1173a4–5). This view can be contrasted to that of Cooper (1987), who holds that there is a *telos* higher than individuals, which is that there be a permanence to all kinds of animals and plants. Against this Johnson argues that 'perpetual existence is not explanatory enough'; it is a fact and not a reason why. The reason why animal parts are functional or that animals mature is because it benefits them individually or as a particular kind.

[3] Boylan (1984) argues that Aristotle's teleology is centred on a principle of 'the fittest survive'. I add to this (ii) and (iii), which are not directly linked to survival but are evident concerns which shape his account of lesser animals in the *GA* and elsewhere.

[4] All of the senses are not just for the sake of being but for the sake of doing well (*de An.* 434b23–5, 435b20–1).

[5] Johnson (2005) lists 'life, survival, activity and flourishing of the individual living specimen' as the main living goals (176). He further notes how survival, reproduction and sensation are directly beneficial to the individual (92–3), while protection of those three is of indirect benefit (191).

[6] This is most plausible in human beings. Aristotle seems to think that having children aids human flourishing by developing the virtues. See Scott (2000). Also *GA* 753a15–17 suggests that birds that have not been able to sit on their eggs 'deteriorate in their condition as though they were being deprived of their natural endowment (ὥσπερ ἑνός τινος στερισκόμεναι τῶν συμφύτων).'

Although this teleology does not have an overarching goal or end, it does have an order. The order is centred on human beings as the best type of animal, serving as a paradigm. As Aristotle says in the *GA*, 'we must begin first of all with the animals that come first. These are the perfect animals . . . and the first of these is the human kind' (*GA* 737b26–9). 'Nature' is sometimes used by Aristotle in a normative sense to indicate a hierarchy of value. Since being a human male is the most valuable animal existence achievable, the human male can be said to be the most 'natural' sublunary living being. Other animals are compared negatively: they are less natural.[7] When using 'natural' in this way, non-human animals can be called 'unnatural' without this meaning that they do not occur always or for the most part as they are, healthy and sound. Thus Aristotle maintains that in their own right lobsters are perfectly lobster-like, while at the same time maintaining that to be a lobster is not an ideal way to live. If we think of teleology in terms of the *telos*, the completion or perfection achievable in nature, then in one sense all animals that are fully grown, whether male or female, are *teleia* (complete/perfected) (*GA* 737b11). On the other hand, females and animals other than humans are considered to be incomplete or imperfect.[8] The statement that 'nature does nothing in vain' does not mean that, for Aristotle, nothing natural can be incomplete without qualification.[9] Many natural items are incomplete in the sense articulated here. Furthermore, it becomes clear that this incompleteness is linked, for Aristotle, to animals' material natures – the fact that each type of animal is made up of certain materials.

7 Humans are the most 'natural' animal (*PA* 656a7–14; 686a27; *HA* 608b7; *IA* 706a19, b10, *HA* 494a26-b1; *Resp.* 477a21–33). Although we might feel inclined to reject Aristotle's normative ordering of animal varieties, on reflection it is worth considering how common this assumption is and how it continues to inform biological and cosmological research, even when our worldview offers little in the way of theological underpinning for it. Modern study in these fields often chooses to characterize the human race as the pinnacle of evolutionary success, not just in terms of propagation but also in terms of cognitive sophistication. The idea that humans are special persists despite the fact that some prehistorical species were more successful in strictly evolutionary terms. See, for example, Rees (2010) 12: 'Humans are more than just another primate species: we are special . . . '

8 Aristotle's emphasis on human biology as a normative assessment differs from his zoological practice, where he comes to treat all types of animal as good ends in themselves. He 'sees beyond this limited anthropocentric approach', Freeland (1994) 171. See also Pellegrin (1986) 92.

9 Johnson (2005) says that according to Aristotle 'natural things are not incomplete, failed, aborted, mutilated, worthless or useless' (82).

The materials composing an animal tend to accord with the region it inhabits and what it eats, its body being predominantly watery, earthy or airy (*HA* 590a14–596b20).[10] Learning that an animal has a watery nature can help us to understand certain aspects of its life, such as the type of food it likes and where it thrives.[11] This set of facts will aid our understanding of certain aspects of its formal existence. But the wateriness of a particular animal is not just part of its essence – it is also a brute fact about that animal. The ways in which water acts and reacts shape how this animal will cope with its environment, how it will survive and how it will reproduce.[12] This state of affairs means that some of the general goals of living nature, including survival (i), flourishing (ii) and generation to type (iii), are not possible in all animal types due to their material natures. Very generally, coldness and wetness or wateriness impede completion in its various manifestations.[13]

Aristotle's desire to discuss generation in many different animals stems from the belief that animal kinds are, or at least ought to be for the most part, arranged appropriately.[14] Close study is designed in part to reveal such truths in nature, as Aristotle makes plain in *GA* II.

[10] 'We say that plants belong to the earth, aquatic creatures to the water, and land-animals to the air, but the more and less and the nearer and further make a surprisingly great difference. As for the fourth tribe, we must not look for it in these regions, although there wants to be a kind corresponding to the position of fire in the series . . . fire is always in some other one of the substances, for the object which is on fire appears either as air or smoke or earth' (*GA* 761b13–22). This can also be expressed in terms of their relative heat, coldness, dryness and fluidity. See the beginning of *GA* II.

[11] Each type of animal is to be analysed, according to Aristotle, in terms of four major kinds of *differentia*: parts, activities, habits and ways of life (*HA* 487a11–12). Lennox in Aristotle (1996) equates the material nature of animals with their food, writing: 'certain biological explananda are explained solely by the elemental material potentials of the basic nutrient of the animal in question' (224 n.12).

[12] In the *A. Po*. Aristotle details the importance of grouping together 'essential' features which define a kind. *A. Po*. 92a6–33, 98b32–7, 99a30-b6. Charles understands this to mean both formal essential natures and material ones (1997b) 29–31 and (2000). Many of the examples mentioned in the *A. Po*. include material features. For instance 'the explanation of longevity for quadrupeds is their not having bile, but for birds their being dry or something else' (*A. Po*. 99b5–6). See also *A. Po*. 98b36–38, Lennox (1996) and Charles (2000). Explanations that start with material givens abound particularly in the *PA*. See Lennox (2001) ch.8, 202.

[13] '[N]ature never fails nor makes anything in vain so far as is possible with respect to each thing' (788b20–2). For some kinds of animal perfection is not possible even in best circumstances. See also *IA* 704b11–17.

[14] See Aristotle's exhortation to study ignoble animals at *PA* 645a7–15. He also wishes to account for the phenomena in all animals in order to defeat opponents who habitually

In all her workmanship herein nature acts in every particular as reason would expect (*GA* 731a24).

We should notice how well nature brings generation about in its several forms: they are arranged in a regular series, thus: the more perfect and hotter of the animals produce their young in a perfect state so far as their quality is concerned (no animal brings forth young that are perfect in size, because they all grow in size after they have been produced), and these young which they generate are living creatures inside them from the outset (*GA* 733a34–33b4).

And since an actual animal is something perfect whereas larvae and eggs are something incomplete, nature's rule is that the perfect offspring shall be produced by the more perfect sort of parent (*GA* 733a2–4).

Broadly, Aristotle seeks evidence of a gradual decline in three aspects of generation, which link up with the general goals of living things.

(a) Higher animals generate more successfully than lower ones (linked to [iii] in Table 7.1).
(b) An animal's ability to complete offspring marks them as better or higher in kind (linked to [iii] in Table 7.1, but not directly).
(c) It is better for animals to exist separately as male and female tokens, which does not occur in the lowest of animal kinds (linked to [ii] in Table 7.1)[15]

Although Aristotle begins to investigate how generation takes place in lower animals with these patterns in mind, he is happy to modify and adapt them. Furthermore, his interest in lower animal natures at many points makes clear his position on the female role in generation and on sex difference itself. The next three sections provide case studies from *GA* III, which illuminate Aristotle's views on gender roles and reveal the continuity and flexibility of his theory of generation.

7.2 Generation in fish and bees

Better animals not only live fuller lives but generally seem to achieve other ends more successfully than less complete kinds

generalize from too little information. Indeed, in order to highlight his own comprehensive approach, he often uses evidence from some of the very lowest animals in order to establish general principles in the *GA* (e.g. 729b25–33 on insects, 740b2, 741a18, 741b30). A truly exceptional class, such as testacea, however, would not be referred to in this way, as I will later show.

15 'In practically all animals which can move about female and male are found separate . . . The reason for this is that the higher animals (τὰ τιμιώτερα) are more self-sufficient in their nature' (*GA* 732a14–22). See Chapter 8.6 and *GA* II.1.

(e.g. [b] and [c]). Since the *GA* focuses on generation, Aristotle is concerned in this context to find an order from more perfect to less perfect methods of generation (a). His methodology begins with a gradual move away from the paradigm. Thus, to start with he seeks the qualities and attributes common to all animals, for instance, the presence of male and female and copulation in motile animals. He then must account for the differences. He often talks of this sort of method.

> The phenomena of generation here are, on the one hand, similar to those which obtain in the animals that walk so that the same statement will serve for all of them. On the other hand, these animals exhibit certain differences not only as between themselves, but also when compared with the animals that walk (*GA* 749a13–18).[16]

In *GA* III Aristotle reminds the reader that he is following a plan laid down earlier on in the work to cover generation in all the major types; these are sometimes divided in terms of locomotion or the presence of particular types of feet, but mostly, since this is a work on generation, they are broken up with reference to generative methods (*GA* 732a26–733b24). Those animals that gestate offspring internally and bring them to live birth – what he calls the external live-bearers (*GA* 737b16–18, 737b26–8) – are detailed first. Next come those that lay internal eggs, which hatch inside, subsequently bringing to birth the hatched offspring, that is, the Selachian fish (e.g. sharks) (*GA* 749a19–21, 754a21–755a6). These fall between the external live-bearers and the egg-layers but sometimes are classed as falling under the egg-laying kind. Egg-layers are next divided into those that lay perfect as opposed to imperfect eggs – which matches up well with the distinction between air animals (i.e. birds *GA* 749a17–751a32) and water dwellers (i.e. fish *GA* 754a21-b19). Falling within the egg-laying kind are also cephalopods and crustaceans (*GA* 757b31–758a5). The production of larvae is the next method of generation, which is inferior to the production of an imperfect egg, larvae being a stage before the egg for Aristotle (*GA* 758a30–759a3). The final generative type to consider are the spontaneous generators (*GA* 761a13–763b1).[17]

[16] See also *GA* 761a13–763b1 and the discussion about the location of the uterus (749a28–34).
[17] These groups are individuated also in terms of their material constituents, that is, where on a hot-cold/fluid-solid continuum they lie. This is expressed comparatively; for

Animal types generally degrade in the above sequence according to generative success (iii). In modern biology, when the scientist deems live-bearing to be better than egg-laying, she does this in so far as she recognizes that bearing live young is a more effective way to protect and ensure the survival of offspring.[18] Aristotle also recognizes, at certain points, that some methods of generation are more successful in this sense. For example, when he refines the account of egg-layers in *GA* III, he makes clear that although all of them are relatively cold and solid (compared to live-bearers) some are colder and more solid than others within the division. Egg-layers are then further divided into those that lay hard-shelled, double-coloured eggs (the hotter and more fluid ones) and those that lay soft-shelled, single-coloured eggs (the colder and more solid ones). Those that are hotter and more fluid have the better method of generation of eggs – after all, it is clear that eggs with hard shells give offspring a greater chance of survival. As Aristotle notes, the majority of soft-shelled eggs get destroyed (*GA* 755a32–3).

Whether certain methods of generation are better than others in terms of the survival rate of offspring ([a] & [iii]) is not Aristotle's only consideration when seeking order in nature. He also employs two other normative conditions which are rather alien to the modern biologist. First, he believes that when male and female exist in separate animals this is a better situation (c). Then, he thinks that it is better for offspring to be more complete at birth regardless of any positive effect this might have on fertility (b). Hard-shelled egg-layers are better than their soft-shelled counterparts because they produce 'perfect' eggs, with a differentiation between white and yolk (*GA* 749a14–22). Soft-shelled eggs are monochromic and incomplete, displaying the weakness and imperfection of the kind (*GA* 755b5).[19]

example, egg-layers are colder and more solid than the live-bearers. Animals are broadly and more narrowly distinguished by the more and the less (e.g. *PA* 692b4–9).

18 This would most probably be accounted for in terms of the evolutionary advantages afforded by live birth. See Lynch and Wagner (2005). Modern scientists might not think of reproductive success in terms of the individual but rather choose to focus on the gene itself. See Dawkins (1976).

19 Fish eggs are single-coloured in part because fish are cold and earthy, lacking in soul heat (*GA* 751b20–1).

In *GA* III, Aristotle brings an array of evidence to bear to show that (a), (b) and (c) represent an order in animal kinds. So, for example, he often concentrates on finding that 'better' or 'higher' animals have separate males and females (c).

> In the most perfect of them, the female and male are separate, and we say that these abilities are principles of all animals and all plants (*GA* 763b21–4).[20]

The paradigm of sex separation is displayed in higher, live-bearing, animals in whom both sexes contribute residues concocted from blood and deposited in receptacles fitted for them (*GA* 725b3 – uterus and male genitals). The male semen is hotter, being fully and completely concocted. The female contribution is the menstrual blood (*GA* 728a24). Aristotle clearly takes human beings as the model since other animals do not bleed regularly.[21] And in some 'lower' animals Aristotle believes that menstruation does not occur at all. One example is fish that produce soft-shelled eggs. These animals, then, are a shift away from the paradigm but, as they are mobile and sentient (b), they remain reasonably high on the scale of life, particularly in terms of their flourishing capacities (ii). Aristotle is thus concerned to establish separation of the sexes as evidence of the normative ordering of natural kinds. He is aware that female-only species of fish (reproducing parthenogenetically) have been reported and does not think this impossible.[22] But in order to move forward with the analysis of fish generation, he chooses to concentrate on observed facts that correspond to his expectation that male and female are separate in this kind. He begins by mentioning an observation that supports sex separation, which is that some fish do not have a uterus and therefore cannot be female (*GA* 755b18–22).

[20] ἐπεὶ δ' ἐν τοῖς τελειωτάτοις αὐτῶν ἐστὶ τὸ θῆλυ καὶ τὸ ἄρρεν κεχωρισμένον, καὶ ταύτας τὰς δυνάμεις ἀρχάς φαμεν εἶναι πάντων καὶ ζῴων καὶ φυτῶν.

[21] *GA* 727a22, 775b6–8, 776b27. Dean-Jones (1994) 187 n.133. Only humans and some primates menstruate: Balme in Aristotle (1972) 150. Aristotle obviously knew this to be the case; his explanation of it is that most animals do not have as much menstrual material as humans and therefore only discharge it internally (*GA* 727a22).

[22] In some kinds of fish no males are known (erythrinus and channa: *GA* 741a36, 755b22; *HA* 538a20; Kember 1972) and many types of fish have never been seen to copulate. The way in which Aristotle expresses the situation ('no male has ever been seen') leaves open the possibility that males exist but are elusive. John Maynard Smith (1988) displays a similar style when describing an order of rotifers in which 'no one has ever seen a male' (168).

The discussion begins with the dismissal of the idea, suggested by a rival theorist, that because fish produce vast quantities of offspring, their method of generation is just like plant generation (*GA* 755b8–12). Instead, Aristotle is sure that separate males and females exist in fish – a significant difference from the more basic (hermaphroditic) plant kinds. As proof of this, he cannot depend on the presence of male semen and female menses. Instead, he hopes to find copulation (*sunduasmou*) taking place. Copulation requires contact whereby semen is emitted *into* the female by the male (*GA* 749a15–16, Cf. 733b19–21). Cephalopods and crustacea, less noble animals than fish, have been seen to copulate (*GA* 755b34), which serves as proof that 'one of them is male and the other has a uterus' (755b35–6). However, nobody has ever seen fish copulating. Yet, Aristotle insists that they *do* copulate; they place themselves side by side like dolphins, achieving this so quickly they avoid observation (*GA* 756a36–756b3).[23] As well as the swiftness of the operation, Aristotle thinks that observational evidence is in short supply because those who watch fish (such as fishermen) are not concerned about knowledge for its own sake and so are not reliable observers (*GA* 756a35).

In *GA* I, the fact that male fish sprinkle milt on eggs after they have been laid was used to show that the male does not contribute any matter, in terms of bulk, to offspring (730a18–24). The phenomenon of milt sprinkling reappears to aid with the issue of sex separation in *GA* III, where it is taken to be an activity that happens in addition to (rather than in lieu of) copulation. When copulation fails and fish lay unfertilized eggs (like the wind-eggs of birds), the males come along afterwards and make the eggs fertile (*GA* 756a24–8).[24] The first stage when the male and female animals come into contact is unobserved, but Aristotle has witnessed the male *gonê* touching the female's eggs. In the end, milt sprinkling counts as sufficient proof of the existence of and

[23] In *GA* I, Aristotle says that 'fishes copulate by placing themselves alongside each other' and notes that they emit quickly when they 'come into contact with one another' (718a1–2, 9). Copulation does occur in some varieties of fish, but is very rare. Most fertilize externally. Balme in Aristotle (1972) 135.

[24] This cannot happen in birds because once the eggs leave their bodies, they are perfected, meaning that, after that point, they cannot increase in size.

need for the male. On the basis of this evidence, then, Aristotle concludes: 'Fish are males and females and all fish copulate. Without the male semen no fish of any sort comes to be' (*GA* 756a28–31).

Clearly Aristotle finds it is important to confirm the indispensability of the male in nature – a theme also emphasized earlier on in the treatise.[25] However, this project is complicated somewhat by the existence of animals in which male and female principles are intermingled within the same body. In accounting for asexual generation, one of Aristotle's concerns is to maintain that, in lower animals, the male and female principles of generation still exist and that it is the male principle that starts generation. He notices that in lower animals supposed females (i.e. those which bear young) sometimes seem to generate on their own. These are not exclusively female but are simultaneously male.[26] The early Books are pretty much silent on hermaphrodites. Generally, in animals that can move from place to place, whether blooded or bloodless, male and female exist as separate (*GA* 715a26–9, 732a14–5). This means that the highly locomotive bee presents something new for Aristotle to consider in *GA* III. To begin with, it is difficult to know what to expect in terms of sex separation (c) in this kind. In some respects bees display clear signs of being superior. They not only locomote and perceive (*de An.* 428a11), they also lead ordered and industrious lives, with cleverly planned divisions of labour, indicating intelligence.[27] However, bees are also deficient in so far as they are cold, bloodless, short-lived and do not have an evident heart.[28] In terms of generating, they use the worst strategy by producing larvae (failing to achieve [b]). And

[25] *GA* 722b13–14. Chapter 3.2.

[26] In *GA* I Aristotle says that females are animals which bear young from their bodies (*GA* 716a22). The observations of animals that bear offspring alone, or produce foetations without a male, lead him to revise this initial definition. Now he narrows the scope of the term 'female' to 'that creature which is unable to set the life of another in motion from the start'. If a lower animal does not fit this description, then it must be male as well as female.

[27] For their status, along with humans, as social animals see *HA* 488a9; for their industriousness, *HA* 622b19; their clever division of labour, *HA* 627a22-b22. At *HA* 623b26, he says 'their working methods and way of life show great complexity'.

[28] Despite being lower and colder animals, the thinness of their blood allows for a sharp intellect (*PA* 650b24).

when it comes to separation into male and female, they do not entirely fit expectations.

In investigating bee generation, Aristotle uses two types of facts or evidence, 'special and particular facts concerning bees' and 'facts generally applicable to other animals' (*GA* 759a25–7).[29] He speculates about the possible ways in which bees could generate given these facts (*GA* 759a12–24). The three types of bee – kings, workers and drones – could copulate with their own kind or another kind or they could reproduce without copulating. The results of reproduction will differ in all these instances. If bees reproduce without copulation, they might produce a different kind from themselves. Aristotle presents a final option, which is that bee offspring are fetched from somewhere, a theory favoured by those who think that bees collect their young from flowers (*GA* 759a25–759b1). Quickly dispensing with the last suggestion on the basis of 'facts generally applicable to all animals' (i.e. that none could be generated in that manner), he considers more closely particular facts about bees in an attempt to eliminate other erroneous suggestions. We are ultimately left with his answer.[30] The particular facts are as follows:

- No bees have ever been seen copulating (*GA* 759b22).[31]
- Workers are never produced unless a king is present in the hive (*GA* 759b10, 19).
- Drones are produced sometimes when the king is not present but never without the presence of a worker (*GA* 759b25).

Aristotle concludes then that the king, without copulating, produces two types of offspring – workers and more kings (*GA* 760a4) and that workers, without copulating, produce drones (759b29) and, finally, that drones are infertile (760a35). Thus,

[29] The use of particular facts concerning bees recalls Aristotle's treatment of mule infertility in *GA* II, where he refers to particular facts concerning mule parents, horse and ass kinds (748a15–17).

[30] *leipetai* (759b28; 760a4). Karbowski (2009).

[31] Notice that he regards this fact as informative of bee generation, whereas he did not do so in the case of fish generation. Although Aristotle obviously saw the mating swarm of the queen bee (*HA* 624a24–6, 625b6–18), he probably did not see any actual mating. The entymologist (and evolutionary biologist) David L. Stern tells me that the actual mating that occurs in the swarm is very difficult to observe. It is certainly not obvious that this is what swarming is about.

the bee's ability to reproduce to type (iii) is unusual but apparent. They achieve this not directly but in a clear order; A produces As and Bs then and Bs produce Cs. 'This arrangement has been so well constituted' writes Aristotle, 'that all three kinds remain in existence although they don't all [individually] generate' (*GA* 760b2–3).[32]

There is a way in which Aristotle could fit bees to a more rigid scale of life: by labelling the entire kind as defective or deformed, their method of generation and distribution of the sexes as faulty, and thus its significance is marginalized. He does not explicitly do this, instead opting to characterize them as exceptional and 'divine'.

> With regard to the generation of animals that are akin to bees, such as hornets and wasps, the situation is in a way similar in all of them, but the extraordinary features are lacking, and this is what we should expect, because they contain no divine ingredient as the tribe of bees does (*GA* 761a2–6).[33]

What is divine about bees is their intelligence. However, they do not contain the differentiation of male and female into token animals. So his handling of bee generation reveals no strict commitment to the coincidence of sex separation (c) and higher cognitive abilities (ii).[34]

Before reaching the above conclusions, an interest in identifying male and female distracts Aristotle. It may be that since bees are locomotive and sentient, he assumes that the sexes are separate, an initial assumption supported by the fact of regular variation in bee morphology within the kind. Many say that workers have to be 'female', because they care for the young.[35] Aristotle objects:

[32] When Aristotle initially sets out methods of generation in insects, he does not reserve a place for hermaphroditic generation. Instead, he notes that some insects reproduce sexually with separate males and females, and that others generate spontaneously from no parent (*GA* 715b2–18, 721a2–11). Thus it seems that further reflection on these animals in a more detailed analysis has revealed a method of generation in insects which Aristotle did not originally expect or predict.

[33] 'We see then that the manner in which bees are generated appears to be peculiar, in keeping with their extraordinary and peculiar character' (*GA* 760a5–6).

[34] Green (1992) wants to find Aristotle firmly linking female coldness to her (supposed) mental infirmity. Such common coincidences obviously interest Aristotle but do not show that he thought them entirely necessary.

[35] In speculating about the sex of the worker bee Aristotle is perhaps engaging in a debate to which, as a natural philosopher at the time, he would have been expected to contribute. The same issue arises concerning wasps in *HA*. There, Aristotle writes: 'the stingless are smaller and feebler and do not defend themselves, but those with the sting are bigger and

worker bees cannot be female because they have defensive organs (*GA* 759b2–8).[36] On the other hand, he thinks that they cannot possibly be male because males never busy themselves (*diapo-neisthai*) about the young.[37] These difficulties are eliminated by the realization of the hermaphroditic nature of bees. Workers contain both male and female principles within them, thus making sense of both the propensity to care for young and the possession of defensive organs (*GA* 759b32).

Aristotle's preconceptions regarding gendered social roles are clear in the comments on the worker's sex. However, this section of *GA* III also shows some curious tensions concerning gender, revealing that Aristotle does not comprehensively denigrate female qualities. Like a human king, the 'king' bee's occupation is superior to other members of the tribe and because of this it is freed from menial tasks and taken care of by the other bees. Also like a human king, it is said to punish its tribe (*GA* 760b20), the king's relationship to a subject being like that of a father to his son (*Pol.* 1259b14–16). In the human sphere, we are only metaphorically the children of a king because we do his bidding and are held accountable to him. Aristotle sees the physical set up of one ultimate bee parent (a combination of male and female) as parallel to kingship in the human sphere in these respects.[38] 'The workers

fierce; and some call the latter males and the stingless ones females' (628b5–8). Hesiod is extraordinarily flexible on this score, likening drones to women in one work (*Theogony* 590–602) and to men in another (*Works and Days* 304). See Sussman (1978).

36 See also *PA* 661b28–31 and *HA* 538b16. Aristotle, of course, realizes that many female animals possess defensive weapons, such as the she-bear's claws or the lioness's teeth. His point is not that females never have defensive organs but that they never have them *if the male of their species* does not. Workers cannot be female if drones are male, because drones do not have a sting.

37 Although he asserts this at first, he ends non-dogmatically. In the more intelligent species, males do care for their offspring (753a12–14). Also, the male glanis fish is renowned for his vigilance over his offspring (*HA* 621a20-b2). Howard (2006) writes '[i]n the mid-nineteenth century, Swiss Biologist, Louis Agassiz, reported that there were several species of catfish in North America that did, in fact, exhibit this activity and then went on to show that there is a similar species that behaves in this manner in Greece and the surrounding regions (Black Sea). In 1857 a new species entered the Linnaean classification, Silurus aristotelis with the common name, Aristotle's Catfish.' See also Mayhew (2004) ch.2.

38 Smith (1988) 178–9: 'To watch animal behaviour without using human analogies would be to rob oneself of a major source of insight. Of course, any insight gained in this way must be tested and will often prove to be wrong.'

attend upon the kings because the bees are generated from the kings' (*GA* 760b16–17).

Although the 'king', as with all generating bees, contains both male and female principles within it, Aristotle is tempted to regard it as primarily male. He finds it hard to call the leader bee a queen or, more properly, by a gender neutral term, such as 'leader' or 'ruler'.[39] However, although he wishes to masculinize the hermaphroditic leader, it is its female or feminine characteristics which come to the fore. Aristotle thinks of animals which bear and care for young as mostly feminine. So, with reference to workers, he notes that, although technically they are hermaphroditic, 'as far as generation is concerned they are female' (*GA* 759b29–30). Because their life aims at producing young, the female portion of their nature is most apparent. This dominant femininity is also present in the 'king'.[40] Its body is very bulky (*megethos*, 760b10) because it must produce many offspring out of it, which is why it remains inside (760b9). This description of the 'king' bee brings to mind the secluded child-producing role of the good Athenian citizen wife.[41] The life of the Greek male landowner and citizen, involving himself in intellectual pursuits has been taken to represent for Aristotle the epitome of mortal animate existence. However, when investigating lower animals, the drive to separate male from female activities in this manner is not so apparent. The discussion of the king bee thereby indicates how one cannot assume that Aristotle set out to glorify only male characteristics in nature and to denigrate all female ones.

7.3 Wind-eggs

GA III provides insights concerning Aristotle's views on the content and the nature of the interaction between male and female

39 The ability to label the animal without making it either male or female is easy in Greek, since there is a neuter gender. In the *HA* also the masculine pronoun is used for the leader of the hive (624a26). Aristotle explains that we cannot think of the king bee as female because 'it is wrong to apply the term "female" to creatures where no separate male exists' (759b33). He does not state the converse, that there is no such thing as a male without a separate female.

40 At *HA* 553a29–30 Aristotle notes that some people call the leaders 'mothers' because they produce the young (literally, they are generators; *gennôntes*). Aristotle refers to the main wasps in *HA* as 'the so-called mothers' (628b25).

41 Pomeroy (1975), Lacey (1968), Lefkowitz and Fant (1982) and MacDowell (1978).

contributions to generation. This is best illustrated with reference to his analysis of wind-eggs in birds.[42] In *GA* I and II the discussion of these unfertilized eggs focuses on the shortcomings of the female principle (730a30–3, 737a30–3). Because wind-eggs remain undeveloped and eventually rot, they provide evidence that these female-only products have the lowest level of soul potentially (*GA* 741a15–26). *GA* III takes a longer time over the phenomenon (750b9). Wind-eggs are already differentiated into two colours – white and yellow (yolk). As was explained earlier on in the treatise with reference to fertilized eggs, the white portion is the more form- or soul-like part.

> The secretion[43] is from the blood, the blood is the matter of the body – the hot part is closer to the shape (*morphê*) of the developing animal; the other, earthier part, is present to build up the body and is further away [from the shape]. That is why in the case of all double-coloured eggs the young animal gets its principle (*archê*) of generation from the white, because hot substance is the place where the soul principle is to be found (*GA* 751a35-b6).[44]

In the *GA* II description of fertilized eggs, Aristotle was inclined to identify the soul-like white portion of the egg with the male contribution. In his more detailed account in *GA* III this is no longer the case. While he still maintains that 'once the white and the yolk have been separated from each other, they already possess the principle that comes from the male'[45] (*GA* 757b13) and that when this occurs the male is in the white,[46] in the unfertilized egg the white is from the female animal. 'The double colour', he writes, 'is not due to the male and the female [principles], the white being male and the yolk female: both are from the female' (751b26–7). This means that the living potentials within the female contribution come in two grades: one that is closer to soul and the other that is less concocted (*GA* 751b6–7).[47] Rather

42 The equivalent of wind-eggs also occur in some varieties of fish (*GA* 756a18–21).

43 This indicates the female secretion which is to constitute the egg.

44 The white portion is closer to form and 'contains in itself the soul heat (ἔχει δ' ἐν αὐτῷ τὴν θερμότητα τὴν ψυχικήν)' (*GA* 752a2). See also *HA* 561a24: 'the principle (ἀρχή) of the chick is from the white'.

45 τὴν ἀπὸ τοῦ ἄρρενος ἀρχήν.

46 At *HA* 559b6–10 Aristotle implies that the male semen becomes the white of the egg.

47 Aristotle maintains that wind-eggs appear without any copulation having occurred, disagreeing with some who say that they are 'relics of an earlier impregnation'

than presenting a problematic contrast with Aristotle's earlier theory, this account meshes well with it. So, for instance, we learn early on that there are two portions of menstrual blood, the purer one and the growth-promoting one (*GA* 730b3–4, 739a7–13, 744b33–6).[48] The purer portion is like the white of the egg, the portion of the female contribution which has more of the soul principle in it.

Although it has more and less refined portions, Aristotle insists that the wind-egg cannot become an animal without male input (*GA* 730a32, 737a33, 741a15, 31, 750b25, 751b25).[49] Wind-eggs, like other female productions, only have potential nutritive soul (741a19–25). The wind-egg is complete in size and has some sort of differentiation, but it never has actualized soul and never becomes complete or finished.[50] The nutritive soul is actualized only when the foetations 'begin to draw nourishment to themselves like foetations that get separated from the mother' (*GA* 736b10–11). Once this occurs, male and female contributions

(*GA* 751a11–12). This is worthy of note: he could have claimed that the male was involved, as others did, which means that he does not attempt to fudge the obvious fact that the female contributes to soul in some way. See also *HA* 559b21–3: 'Some people allege that wind-eggs are the remains of eggs previously produced by copulation. They are wrong, because we have sufficient observations to establish that chickens of the domestic fowl and of geese lay wind-eggs though they have never copulated.'

48 The difference between these two must be one of degree of concoction if one considers that in rare cases, such as superfoetation, the nutritive material is capable of forming a second offspring (*GA* 773b8–25).

49 As it turns out, intercourse has to happen numerous times before the female conceives (*GA* 751a17, 34).

50 Near the beginning of *GA* IV Aristotle reinforces the idea that females are unable to 'set' offspring (765b15), meaning that on its own the female contribution cannot begin the process that will lead to the development of a sentient animal, differentiating (διορισθῇ) and bringing the parts to their completion (741a26, 758b36). We now know that female animals can on occasion do more than Aristotle allows. In some cases, the male contributes a mere 'stimulus'. 'Parthenogenesis, or the development of an individual from an unfertilized egg, is a type of reproduction occurring naturally in various invertebrates. Artificial stimuli (physical or chemical) may initiate embryonic development in the unfertilised eggs of certain animals that normally reproduce bisexually. This fact seems to indicate that some impetus is given to the egg by the sperm cell, an impetus that is quite apart from the mere contribution of protoplasm. There is, also, a further implication: that the unfertilized egg itself possesses the potential capacity for development, at least up to a certain stage.' Romanoff (1960) 126. See also Smith (1988) 169–71 where he explains that parthenogenetic species require the stimulus of male semen from a male member of another species. Although the jump-start is necessary in these cases, the male makes no genetic contribution to the offspring.

no longer exist as such. It is the embryo, or fertilized egg, that has potential and actual soul.[51]

Aristotle's discussion of wind-eggs reveals that the white in double-coloured eggs is already present before the male contribution does its work; the work then continues within that hotter portion, where the principle (*archê*) resides. Thus, after conception, what male and female both contribute must come together there. Similarly, in higher life forms, since there is a hotter portion of menstrual blood which contains more of soul and the principle of life, this combines with the male semen. It seems reasonable, then, that, post-conception, the female's purest contribution can shape parts and develop them in the way they would have done if they had remained internal to the female animal's body and act in a reciprocal fashion to the male contribution.[52] That the female contribution contains only potential nutritive soul need not imply that it cannot contribute to the formation of the embryo.[53]

7.4 On the margins of life: the phenomenon of spontaneous generation

In attempting to account for animals that come to be spontaneously without any parents Aristotle faces many challenges. Animal generation normally indicates the pre-existence of form and the attempt to continue that form. Any matter that comes to be an animal is usually matter that has been worked up by soul to be in a potential state to come to life. Both form and matter in these senses are absent in spontaneous generation. More generally, the phenomenon challenges a teleological outlook: the fact that generation of a type of living thing happens always or for the most part normally indicates goal-directedness.[54] If nature does not operate due to chance or spontaneity (*Ph.* II.8), can spontaneous generation even count as natural, according to Aristotle?

[51] It is misleading to imagine stark separations in soul functions between male and female. See Chapter 5.3.

[52] Both constitute a set of *dunameis* – potentials or powers – further elaborated in *GA* IV on sex difference and heredity. See Chapters 8–9.

[53] This will be crucial to our understanding of hereditary resemblance. See Chapter 9.

[54] Balme (1962a) and Hull (1967–8).

As is apparent already in *GA* I, 'testacea'[55] occupy an anomalous position. The lack of differentiated sexes accords with their immobility, being attached to rocks or the earth like plants (*GA* 715b16–19; *HA* 621b11–12). With no separate males and females, they cannot copulate and generate in that manner, but neither do they seem to contain male and female within themselves and generate out of themselves like plants.[56] They perform neither the function of plants, nor that of animals, because they are unable to generate; somewhat unhelpfully, Aristotle offers that 'being in both groups they perform the function of neither' (*GA* 731b8). Next, in *GA* II when Aristotle divides living things with respect to reproductive strategies (732a26–733b15) he does not have a place for testacea – they are not live-bearing, egg- or larvae-producing. Indeed, they simply do not *re*produce. New examples of the testacea kind are not generated from the older versions; instead, they are generated from the environment. *GA* III attempts to provide an account of this phenomenon: testacea are produced out of mud and sea water under the influence of the sun's heat (762a8-b18; Cf. *HA* 547b18–23). The putrefaction and frothing up of mud results in an earthy bubble forming around *pneuma*, which becomes ensouled.

Reference to paradigm instances of generation, in which male and female principles interact in a specific way in order to produce a replica of parents, structures Aristotle's attempts to uncover the mysteries of spontaneously generating types. However, the model is found to be of limited use. Earlier it was noted that every instant of natural generation requires male and female principles. Aristotle even went as far as to suggest that things are alive because they share in these two (*GA* 732a12–3). Such general principles ought to be apparent in all instances of generation. However, the initial attempt to locate male and female principles in instances of spontaneous generation (*GA* 762b2–6) does not go

[55] 'Testacea' is used here to translate τὰ ὀστρακόδερμα – 'potsherd skinned ones'. What Aristotle places in this group does not correspond to any modern classificatory group. He means molluscs with hard shells that usually (but do not always) occur in the sea. Some examples are oysters, clams, barnacles, limpets, mussels, snails and sea-urchins (*HA* 527b35–528a6).

[56] Testacea are non-copulating (*GA* 720b8–10). Separation of the sexes and subsequent coming together for copulation is impossible for immobile kinds.

to plan. In *GA* II we were told that the male principle is both the efficient and the formal cause of generation. As efficient cause the male brings about substantial change through the agency of his currently operative psychic functioning.

> As semen is a residue, and as it is endowed with the same *kinêsis* as that in virtue of which the body grows through the distribution of the ultimate nutriment, when the semen has entered the uterus it sets the residue produced by the female and imparts to it the same *kinêsis* with which it is itself endowed (*GA* 737a18–23).

As the formal cause, the male principle ensures offspring are the same in form or type, as Aristotle also explains in his *Metaphysics*.

> And in general both that from which and that in accordance with which they are generated, is nature; for the thing generated, e.g. plant or animal, has a nature. And that by which they are generated is the so-called formal nature, which has the same form (*homoeidês*) as the thing generated (although it is in something else); for human generates human (*Metaph.* 1032a21–6).

The male principle originates in a male animal living its specific life; this guarantees that a creature like itself develops.

Aristotle is at pains to locate this male principle in spontaneous generation. He equates male heat with the heat in the environment at *GA* 762b14 and earlier when he writes: 'with those animals that are formed spontaneously, the formal cause responsible is the *kinêsis* and heat of the climatic conditions' (*GA* 743a35). However, this sort of heat is normally a mere instrument.[57] The sun's heat, which froths up the mud and sea water, is not from a previous animal's soul. Thus, it does not seem that there is any clear efficient or formal cause of testacean generation.[58]

If there is no father or formal cause, where do testacea get their forms from? We might suspect that the new testacean does indeed have a form since it is alive – soul being form in living beings.

57 If the male contributes only heat then the female would sometimes, of course, be able to generate on her own. Female birds help to concoct their offspring by sitting on the eggs (*GA* 752b16–17). The weather assists in the concoction of reptile eggs (752b30–1). Finally, in hot weather offspring mature faster (753a17–19). If the female's own body heat or the heat of the weather were sufficient efficient causes, what use would males be?

58 It is incorrect to say that 'the sun was seen to be the efficient cause and the formal cause of these spontaneously generated organisms' (Dudley 2012, 185). Strictly speaking, Aristotelian efficient and formal causes must be goal-directed, which is not the case with the sun's heat. See Chapter 6.4.

Aristotle also notices that the same types are produced in certain areas – such as barnacles, oysters and clams – seemingly implying the presence of specific forms.[59] However, rather than types occurring due to the presence of formal causes, instead they correspond to the matter which is present.

> The animal that is constituted differs as to being a better or worse kind – by differences in the enclosure of the soul-principle. The causes of this are the places and the body which becomes enclosed. There is a lot of earthy stuff in the sea. For this reason, the nature of testacean kinds come to be constituted from such materials, and the earth hardens all around in a circle and congeals in the same way as bones and horns do (for these are not capable of being melted by fire), and the body which is alive is enclosed inside (*GA* 762a24–34).[60]

A spontaneously generated animal has a certain morphology because of the place (*topos*) it develops in and the body (*sôma*) – the material stuff – it is made up of. Aristotle's survey of spontaneously generated kinds other than testacea in his *Historia Animalium* reinforces the idea that the nature of the material determines the spontaneously generated animal's structure – horseflies are produced out of timber, bud eaters in the stalks of cabbages, vinegar flies in vinegar, red furry larvae from the putrefaction of snow, and so on. (*HA* 552a29–552b11).[61] This is also the case with regard to testacea.

> Speaking generally, all the testacea arise by spontaneous generation in mud, though they exhibit differences according as the mud differs: in slimy mud oysters grow, in sandy mud cockles and the others I mentioned; on the eroded hollows of rocks the tethya, barnacles, and the commoner kinds, such as limpets and nerites (*HA* 547b18–23).

59 I follow Kember's translation (1972) 172–3 and its emphasis on the importance of *kata logon*. Animals are produced, usually, according to some pattern.

60 αἱ μὲν οὖν διαφοραὶ τοῦ τιμιώτερον εἶναι τὸ γένος καὶ ἀτιμότερον τὸ συνιστάμενον ἐν τῇ περιλήψει τῆς ἀρχῆς τῆς ψυχικῆς εἰσιν. τούτου δὲ καὶ οἱ τόποι αἴτιοι καὶ τὸ σῶμα τὸ περιλαμβανόμενον. ἐν δὲ τῇ θαλάττῃ πολὺ τὸ γεῶδες ἔνεστιν· διόπερ ἐκ τῆς τοιαύτης συστάσεως ἡ τῶν ὀστρακοδέρμων γίνεται φύσις, κύκλῳ μὲν τοῦ γεώδους σκληρυνομένου καὶ πηγνυμένου τὴν αὐτὴν πῆξιν τοῖς ὀστοῖς καὶ τοῖς κέρασι (πυρὶ γὰρ ἄτηκτα ταῦτ' ἐστίν), ἐντὸς δὲ περιλαμβανομένου τοῦ τὴν ζωὴν ἔχοντος σώματος.

61 See Gotthelf's index under 'spontaneous generation, materials of' in Aristotle (1991). Animals are generated in wool, from cheese, from books, from figs and in vinegar (*HA* 557b1–13, 26–31, 552b5). All of these are different in character. At *HA* 552b7–8 animals are said to come to be from snow, which 'after lying for some time turns slightly red, hence the larvae it produces are red too'.

Thus the morphology of varieties of testacea are linked to the materials out of which they arise. Form (*eidos*) in Aristotle's biological ontology is a much more robust concept than shape, and in living beings requires the ability to self-perpetuate (*GA* 735a17–19). As testacean kinds lack this, it makes sense to refrain from referring to the morphology of different testacea as their 'forms'. There fails to be any account, then, of how they acquire any form or soul.

Having discovered no evidence of a male principle, the search begins for a female principle in testacean generation.

Anyone who wishes to follow the right line of inquiry might well inquire what it is which, as it takes shape, corresponds in the case of these creatures to the 'material principle'. In females, of course, this is a residue produced by the animal, a residue which potentially is such as the parent is from which it came, and which is perfected into an animal by the principle from the male imparting *kinêsis* to it (*GA* 762a37–762b4).

In the case of normal (non-spontaneous) generation it is crucial that the female contribution has the same character as the animal it comes from. The male principle cannot act on matter that is not of a very particular sort.[62] If the male alone could determine the type of creature generated, then males could deposit semen in the female's stomach and produce animals directly from nutriment.[63] However, Aristotle does not think that this is possible. Without the female contribution, which resides exclusively in the uterus, no amount of heat or concoction will be able to produce the kind of creature required.[64] In cases of spontaneous generation there is, of course, no pre-existent animal to contribute specific materials. Thus, there is no material cause, strictly speaking.

[62] The female contribution contains all the potentialities for the type of animal it is to become (*GA* 737a24, 738b2–4, 740b19–21).

[63] This is why female fish cannot become pregnant by swallowing male semen (*GA* 756b5–8). The stomach concocts whatever enters it (756b10–12). The uterus is also a locus of concoction. On the womb as an oven which concocts offspring see Herodotus *Histories* V.92 (Herodotus 1920, 113); *GA* 739b9–12: The hot womb draws up the male semen.

[64] Freudenthal (1995) believes that the heat determines the specificity of the offspring. If this were the case, then the male semen of an animal could produce life from any material.

Where, then, are we to locate the correct causes of generation in testacea? These animals are so far away from the paradigm that the usual causes are absent. Aristotle thinks that perhaps we need to consider the role of external matter and can learn something about this from rare cases in which testacea manage to reproduce non-spontaneously from side-shoots.[65] These cases are more in keeping with the general account and serve to reveal something about the materials out of which testacea are constituted, which, in turn, will help to explain their spontaneous generation. The seed-like residues, or side shoots, differ very little from materials that just happen to be lying around on the seabed, as Aristotle explains.

It is reasonable to suppose that there is a surplus portion of residue close by each of the original stock, from which each of the side-shoots springs up. And since the residue is a substance possessing one and the same potential as the nutriment of which it is the residue,[66] it is probable that the stuff produced by the 'honey-combers' is similar to the substance out of which they were originally constituted; hence it is reasonable to suppose that it too gives rise to young ones (*GA* 762a2–8).

Because testacea are cold creatures, they do not concoct very well. Thus, their bodies differ very little from the stuff out of which they were originally composed. However, this is not the whole story: in other animals, internal heat concocts nourishment to bring about generation; in the case of spontaneously generated testacea *external* heat appears to play this part.

Nutriment for testacea is water and earth, and the nutriment of animals is things that are formed from these. So as the heat in animals works things up from food, in testacea, the heat of the surrounding weather conditions, by concocting, constructs and constitutes them from sea water and earth (*GA* 762b13–15).

The heat testacea expend on changing the food they consume to convert it into their body is so slight and weak that the sun can easily mimic this, creating new bodies out of their sea-water food without initiation by soul or form. As Aristotle explains: 'with those animals that are formed spontaneously the cause responsible is the *kinêsis* and heat of the climatic conditions' (*GA* 743a35–6).

65 See also *HA* 546b18–547a3, a passage which makes it unclear whether the mucus emitted by honey-combing purpuras contributes to the offspring or not.
66 ἐπεὶ δὲ παραπλησίαν ἔχει τὴν δύναμιν ἡ τροφὴ καὶ τὸ ταύτης περίττωμα.

The difficulty with this account is that in normal cases heat is merely instrumental. The animal's soul must concoct nourishment; it is never heat itself which generates. How, then, can it be that heat, detached from goal-directed psychic processes, could possibly create living beings with souls?

The oddities of Aristotle's account of spontaneous generation are sometimes used as support for the view that *pneuma* is the material basis of life for him.[67] There is no denying that he leans heavily on the nature of *pneuma* in his description of the phenomenon.

> Animals and plants come to be in earth and in water because water is present in earth, and *pneuma* is present in water, and heat of soul is present in all this [*pneuma*], so that in a way all is full of soul. It is on account of this that things are quickly constituted once they become enclosed. As the corporeal fluids are heated up, it is enclosed and becomes, as it were, a frothy bubble (*GA* 762a19–24).

Pneuma is not literally alive for Aristotle since no soul or life exists until it has been enclosed.[68] However, there would seem to be some soul-like stuff in the *pneuma* once it has been enclosed.

> The enclosed or separated off part of the soul-principle within the *pneuma* makes a foetation and establishes *kinêsis* in it (*GA* 762b17–18).

Aristotle's attempts to elucidate spontaneous generation as the closing off of *pneuma* suggest to some a commitment to hylozoism and to others that there is a purely physical explanation in all cases of generation, that is, 'a form-free causal chain'.[69]

There are good reasons, however, to resist assigning any broad-ranging hylozoist or mechanistic slant to Aristotle's embryology on the basis of his discussion of spontaneous generation, the main one being his obvious discomfort with the phenomenon. Testacean generation is treated as anomalous, not fitting any of the usual expectations when it comes to animate generation.[70] Even in his

[67] Preus (1975) 88–9, following Solmsen (1957) 122, takes it that there is a soul matter – namely *pneuma* – which the male contributes to offspring. See Chapter 6.3.

[68] A 'closing off' occurs in spontaneous generation (ἐμπεριληφθῇ, ἐμπεριλαμβάνεται *GA* 762a23; περιλήψει, περιλαμβανόμενον; 762a26–27; περιλαμβανομένου 762a33; ἐναπολαμβανόμενον 762b16) before living can begin. 'The body that has life becomes enclosed' (ἐντὸς δὲ περιλαμβανομένου τοῦ τὴν ζωὴν ἔχοντος σώματος) (762a31–2).

[69] Lewis (1988) 64. See also Chapter 6.3.

[70] The entire discussion displays Aristotle's perplexity and consternation rather than any confident theorizing. It seems hasty to conclude, then, that 'Aristotle was the foremost

very first attempts to place them, Aristotle treats testacea as entirely exceptional. Although earlier testacea are said to be neither plant nor animal, here they are likened to animals, presumably because they have evident sensory capacities.[71] Despite this advantage, testacea ultimately suffer when compared with plants. Plants lack sensation (a key flourishing capacity [ii]), sexual differentiation (c), and their method of nourishment is faulty. In terms of nourishing, they are turned the wrong way around, with their 'mouths' in the soil (*IA* 705a31-b9, *de An.* 416a6); they also cannot prepare their own nourishment, instead depending on environmental heat to do this (*PA* 650a21–3). In this respect, testacea are similar, since they also allow the heat in their surroundings to prepare their nourishment. Plants still trump testacea, though, because they can generate. For Aristotle, a normative assessment of generative ability does not only include (iii) reproductive success but also how generation displays (ii) flourishing capacities. The ability to complete offspring is an achievement which indicates flourishing. Testacea cannot generate (iii) but also they have no share in proper flourishing (ii). Because they come about as a consequence of the action of the elemental powers, they are completely passive in the process that brings about their own existence.[72]

Every single instance of testacean generation is an accident. The regularity with which they come about makes them natural in the sense that certain material conditions tend to coincide.[73] This pulls against Aristotle's usual account of generation. The testacean is

proponent and the great systematizer of the ancient theory of spontaneous generation' (Dudley 2012, 331).

[71] Some varieties display the lowest sense, taste and smell, but it is unclear whether they can hear (*HA* 535a14–25). The ability to see, on the other hand, is definitely absent as none possess eyes (*HA* 491b27–8). Aristotle laments that plants cannot experience sensation due to the fact that they are too earthly and cold to possess sense organs (*de An.* 424b1–2, 435a20-b3). Testacea are more watery and plants more earthy. Aristotle's view that water is a better element than earth and better able to support life (*zôtikôteron*) might also help to explain why testacea are sometimes deemed better than plants (*GA* 761a26–33).

[72] Spontaneous generation is a 'case of necessary processes incidentally generating something that has an end' (Johnson 2005, 200–1).

[73] Lennox (1982) 227: 'Aristotle's view of spontaneous processes ought to be contrasted, not with (or merely with) *regular* and *orderly* processes, but with those that exemplify formal replication.' This could also be said of the regular production of incomplete offspring in some species.

such an unusual sort of living being that it leads the biologist to the uncomfortable question of whether form or soul properly exists in it at all.[74] The fact of enclosing something in a bubble hardly seems an adequate way to explain the beginning of a living being in Aristotelian science. Male semen itself is constituted of many bubbles containing *pneuma*, yet these do not become many small animals.[75] In fact, stirring creates bubbles in inanimate stuffs, such as oils. It is not clear what prevents frothy oil from being alive, if life occurs when bubbles do.[76] This account is unsatisfactory, on Aristotle's own terms, and so it would be odd to conclude that we can learn something about his attitude towards all animal generation from it. Although he says the soul principle emerges in the enclosed *pneuma*, this counts as a somewhat enfeebled attempt to describe what is taking place rather than any positive endorsement of the idea that all generation is down to a special material or a mechanical process.[77]

If, in attempting to find a theory of psychic *pneuma*, a reader dips into *GA* III on spontaneous generation, she may leave with the distinct impression that Aristotle was some kind of hylozoist. This reaction is tempered, however, when the context of the discussion is taken into account; testacean generation is not part of Aristotle's account of the role of *pneuma* in generation. It is an attempt to tackle something untoward and anomalous, which is left to the very end of the section. He is unable to resolve the various difficulties it presents. More generally, although *GA* III sees Aristotle attempting to fit his general theory of generation and his normative teleological expectations to the available empirical facts about lower animals, when the theory conflicts with these facts, he is not rigid. Indeed, he acknowledges psychic influences in the female wind-egg, and the femininity of the 'king' bee and

74 They lack proper form since form implies that an animal is complete and an animal cannot be complete until it is able to generate another like itself (*GA* 735a17–19). Testacea never reach that stage and so remain perpetually incomplete. See Connell (2001) 314.

75 *Pneuma* is enclosed in the male semen at *GA* 736b36.

76 Bubbles (αἱ πομφόλυγες) at *GA* 735b12 and 736a17.

77 It seems that Aristotle would have been much more comfortable with the modern realization that semen or seeds from parents, undetectable to the naked eye, are in fact responsible for generation in such instances, welcoming the rejection of living beings coming to be out of the non-living.

that spontaneous generation does not fit the norm. Thus, we can see that his theory of generation is flexible and complex: the female principle contains soul and some of the materials in the natural world have a strong potential to become ensouled under extraordinary conditions. This is not out of line with Aristotle's ideas of material substance, along with his hylomorphic account of generation. But these theories ought not to be simplified and generalized or else large portions of the *GA* would have to be dismissed.

We can now see how important it is when reading the more detailed portions of the *GA* to differentiate between instances when the intricacies of generation in lower animal allows Aristotle to refine and expand his theory and those which stand outside his general account. His discussion of spontaneous generation is the latter variety, so rather than help the reader to understand his theory, as the elaboration of the female contribution of wind-eggs does, it is misguided to generalise from it. Testacea are exceptional in a way that is beyond the usual exceptions, so that very few specific principles of explanation will help to elucidate their completely incongruous method of generation.

8

ARISTOTLE ON SEXUAL DIFFERENTIATION

8.1 One or two sexes for Aristotle?

According to the popular one-sex model, there can be, for Aristotle, only one true sex, the male.[1] On this understanding the female is an anomaly, which fails to be properly human (or any kind of animal).[2] She is a negative of form, the privative to his positive, the defect to his norm, fundamentally disabled. There are two difficulties with maintaining this account of his views. The first is that it clashes with a very general feature of Aristotle's work on male and female in nature. The second is that it does not do justice to his much more specific description of sex difference in the *GA*. The first problem is this: these interpretations have to posit that, according to Aristotle, female anatomy and physiology is useless, that is, non-functional. This is, however, a ridiculous and unsupportable position for any zoologist. Aristotle of course recognized the functionality of the female. In light of this, the charge may then shift to the thought that he is not entitled to do so. This consideration can be left aside as irrelevant to textual interpretation; instead the focus of this section will be on a bigger challenge for the one-sex interpretation: the fact that in Aristotle's writings there is much to suggest that a two-sex model is in place.

1 This view has been very influential. See Flemming (2000) 119: 'Aristotle constantly measures the female against the male, ranks them on a single scale of ability, heat and so forth ... ' Sissa (1992) 65: 'Female characteristics are described in two ways [only]: by analogy with the male body and by comparison with the male body'. Whitbeck (1973–4) 56: 'Aristotle repeatedly expresses the view that what is essential to being female is a lack of certain attributes of the male. Aristotle was one of the first to build a detailed biological theory upon this idea [of female lack]'. See especially Laqueur (1990) 28–35. King (2013) provides a compelling critique of the one-sex narrative in the history of medicine.

2 Irigaray (1985b) 165: 'Woman, for her part, remains an unrealized potentiality – unrealized, at least, for/by herself ... And in her share of substance, not only is she secondary to man but she may as well not be as be. Ontological status makes her incomplete and uncompletable. She can never achieve the wholeness of her form. Or perhaps her form has to be – paradoxically – as mere privation?'

The content of *GA* IV in particular displays a tendency to treat femaleness as something distinct from and opposed to maleness. Despite the female being the result of a failure in generation, it cannot be reduced to or defined solely in terms of maleness; for Aristotle there must be two sexes.

In many ways, it is understandable why the one-sex hypothesis emerged. First of all, *GA* I and II tend to describe the female as lacking a male ability. The female is weaker (726b31–3), colder (728a22) and more 'infertile' (728a18). Women are even likened to boys, that is, undeveloped men.

> [A] boy actually resembles a woman in physique, and a woman is as it were an infertile male; the female, in fact, is female on account of an inability (*adunamia*) of a sort, viz., it lacks the power to concoct semen out of the final state of nourishment (this is either blood, or the counterpart of blood in bloodless animals) because of the coldness of its nature. Thus, just as lack of concoction produces in the bowels diarrhoea, so in the blood-vessels it produces discharges of blood of various sorts, and especially the menstrual discharge (which has to be classed as a discharge of blood, though it is a natural discharge, and the rest are morbid ones) (*GA* 728a18–25).

The female system is unable to produce the 'hotter' male contribution – residue in its final form (*GA* 726b7–8). The female, being colder, discharges the blood-like menses (*GA* 726b31–727a2), an 'excessive' residue which is compared to morbid diarrhoea.[3] Aristotle also describes an instance of a female being generated instead of a male, as a sort of failure of generation (*GA* 767b6–8).[4] Second, Aristotle's ideas about the differences between the sexes are quite different from those found in many contemporary accounts. In the Hippocratic work *On Diseases in Women*, for example, women's diseases are said to be distinct from those of men because women are an entirely separate sort of animal with a completely different constitution.[5] The contrast between this robust 'two-sex' model and Aristotle's own account could lie behind the inclination to regard him as a one-sex theorist.

[3] This comparison cannot stand since menses are not unnatural like diarrhoea.
[4] This passage will be analysed in greater detail in Chapter 10.
[5] *Mul.* I.62, *Loc. Hom.* 47, Dean Jones (1994).

Despite its initial attractiveness, the passages on female inability noted above cannot be taken as conclusive evidence for a one-sex reading of Aristotle for two main reasons. First, they must be understood within the context of the *GA* as a whole, where many more particular characterizations of the female and the female role exist. In *GA* IV there is a positive propensity and drive towards the production of females and also a robust opposition of male and female principles in the process of generation. The text also shows that Aristotle, in objecting to rival views, commits himself to the functionality of the female. Second, the one-sex reading depends upon three other claims, each of which is undermined by close readings of Aristotle's biology. The first of these is the idea that the female for Aristotle is characterized solely as a privation of maleness and that there is no positive or separate account of the female condition. The second is that the birth of a female animal is non-telic; it is only an accident in development and cannot have any determinate or positive cause, which means that it cannot be linked to teleology. Tied to this is the conviction that Aristotle held to monotelism: the view that any natural process, such as the generation of a new animal, can only have one aim. And finally, the third claim is that the form of a kind is essentially masculine. What it is to be a dog, for example, is taken to be the same as what it is to be a male dog. I will come to challenge all of these assumptions.[6]

8.2 Male and female animals are functional wholes

GA IV.1–2 takes the reader back to the types of disagreement with opponents we met earlier on in the work. Rival theorists look to the material conditions at conception or the material parts which exist in semen to explain sexual differentiation rather than attending to unified form and function. For instance, Democritus thinks that offspring become male or female 'according to which of the two parents' semens prevails, the semen, that is to say, which has come

[6] I do not deny that Aristotle's ideas are deeply and problematically sexist – however, I think that when they have been simplified, this is detrimental both to scholarship on Aristotle and to those who wish to make certain points using his ideas. When critics are not entitled to such points they damage their own credibility.

from the parts wherein male and female differ from one another' (*GA* 764a10–12). Aristotle contends, contra this view, that the material parts cannot explain how animals come to be male and female; instead the end product and the 'substance' (*ousia*) are primarily explanatory (*GA* 764b36–765a3). Along the same lines, Aristotle also disagrees with Empedocles, who posits that hot conditions in the mother's uterus result in male embryos, cooler conditions in female ones (*GA* 764a2–4). For Aristotle, Empedocles is rather careless (*rathumoteron*) in his assumption that hotness and coldness alone is the difference between male and female animals, seeing the whole of the genital parts differ greatly (*GA* 764a13–16) and offers the following proof against his account.

> Supposing that once the animals have been fashioned, and one has got all the parts of the male and the other all the parts of the female, they were to be put into the uterus as though it were into an oven, the one which has a uterus into a hot oven, the one which has no uterus into a cold one, then it follows that the one that has no uterus will turn out a female and the one that has a uterus a male. This is impossible (*GA* 764a16–20).[7]

Temperature alone is not sufficient to explain the difference in the sexual parts.

Empedocles could perhaps defend himself by connecting temperature to the development of generative parts.[8] First, he could say that half of each of the parents' generative parts exists in the seminal mixture, having been halved before entering the semen (*GA* 722b11–13). If half the parts of the offspring come from the mother, half from the father, when they meet in the uterus heat can then determine which half gains mastery over the other. For instance, half of the female's head could gain mastery over half of the male's head. The half of the male's head, which is present in the uterus, changes into another half of the female's head and the offspring gets one head just like the female's. The hotter the uterus, the more likely the male halves are to gain mastery, especially when it comes to the

[7] By mentioning the female organ that the male lacks, this is one of the rare occasions when he reverses his usual male vantage point. That males have no uterus is also noted at 738b36–739a2.

[8] It has been suggested that Aristotle purposefully misrepresents Empedocles' theory. See Cherniss (1935) 278: 'This criticism simply neglects the use which the theory made of temperature as the determining agent in the struggle of the germs for dominance . . . '

generative parts.[9] Despite this refinement, Empedocles' theory is still deficient. This is because by concentrating on male and female generative parts, he fails to explain how these parts come together to form a unified whole.

> To state the matter in this way is attempting to lay hold of the cause from too great a distance, and we ought to come as closely to grips as we possibly can with primary causes (*GA* 765b4–7).

The principles or primary causes of sexual differentiation, which are not clear to us at first (*Ph.* I.1), are teleological – the ends served by these parts.

> It is impossible that a complete talė of parts is secreted off from each of the parents and that a male or female embryo is formed according as one part prevails over another part. Considering the matter generally: to hold that the superiority of one part prevails and that this is what makes the embryo female is certainly better than saying that heat alone is the cause . . . but the fact that at the same time the conformation of the pudendum as well is different requires an explanation to show why these parts are always of a piece with each other (*GA* 764b18–24).

Aristotle contends here, against theories such as that of Empedocles, that being male or being female does not simply amount to having either a uterus or a penis/testes. The whole body of each sex is 'of a piece' – the female also has breasts, softer flesh, and so on, and the male has its other bodily male characteristics. '[I]t is fantastic to imagine', writes Aristotle, 'that these parts alone can be formed, without the whole body also having undergone a change' (*GA* 764b28–9).

Aristotle asks why one part follows on another: for instance, if the offspring gets a uterus, how does it also manage to get the other requisite functional female parts? And why are these always (or almost always) connected together as a whole (*GA* 764b24–5)? If close proximity is the sole cause, then all parts that are near should attach together and a replica of the female will result (764b28). But this does not occur – for we often find females that resemble their fathers and males that resemble their mothers (768a30–1). Any explanation of sexual differentiation must be based on the

9 Michael of Ephesus (1903) 167, 81v.31–35 helps to make sense of Empedocles' 'talé' theory in this manner.

end product or 'substance' (*ousia*) that results. Because male and female generative parts are immediately obvious to us, this is what many theorists of sexual differentiation start with.[10] On closer inspection, these parts turn out not to be *principles* at all.[11]

> [T]his difference of the sexual parts as between male and female is a contingent phenomenon;[12] we must not look upon it as being a principle or a cause (*GA* 764b36–765a1).[13]

It is not the parts that dictate the sex of the body, but the principle of maleness or femaleness which determines how the whole body is structured.

Empedocles' mistake is not to say that hot and cold affect embryonic development, but to fail to explain how each animal comes to be a functioning whole with male or female parts. Aristotle proceeds the other way around, from function to parts to temperature. And, by doing so, he must concentrate on the functionality of male and female. The robust independent status of femaleness in this context emerges even more strongly from a closer study of the causes of sexual differentiation and Aristotle's explanation of how differently sexed bodies are formed *in utero*. The next two sections set out these views as presented in *GA* IV.1–2.

8.3 The chemistry of sex difference

Aristotle already mentions the most plausible cause of sexual differentiation in *GA* I when arguing that it is not due to the presence of generative parts in parental semen.

10 *GA* I.3–16 give a survey of the generative parts of animals. As I have said, Aristotle thinks it is an obvious fact about the world that males and females have different generative organs. See Preus (1975) 108.

11 Deslauriers (1998) 147: 'The differences in capacities – concocting capacities as it turns out – are less observable (less knowable to us) but more fundamental in the sense that these capacities explain the differences in organs (and are therefore more knowable in themselves)'.

12 Συμβαίνει. This term can also be translated as 'accidental' as opposed to 'essential'. See Peck in Aristotle (1942) 380–1, note *d*.

13 Another reason why a focus on the dominance of sexual parts in embryogenesis is a flawed approach is that generative parts are among the last to form. Thus, they cannot possibly be the primary principles and causes of sexual differentiation.

[B]oth women and men are observed to change: not only do the infertile become fertile, but also those who have borne females bear males, which suggests that the cause is not that the semen is or is not drawn from the whole of the parents, but depends upon whether or not that which is drawn from the woman and from the man stand in the right proportionate relation (*summetron*) to each other (*GA* 723a27–31).

In rejecting pangenesis, Aristotle accepts that sexual differentiation is due to the nature of the blend of generative residues. A relevant balance (*summetria*) between male and female contributions causes sexual differentiation. Before fully explaining how this happens in *GA* IV, Aristotle emphasizes the need for *summetria* in *any* instance of generation.

Male and female differ generally with regard to each other in respect of the generation of male and female offspring on account of the causes which have been stated. At the same time, they must stand in a right proportional relationship (*summetrias*) to one another, since everything that is formed either by art or by nature exists in virtue of some due proportion. Now if the hot is too powerful it dries up fluid things; if it is very deficient it fails to make them set; what it must have in relation to the object which is being fashioned, is the mean proportional, and unless it has that, the case will be the same as what happens when you are cooking: if there is too much fire it burns up your meat, if there is too little it will not cook it – either way what you are trying to produce fails to reach completion. The same applies to the mixture of the male and the female: they require the right proportional relationship (*summetrias*), and that is the reason why it happens that many couples fail to effect generation with one another, but if they change partners they succeed (*GA* 767a14–26).[14]

We already know that the female contribution, as a residue of the nutritive process, has to be of a very particular type in order to respond appropriately to the male action. Aristotle now elaborates on this: in order for generation to occur at all the female matter must accord with that of the male – a balance must be set up between the two. The craft analogies, which lent the idea that the efficient cause does most of the forming, are not prominent here. Instead the present analogy to cooking suggests that male and female are comparable. At the material level, in any instance of successful conception, the female contribution counterbalances the male contribution's heat with cold. The male contribution is

[14] Cf. *HA* 585b9–19.

only hot accidentally (*allotrian*), being hot in the same way as blood and heated by the same agent (i.e. the heart).[15] If it does heat something else up, it will become cooled down in turn. If the female contribution cools down too much and the male contribution does not heat enough, generation will not occur. To go to the other extreme, if the male contribution heats up too much and the female contribution does not cool down enough, generation will not occur either because the potential *conceptus* will be burned up and dried out.[16]

When the term *summetria* appears in Aristotle's works it tends to have one of two meanings. It either stands for mathematical proportion or a good or proper balance: a well-proportioned blend.[17] *Summetria* in this latter sense indicates the creation of a unity and is often used of bodily heath.[18] As it was for his predecessors, *summetria* for Aristotle is a state in which the elemental powers within an entity are properly balanced against each other and proportioned to external factors.[19] The use of *summetria* in this passage implies that the sexes exercise a dual influence at conception. Aristotle elsewhere explains that hot and cold are active – *poiêtika* – acting on moist and dry (*Mete.* 378b21–6).[20]

Elements must be mutually active and passive, for they mix and change into one another. But hot and cold, and dry and moist are terms of which the first pair get their natures because they are active, the second pair because they are passive (*GC* 329b23–7).

We have distinguished in the elements four causal factors whose combinations yield four elements: two of the factors are active, the hot and the cold, two are passive, the moist and the dry (*Mete.* 378b10–14).

15 747a19: heat is a mere supplementary acquisition (ἐπίκτητος) of semen. Hipp. *Aph.* 5.63, L IV 556–7, also holds the view that semen is not hot of its own nature.

16 *GA* 772a12.

17 For instances of the former use, see *A. Po.* 71b26, 89a30; *Metaph.* 1012a33. For the latter use see *Ph.* 246b3–10, *EN* 1104a10–26; *PA* 652b35; *Top.* 139b21; *A. Po.* 78b20.

18 It therefore has a lot in common with a more familiar ancient term for proper blend – *krasis*. *Krasis* is not just any mixture but a properly proportioned one. *Eukrasis* is the optimal blend – the good blend which produces the very best results. These terms and concepts are used extensively in Hippocratic humoural theory. See Lonie in Hippocrates (1981) 130 and Vlastos (1947).

19 See in particular the stance of Alcmaeon of Croton in Tracy (1969) 22.

20 At *GA* 743a37 heating and cooling are both called *dunamis* – an ability. Peck in Aristotle (1965b) lxix: 'The two passive substances thus tend to serve as "matter," and the two active ones as instruments of "form" and "movement"'.

Male and female contributions, then, are both active. Yet we know from *GA* 729b15 that the male principle is the *poiêtikon*, the active principle, and the female, the *pathêtikon*, the passive principle. Aristotle seemingly endorses contradictory propositions: that the female principle is passive (*pathêtikos*) and that it is also active (*poiêtikos*). However, when properly contextualized, the problem dissolves. When Aristotle says that the female is passive in *GA* I he is discussing generation in terms of substantial change. On this general level, the female principle has the passive potential and the male one the active potential to develop into a new substance. Here in *GA* IV the discussion is more detailed and involves the necessary qualitative changes occurring in each instance of generation. In this context, there is more than one active cause. Thus the female principle counterbalances the male one by the active power of cold, which opposes active heat, resulting in the correct balance.[21]

Although it makes sense to think of the coldness of the female fluid as actively counteracting the heat of the male semen, Aristotle equivocates on whether cold is really an active agent in generation. On the one hand, he indicates that it is when he writes that some things are set by cold in foetal development (*GA* 743a11, 734b31). On the other hand, he later reverts to talking of cold as the privation of hot (743a37).[22] Throughout Aristotle's works hot is more form-like, active and important than any other physical power. Indeed it is a sign of superior *ousia* – substance or being. Despite acknowledging that in a particular instance, when the mixture is too hot, generation cannot take place, Aristotle would never describe the male animal (even in such cases) as

[21] Cooper (1988) 24 admits that '[t]he heat of the [male] semen must be in the right proportion to the moistness (and cold) of the menstrual fluid'. Yet, directly afterwards he disregards the cold aspect of the female contribution and denies that it has any active role, stating 'wetness, unlike heat, certainly does not connote movement in Aristotle's physical theory'. But coldness can connote *kinêsis*, the Greek term here translated 'movement', which literally means change.

[22] Düring complains that '[i]t is not easy to see how Aristotle could acquiesce in this obviously inconsistent mode of reasoning: cold regarded as a passive quality of matter as well as active elementary force, by the operation of which matter is transformed. It is a source of perpetual confusion' (Aristotle 1944, 81). This confusion is not confined to the *GA*. Cold is the privation of hot at *Mete*. 384b26–28 and *GC* 318b15–19 and not a privation of hot at *PA* 649a19–20.

'too hot',[23] whereas he often states that the female is 'too cold', or deficient in heat (*GA* 726b33, 728a22, 766b17). He also emphasizes the close connection between coldness and matter.

It must be understood that matter is a sort of coldness. For as dry and moist are matter (being passive), and find their principle embodiments in earth and water which have cold as a defining characteristic, it is clear that all bodies that are made of either element alone tend to be cold (*Mete.* 389a29–389b2).[24]

Summetria in generation requires that the female contribution's coldness be active and act reciprocally to the male contribution's warmth. However, Aristotle's natural science dictates that the hot or a heating capacity is superior to coldness or a cooling capacity. Added to this, factors that combine to produce a symmetric unity need not be equal. This is clear, for instance, in the *Nicomachean Ethics* where symmetrical friendships include friendships between unequal parties (1158b20–8).[25] In generation, when male and female must combine to create a well-proportioned blend, the female remains the lesser party in this arrangement. Thus, the coldness of the female, despite its counteractive power, is still inferior.

So far we have learnt that generation can only occur when a *summetria* does. Aristotle employs this concept to further explain differences between generative outcomes, such as why one animal will become male and another female. The answer to this question

[23] Lucrezia Marinella, a Renaissance scholar, using her knowledge of Aristotle's biology, deemed men to be *too* hot, using Aristotle's own terms (Marinella 1999). MacLean (1980) 34: 'Nearly all other [than Hipp. *Mul.*] ancient texts can be interpreted to indicate that woman is colder and moister in dominant humours, but after 1580 this is no longer assumed to be a sign of imperfection. Woman's temperature is functional; her colder metabolism causes her to consume ("burn up") food less fast, thus leaving residues of fat and blood which are necessary for the nutriment of the foetus and for the eventual production of milk . . . Such functionalism is, of course, subjacent or even sometimes explicit in ancient texts; but it had become overlaid by the conclusions it had engendered.'

[24] δεῖ δὲ λαβεῖν τὴν ὕλην ψυχρότητά τινα εἶναι· ἐπεὶ γὰρ τὸ ξηρὸν καὶ τὸ ὑγρὸν ὕλη (ταῦτα γὰρ παθητικά), τούτων δὲ σώματα μάλιστα γῆ καὶ ὕδωρ ἐστίν (ταῦτα γὰρ ψυχρότητι ὥρισται), δῆλον ὅτι πάντα τὰ σώματα ὅσα ἑκατέρου ἁπλῶς τοῦ στοιχείου, ψυχρὰ μᾶλλόν ἐστιν.

[25] The lesser party must compensate with a greater proportion of love. In the *Magna Moralia* Aristotle explains that unequal friendships exist between opposites, who are like fire and water; friendship then provides the correct balance between the opposing powers (1210a11–22).

depends on the state of the blend of the parents' contributions and the blend of the surrounding environment.

[T]he mixture[26] of male and female . . . requires *summetria*, and that is the reason why it happens that many couples fail to effect generation with one another, but if they change partners they succeed; and also that these impediments occur sometimes in young people, sometimes among those who are older, both with regard to failure and success in generation and also with regard to the generation of male and female offspring. Also, one country differs from another in these respects, and one water from another, on account of the same causes, for the quality of the nourishment especially and of the bodily condition of a person depends upon the blend of the surrounding air and of the foods which the body takes up, and especially upon the liquid nourishment. Hence hard, cold water in some cases causes barrenness, in others the birth of females (*GA* 767a23–35).

For Aristotle, sexual differentiation is the result of a difference in the initial blend of generative residues. One *summetria* results in a male, another in a female. The way in which not all *summetrias* are the same is clearest when he compares pleasure and health in his ethical works.

Why should not pleasure be like health, which is definite although it admits of degrees? For health is not constituted by the same *summetria* of elements in all persons; nor a single *summetria* always in the same person, but it may be relaxed and yet persist up to a point, and it may differ in degree (*EN* 1173a24–9).

Two key points can be gleaned from this: a perfectly successful *summetria* need not be exactly the same in each instance. Because individuals have slightly different constitutions, health and pleasure can supervene on different *summetria* in different people. In the *GA* context, this idea helps Aristotle to explain why certain couples are better at generating than others. For example, the combination of a slightly hotter female and a slightly cooler male will result in generative success if they combine with each other, although when they come together with more run-of-the-mill partners they may fail. Another, separate, line of thought is

26 μίξει. Although this term can refer to sexual intercourse itself, thus the implication would be that intercourse between individuals must be, in some sense, proportional or balanced, this does not appear to be what Aristotle wants to say here. It seems that the term refers, rather, to the actual mixing of male and female fluids in the female's body. Aristotle's preferred terms for copulation are ὀχεύονται (*GA* 756b1) and συνδυασμὸν (755b35), which he uses exclusively not only in the *GA* III discussions of lower animals but also in the *HA* V.1–8.

also apparent here. *Summetria*, he thinks, allows for degrees of success since not all *summetrias* are equal. The reasons for degradation are the same as those for variation: the constitution of an animal's generative residue will differ at any given point (*GA* 725b26–30), which helps explain why a certain individual is more fertile at some moments in their life than at others.[27] Because some combinations of generative residues from parents are better than others, Aristotle can thereby account for various more negative results, for instance, deformity and the failure to generate. And the first step towards these failures is the production of a female (*GA* 767b8). Aristotle separates generative failures from the generation of females, indicating that the latter is not a complete failure (*GA* 767a27–8, 35). However, an animal which is female is an indication of non-ideal conditions during conception, either in the initial mixture or in the environment surrounding it.[28] Although there is a *summetria*, it is somewhat relaxed. A better blend results in a male; a slightly worse, or colder one, in a female.

8.4 The physiology of sex difference

From Aristotle's works it is possible to find an account of what happens from the point at which the *summetria* results in a male or female, to the development of their differing bodily condition and parts.[29] Before looking more closely at this, it will be helpful to review a few general points about the use of heat and the hot in Aristotelian biology. For Aristotle, something is essentially hot if it possesses its own (*oikeian*) heat, whereas it is accidentally hot if heat has been added to it by something else (*allotrian*) (*PA* 648b36–649a10). Because an animal possesses its own heat, it is essentially hot, does not lose its heat and is able to heat up other objects. And the hotter this animal is, the more able it is to

[27] This may be due to diet, aging, the state of health and also external weather conditions. This is why Aristotle is interested in what the winds are like at the point of conception. See *GA* 766b35–767a13, 767a29–35; *Pol.* 1335a36-b1.

[28] Parents who are too young or too old are particularly likely to have imperfect, small or female offspring. See also *GA* 766b29–32, *Pol.* 1335a12–15, 1335b29–32.

[29] For Aristotle, it is crucial that the body of any animal, whether male or female, be 'of a piece' (764b25). This idea must come first in any explanation of the development of its parts, in contrast to the account of someone like Empedocles. See Chapter 8.3.

bring about that heating (of other objects). An animal's heating capacity is contained in its heart, which concocts all the incoming nutriment and internal blood (*Juv.* 469b10–17). The hotter an animal is, the more capable its heart is of concocting. In contrast, the separable parts of animals, including blood and residues, are only hot accidentally and quickly lose that heat when they are no longer in contact with the heating agent (*PA* 649b23–8; *Mete.* 389b9–12). This is true of blood and the male and female generative residues.

Aristotle notes that the *logos* – definition or account – of male and female differ. The male concocts the ultimate nourishment, the female does not (*GA* 766a31–4). He then notes that in terms of the material mode of explanation, this difference lies in the heart.

The reason for this [sex difference] is in the principle and in the part [of the body] which has the principle of natural heat. Therefore, of necessity in blooded animals a heart is formed, and the animal will come to be either male or female (and in other animals where male and female exist, the analogue to the heart). So, this is the principle and cause of male and female and this is where it is located (*GA* 766a35–766b5).

In a hotter material condition at conception, a heart develops which is able to concoct the residue of blood to its ultimate stage (*GA* 766b14).[30] The presence of this male heart also explains, on a material level, how the embryo develops male characteristics and parts.[31] Once the heart is formed, the embryo begins to nourish itself, since the heart is 'the first principle of the animal or organism from the moment when it begins to need nourishment' (*GA* 740a19–21). The nutritive process begins with the production of blood (*Somn.* 456b3f.) and the blood produced by a male embryonic heart will be hotter than female blood (*PA* 648a28).

Early on in embryonic development blood vessels are composed out of blood coming forth from the heart.[32] In an adult, blood

[30] Reiche (1960) 65.

[31] We must keep in mind that the heat of the heart is not soul and that soul cannot be reduced to heat. Male animals are hot in order to serve their function (*GA* 765b9–19).

[32] Balme in Aristotle (1972) 156–7 suggests that the formation of blood-vessels was first gleaned from looking at bird eggs because early on they develop a 'net' of observable blood vessels. *HA* VI.3 details the development of a chick.

maintains the being of the parts, but in embryos, blood must also *form* these parts.

> Beginning at the heart, the blood-vessels extend all over the body. They may be compared to the skeletal models which are traced out on the walls of buildings, since the parts are situated around the blood-vessels, because they are formed out of them (*GA* 743a1–4).

The blood vessels that form in the male embryo will be of a character that corresponds to its hotter blood; they will be tougher and stronger than the female variety (*GA* 727a16–17).[33] Next, flesh is formed from the overflow of blood from the blood vessels.

> The formation of the uniform parts is effected by the agency of cooling and heat; some things are set and solidified by the cold and some by the hot ... As the nourishment oozes through the blood-vessels and the passages in the several parts (just as water does when it stands in unbaked earthenware), flesh, or its counterpart, is formed (*GA* 743a4–11).

The male blood solidifies into male flesh, which is more sinewy and articulated than the female variety.[34] Finally, the residue of the male's blood, because the animal is hotter, will be the ultimate residue, male semen. This residue requires a suitable receptacle, which is why the male genitals form.

> And nature assigns to each of the residues the part which is fitted to receive it. Semen is a residue, and in the hotter of the blooded animals, i.e., the males, this is manageable in size and quantity, and therefore in males the parts which receive this residual product are passages (*GA* 766b18–22).

Although the new animal will not be able to generate until puberty, it is potentially able to produce the generative residue from the very beginning – this is contained in the definition (*logos*) of what makes it male.

[33] Galen (*Sem.* II.5.13–14) criticizes Strato, the second leader of the Lyceum after Aristotle, for saying that female and male differ by veins and arteries (1992, 182–3). Strato was no doubt following Aristotle in this.

[34] These differences might be referred to now as 'secondary sexual characteristics'. *HA* 538a22–538b24: 'in all animals the upper and front parts of the males are better, stronger and more fully equipped than the female'. He goes on to say that the female is less sinewy, less articulated, has looser flesh, is more knock-kneed, has thinner calves and more delicate feet. She will have no defensive organs or the male's will be bigger and better. The material causes of these differences can be traced back to embryonic development.

In cooler material conditions at conception, a female heart develops. The heart begins to nourish and blood is produced which is less concocted or cooler than the male's – so that finer blood vessels (*GA* 727a16) and looser flesh result (*HA* 538b7). The female embryo will produce a less concocted generative residue requiring a receptacle different from that of the male.

> In females, due to the plentiful amount of unconcocted blood-like stuff (for it is less pure), some receptacle part must be present and this is unlike [the male receptacle] and has a considerable size. It is on account of this [the quantity of the fluid residue] that it has the nature of the uterus (*GA* 766b23–5).

On a material level, differences in heart, blood and blood vessels, brought about by the cooler conditions at conception, result in the formation of the uterus.

> It is reasonable to suppose not that the blood-vessels have been formed to be of a particular character on account of the uterus, but rather the uterus has been so formed on account of them, since although each is a receptacle of blood in some form, the blood-vessels are prior to the uterus (*GA* 764b32–5).

Aristotle thinks that the offspring is completed as male or female 'when it obtains the part by which female differs from male' (*GA* 766b5–6).

This explanation contains six developmental and causal stages. It begins from (1) the material conditions at conception (e.g. hard and cold water produces female; *GA* 767a34–5), which then determines (2) what kind of heart develops, which then explains (3) the nature of the blood, which explains (4) the type of blood vessels (and flesh) created out of that blood and finally (5) how much blood will be left over and how concocted it will be, which results in (6) fitting receptacles for the residue coming to be (i.e. the generative organs).[35]

So we have seen that, on a material level, the differences between male and female animals originate in the heating capacities of their variant foetal hearts (*GA* 764b32–765a1). But this is not the only explanation Aristotle provides of the differences between the two – the account is not complete even from a

[35] Note how Aristotle aims to refute the pangenetic thesis that the sexual parts are principles and causes of sex difference.

material perspective because it relies on other principles. Although the receptacles form in male and female animals because they are appropriate to receive the generative residues (uterus for female, vessels for male), these residues are not present in the animal when it is an embryo or even when it is born. They come about later on when the animal reaches sexual maturity. Thus, it is not the presence of the material (i.e. male and female generative residues) which directs the development of the sexual parts; the cause of (6) is not (5). (6) occurs because there is some principle in the process that includes the specifications that will be appropriate for fully formed and functional males and females. The principles of maleness and femaleness must be present.

8.5 Sex opposition: the female as simultaneously successful and failed

The female body serves a function in generation and the way in which it develops *in utero* ensures that it is 'of a piece' as a complete functional whole. However, for Aristotle, the development of a female body is also, on some level, a failure: the cooler situation at conception occurs because the male principle *fails* to fully master the matter (*GA* 766a19). How can it be that the female is both functional and a failure?[36] The inconsistency, or at least awkwardness, of holding both views simultaneously could lead one to advocate a one-sex reading of Aristotle, disregarding the places where he speaks as if femaleness were robust and functional. However, the one-sex model rests on many assumptions which do not accord with large stretches of the text of the *GA*.

The one-sex interpretation must try to maintain that Aristotle only views the sexes as positive and privative. The male animal is the norm or positive condition, the female animal the defect of this

[36] Some scholars worry that so frequent an occurrence as female birth ought not to be characterized as unnatural (since in *Ph.* II Aristotle says what is natural is what occurs always or for the most part). However, this is not as problematic as it might seem since for Aristotle frequency of occurrence is not a sufficient condition for naturalness; many frequent occurrences, such as spontaneous generation of testacea, are unnatural. Also, as Nielsen (2008) points out 'always or for the most part' does not have to indicate a numerical or statistical frequency. Aristotle also uses the phrase in a more normative sense: that is, what should ideally happen.

norm, or the privation of it. Even without considering *GA* IV.1–2 in detail, this view is initially troublesome as a reading of Aristotle's metaphysics. If a female animal is 'a mere privation', as Irigaray puts it, then surely it could not be alive, much less move around, sense, communicate, and so on. Aristotle defines privation as:

> that which has not a certain quality and that which might naturally have it but has not got it, either in general or when it might naturally have it, and either in some particular way, e.g. when it *completely* fails to have it, or when it in any degree fails to have it (*Metaph.* 1046a32–5).

The production of a female is an instance of privation of maleness because it fails to have the ability to concoct male semen. In another way, though, it does not make sense to view the female condition as a privation. Aristotle elsewhere explains that if something has a privation of some faculty, the faculty is one that it *should* have at that particular time (*Cat.* 12a28–31). It hardly seems correct to say that the female animal which produces the needed matter for the next generation *should* produce the male residue and *should* be male. Aristotle says that men without the ability to reproduce (i.e. eunuchs) become like women (e.g. *GA* 766a25–6, 784a11), which makes it seem that female is the privation of maleness. However, he must be speaking loosely in these instances, since eunuchs do not have *uteri* and cannot provide the reproductive function that women can.

If the one-sex interpretation insists upon the female condition being a privation of the male one, then it must also ignore the other ways in which Aristotle opposes the sexes. Recall that for Aristotle there are four different types of opposition:

O1. correlatives (*ta pros ti*).
O2. contraries (*ta enantia*).
O3. privative vs. positive (*sterêsis kai hexis*).
O4. affirmation and denial (*Cat.* 11b15–20).[37]

In his chemical account, as detailed earlier, male and female are opposed in terms of active hot and active cold. Also, it is

[37] From Chapter 1.2.

evident in *GA* IV.1–2 that the female has an 'ability' in opposition to the male one.

There must be an organ both for the male and for the female; hence the male has his genitals and the female has the uterus. Nature gives each one its organ simultaneously with its ability, since it is better done thus. Hence each of these regions of the body gets formed simultaneously with the corresponding secretions and ability, just as the ability to see does not get perfected without eyes, nor the eye without the ability to see (*GA* 766a3–9).

The female condition is characterized then as an ability in its own right, analogous to the ability to see: a function requiring an organ. In the case of sight, the organ required is the eye; in the case of the female function, it is the uterus.[38] For Aristotle, the uterus comes to be in the female because this animal needs it in order to perform the function of discharging menstrual fluid (and gestating). Thus are the sexes contrapoised as two abilities, like contraries (*Cat.* 12a1–25).[39] Aristotle does not see females as only failed males but also as functional generators. This helps explain why each time female infirmity comes up, he qualifies his pronouncement. The woman is 'as it were' (*pos* – sort of, but not really) an infertile man, and the female has 'an inability of a sort', that is, not total incapacity.

Key passages in *GA* IV.1–2 on the causes of sexual differentiation also provide support for this way of opposing male and female. As Aristotle explains:

[I]t must be laid down that, assuming the extinction of a thing means its passing into its opposite condition, then also that which does not get mastered[40] by the agent which is fashioning it must of necessity change over into its contrary

[38] One of Aristotle's favourite examples of the opposition of positive to privative is sight as opposed to blindness. See *Cat.* 12a26. Male and female do not fit with this type of example, as is clear from the above passage. However, it may be that the female, although it possesses an organ and has an ability, is less able than the male, just as blue eyes are less good at seeing than dark ones. This was suggested to me by an anonymous reader.

[39] This is also supported by Aristotle's discussion at *GC* 323b18–324a9 of the way in which contraries are involved in reciprocal action, as happens when male and female residues combine to create *summetria*: 'agent and patient are in some sense identical, but in another sense other than (i.e. unlike) one another . . . contraries exhibit this character' (*GC* 324a4–6).

[40] Mastery is a crucial concept which Aristotle uses throughout *GA* IV. He did not invent the idea, which is prominent in certain Hippocratic works, as well as in the work of Democritus and Empedocles, as Aristotle himself mentions (See *Genit.* IV, *Nat. Puer.*, *Morb.* IV, *GA* 764a10, 764b27). Van der Eijk (2007).

(*enantiou*). With these as our premises it will perhaps be clearer the reasons why one embryo becomes male and another female (*GA* 766a14–18).[41]

A female is produced, then, when the male fails to master and so must change over into its contrary. His explanation runs more fully as follows.

When the *archê* [male][42] is failing to gain the mastery and is unable to concoct owing to deficiency of heat and does not succeed in bringing it over to its own proper form,[43] but instead is defeated in this, then of necessity it changes into the contrary (*tounantion*). Now the contrary of the male is the female and it is opposite in respect of that whereby one is male and the other female. And since it differs in the ability it possesses, so also it differs in the organ it possesses. Hence this is the state into which it changes over (*GA* 766a18–24).[44]

The cause of the formation of a female, as these passages show, is the failure of the male contribution to gain mastery. The extinction of a thing means the passing of it into its contrary; thus, the extinction of male results in its contrary, female (*GA* 766a14–18). When the male factor has failed, the female factor has succeeded; if it does not get mastered by 'the agent', that is, the male, then it must be mastered by the other sex, that is, the female. The female masters and makes the offspring female. There is something positive pushing for femaleness – an opposite power, related to an organ and a function which 'masters' or 'defeats' the male principle.[45] Aristotle

41 δὲ πρὸς τούτοις ληπτέον ὅτι εἴπερ ἡ φθορὰ εἰς τοὐναντίον, καὶ τὸ μὴ κρατούμενον ὑπὸ τοῦ δημιουργοῦντος ἀνάγκη μεταβάλλειν εἰς τοὐναντίον. τούτων δ' ὑποκειμένων ἴσως ἂν ἤδη μᾶλλον εἴη φανερὸν δι' ἣν αἰτίαν γίνεται τὸ μὲν θῆλυ τὸ δ' ἄρρεν.

42 Coles (1995) 70: '[T]he principle (*archê*) referred to in the first line must be taken as the male genital'. He goes on to say that *to arren* and *to thêlu* refer 'to the male and female sex organs respectively' disregarding the fact that *to thêlu* means teat (71). I would argue that the penis is not essential to maleness. For one thing, the heat of maleness is contained in the heart. Semen is not produced in the old, the young or the sick not because there is anything wrong with their sexual organ, but because of their coldness (*GA* 766b27–31, 725b19–21). Also, certain animals which Aristotle considers to be male lack penises, such as insects.

43 Cooper (1988) 21 translates *idion eidos* 'own peculiar form'. Some think that this phrase indicates that Aristotle is dealing with individual forms (Nielsen 2008, 400). We must recognize, however, that the subject matter is sexual differentiation and not heredity, and so the focus is on sub-types and not individuals.

44 ὅταν γὰρ μὴ κρατῇ ἡ ἀρχὴ μηδὲ δύνηται πέψαι δι' ἔνδειαν θερμότητος μηδ' ἀγάγῃ εἰς τὸ ἴδιον εἶδος τὸ αὑτοῦ, ἀλλὰ ταύτῃ ἡττηθῇ, ἀνάγκη εἰς τοὐναντίον μεταβάλλειν. ἐναντίον δὲ τῷ ἄρρενι τὸ θῆλυ, καὶ ταύτῃ ᾗ τὸ μὲν ἄρρεν τὸ δὲ θῆλυ. ἐπεὶ δ' ἔχει διαφορὰν ἐν τῇ δυνάμει, ἔχει καὶ τὸ ὄργανον διαφέρον· ὥστ' εἰς τοιοῦτον μεταβάλλει.

45 I will argue for this idea more fully in the next section on hereditary resemblance. Because the individual male parent is said to contribute *dunamis* of maleness, this

thus assumes some positive cause for the development of a female animal, a cause which pulls against the drive to form a male animal.[46] What opposes the male in terms of function and ability has overpowered it in order to establish an ordered female anatomy. The idea of female as counterpoised and functional is not novel to *GA* IV. Even at the very beginning of the work, Aristotle includes in his definition of femaleness a sense in which it fulfils a function: the female is that which generates in itself (*GA* 716a15) – it is a functioning generator.[47] The simplicity of the one-sex interpretation obscures the insight that, for Aristotle, the sexes can, and ought to, be opposed as contraries also and not only as positive and privative.[48]

8.6 Females and teleological explanations

In addition to the idea of female as privation, the one-sex interpretation is sustained by another related idea: female birth as entirely non-telic. For the one-sex hypothesis, when a male comes to be, the process of generation has reached its proper end (*telos*). On this reading there is only one possible positive outcome of any generative process – the production of a male.[49] The first

implies that the individual female parent contributes a *dunamis* for femaleness. See also Henry (2006b) 435–6.

46 Sexual differentiation is accidental in the sense of not being part of the essence of the kind but it is not accidental in the sense of being random. Any offspring which is not male, will not become just any random thing, it will become a functional female. The female condition is no mistake and happens due to powers within semen.

47 The presence of organs indicates a function (*GA* 716a17–34).

48 Contraries and privation are not mutually exclusive (Ross 1924, vol. 2, 287; *Metaph.* 1055b14, 1004b27, 1011b18). At their extremes, contraries are privations of the opposing contrary. But male and female are not these extremes.

49 See Balme (1987b) 292, Furth (1988) 115–17, Dean-Jones (1994) 196–7, Cook (1996) 62 and Freudenthal (1995) 24. Van der Eijk (2007) 422: 'on peut regarder l'occurrence de la femelle comme le réultat d'un manque, d'une absence des propriétés mâles' ('we can look at the occurrence of the female as resulting from a lack, an absence of male properties'). The broader implication of this line of interpretation is that the ideal result of the reproductive process will be the continuation of an all-male race. Lange (1983) 11–12 holds that for Aristotle, the female 'is instrumental to species eternity, and potentially rather than actually human. The actuality of her human potential is the incubation of a male child.' Lange comes to the bizarre conclusion that 'the female [animal] is no more than a material cause of the animal. Reading "male" for "person", what else are women but "matter set in motion by and for the soul of the unified male, for the ends of the (male) species"?' Nielsen (2008) writes that, according to Aristotle, the

problem with this reading is that Aristotle himself characterizes the production of female as telic in one sense. As he explains, something can occur both due to necessity and also to serve a higher purpose and be 'for the better'.[50] Earlier on he explained the natural occurrence of menstrual blood in this way.

> [T]he production of [menstrual] residue by females is, on the one hand, the result of necessity, and the reasons have been given: the female system cannot effect concoction, and therefore of necessity residue must be formed not only from the useless nourishment, but also in the blood-vessels, and when there is a full complement of it in those very fine blood-vessels, it must overflow. On the other hand, in order to serve the better purpose, the end, nature directs it to this place and employs it there for the sake of generation, in order that it may become another creature of the same kind as it would have become, since even as it is, it is potentially the same in character as the body whose secretion it is (*GA* 738a34–738b4).

Here it seems that the chemical process (the materially necessary interactions of elemental powers in the body) occurs without being goal-directed and connects to a teleological framework because its results happen to be beneficial.[51] Due to the coldness of the female animal's heart, it does not produce highly concocted and scanty male residue, but of necessity produces a bulkier, bloodier counterpart. However, this eventuality is also 'for the better'. Similarly, due to a certain degree of coldness at conception, a female heart forms in the embryo rather than a male one. But again this is 'for the better'.

> The first beginning of this deviation is when a female is formed instead of a male, though this indeed is a necessity required by nature, since the race of creatures which are separated into female and male has got to be kept in being; and since it is possible for the male sometimes not to gain the mastery either on account of youth or age or some other such cause, female offspring must of necessity be produced by animals. As for monstrosities, they are not necessary so far as the for the sake of which and the final cause is concerned [i.e. in contrast to the female] (*GA* 767b8–14).

Females, unlike monstrosities or failures in generation, do not come about solely due to material necessity separate from natural

female 'is unable to participate in the eternal individually, and only stands as handmaiden to the process of male self-perpetuation' (403).

[50] These cases are discussed in greater detail in Chapter 10.1.

[51] However, this analysis may be too chance-like. Talk of nature 'using leftovers' seems to indicate a more goal-directed account; Leunissen (2010) on 'secondary teleology'.

goals. Aristotle hereby clearly acknowledges a place for the female in his teleological explanation of generation: a female is 'for the better' so that the kind may continue to exist.

The one-sex idea of female as failure in the sense of failing to reach an end (*telos*) depends on Aristotle's natural science being monotelic, there being only one end to which any natural process can aim. And if this end is not achieved, the inference is that the process has entirely failed. But the *GA* on the female indicates a different approach. Femaleness is a sort of second-ranking goal for an individual animal.[52] A metaphor may help us to understand how the female can be simultaneously a success and a failure. Imagine that you are making a cake and that you aim for it to be beautiful – instead it comes out a bit lopsided. It is not the case that you now have a privation of a cake – you have a cake which may well be delicious and pleasing, serving the purpose of appeasing a sweet tooth, even though it has not hit the aesthetic ideal you set for it. The failure to reach one goal would seem to have good side effects in terms of another goal. In the case of the cake, it fails to meet an aesthetic goal but instead reaches a culinary one. These could even be causally linked; perhaps whatever made the cake collapse (for instance, the extra butter) also makes it especially delicious. So also the female in failing to produce male semen, produces the highly specialized matter needed for generation.

It is also instructive at this point to consider Aristotle's application of teleological principles to females in light of his views on non-human animals. At the level of each kind, all animals that are fully grown count as *teleia* (complete/perfected).[53] But when ordering animal kinds in terms of relative success, he characterizes some kinds as incomplete compared with humans.[54] In the case of lower animals, their failure to be human is also, in another sense, a

52 One might be reminded of Aristotle's thought in his ethics that the political life is a secondary sort of happiness (*eudaimonia*) because the life of contemplation (*theoria*) is better (*EN* 1178a9–10). Femaleness is similar – a failure to live the best life possible, but a robust and respectable end nonetheless.

53 The foetation is *teleia* by the time it is either female or male (*GA* 737b11–12). See also *GA* 715a21–2: 'among blooded animals, with a few exceptions, the individual when completely formed, is either male or female'.

54 Humans are 'animals constituted most in accordance with nature' (*IA* 706a19, b10). See Chapter 7.1.

success in terms of being the type or kind they are. So the females' failure is also their success.[55]

Another reason to be cautious about ascribing to Aristotle the view that the female is simply a failure (monotelism) is that, as argued earlier, his general goals for animals are tripartite: (i) survival, (ii) flourishing or well-being and (iii) generation to type (see Table 7.1). On this understanding, more than one goal may be in play in any natural process, leaving open the possibility of conflicts between them. In *GA* II.1 Aristotle gives several of these goals as the reason for the development of differently sexed animals.

> It will surely be for the sake of generation that the male and the female are present in the individuals which are male and female. And as the proximate motive cause, to which belong the *logos* and the form, is better and more divine in its nature than the matter, it is better also that the superior one should be separate from the inferior one. That is why wherever possible and so far as possible the male is separate from the female, since it is something better and more divine in that it is the principle of *kinêsis* for generated things (*GA* 732a2–9).

The passage posits that it is best that male and female exist in separate animals for a set of purposes relating to the goal of flourishing or well-being (ii). Given that flourishing is better achieved when the sexes are separated in this way, it follows that the two different sexes will be required in order to achieve another living goal: that of generating (iii).[56] The production of female animals, half the time, will be crucial, and related directly to the goal of generation in any type where the sexes are separated into different token animals (*GA* 767b10).[57]

Aristotle's reasons for thinking that sex separation aids animal well-being belie his sexism, for the benefit only accrues to males.

55 See especially Freeland (1998) on this issue.

56 Aristotle assumes the existence of two sexes, which he sees as an unavoidable part of living nature. Male and female are the *archai* – the starting points – of the inquiry (*GA* 716a5).

57 Aristotle does not link sex separation to the goal of generating to type itself, since the male and female principles do not have to exist in separate animals in order for it to occur (as evidenced in various kinds of hermaphroditic animal as well as in almost all plants). In modern biology also sex separation confers no obvious reproductive advantage and is, in most cases, costly (see Margulis and Sagan 1986). Unlike modern biology, Aristotle does not come from an evolutionist perspective and so does not need to explain how sex separation came about in the first place in terms of reproductive advantage.

Earlier on he suggested sexed animals (of both sexes) live a better sort of life than hermaphrodites.

> In all its workmanship (*dêmiourgei*) nature acts as reason would expect. A plant, in its essence, has no function or activity to perform other than the production of its seed; and since this is produced as the result of the union of male with female, nature has mixed the two and placed them together, so that in plants male and female are not separate. Plants, however, have been dealt with in another treatise; here we are concerned with animals, and generation is not the only function which an animal has – that is a function common to all living things. All animals have, in addition, some measure of knowledge of a sort (some have more, some less, some very little indeed), because they have sense perception, and sense-perception is, of course, a sort of knowledge. The value we attach to this knowledge varies greatly according as we judge it by the standard of intelligence (*phronêsis*) or the class of lifeless objects. Compared with intelligence, it seems as nothing to have the two senses of touch and taste only; but compared with entire absence of sensibility it seems a very fine thing indeed. We should much prefer to have even this sort of knowledge to a state of death and non-existence (*GA* 731a24-b4).

The life of a mature sexed animal is of greater value, thinks Aristotle, than that of the hermaphroditic living being because the latter has no respite from reproduction, and merely exercises the lowest nutritive/generative portion of the psyche. This advantage, however, could apply to either sex. Furthermore, if the key benefit of sex difference is that it allows individuals to avoid the distraction of the drive to reproduce, the fact that there are two different sexes would be inconsequential. Nature could simply require that two or more animals were necessary for generation, rather than one. In the *GA* II.1 passage Aristotle also wants to say that differentiation into two sexes is a good thing because being *male* provides a further advantage in terms of flourishing. Perhaps the point is that flourishing is furthered by the avoidance of the physical burdens of pregnancy, leaving more time for perceiving or thinking – but surely this is not always the case. Certainly within some kinds, there will be nothing between male and female in terms of generative burdens, for instance, in those fish that do not guard their eggs but deposit them externally.[58] We might also note that although the male is freed of pregnancy and the physical

[58] See, for instance, *HA* 543a6, 567b3–5, VI.17.

labours of generation, he is not thereby free from a preoccupation with finding a mate or appeasing procreative desire. This seems to be why Aristotle introduces other ways in which these distractions might be curbed. Higher animals have testes, which make 'the movement of the seminal residue more steady' ensuring that 'their desire [for copulation] shall not be violent or speedy' (*GA* 717a28–32).[59] Overall, then, the passage from *GA* II.1 suggests that the female condition is only a failure in terms of the second sort of goal of sex separation (flourishing), and only in certain kinds of animal, in which males will have additional time for cognitive activities not expended on the tasks of physical generation.[60]

As in the case of menstrual blood, female births occur due to material necessity. These events are tied to teleology in this manner: they are used by nature for the advantage of individual males. In order for males to benefit, there will have to be females in their kind to do the procreative labour they are thereby freed from. This line of thought, apparent in the *GA* II passage, explains the attraction of the view that Aristotle regards females only in instrumental terms.[61] The text, however, does not allow the reader to conclude that this is the only, or even the chief way, in which Aristotle speaks of femaleness. It is a small part of the work as a whole. Furthermore, there is a tension in Aristotle's teleology between what is best for the individual and what is best for the kind: something that persists in modern zoology.[62] The female is caught in the crossfire. Since nature attempts to achieve multiple goals, it is no surprise that instead of a clear and simple picture of the place of the female in nature, Aristotle's female is better in one way and not in another. Even though it is best for the individual to end up male (with regards to [ii] flourishing), it is best for the kind that there be females half of the time (with regards to [iii] generation to type).

59 There are also possible reasons why sex separation is metaphysically superior (and not just beneficial to specific individuals). It is better, according to Aristotle, for the agent to be unaffected by change (*GC* I.7).

60 Even this understanding might be challenged by readings of certain passages in Aristotle's zoology, where he associates females' involvement in generative and parental activities with intelligence (*phronêsis*). See Chapter 6.2.

61 See Lange (1983), Irigaray (1985b) and Nielsen (2008) among others.

62 Consider, in particular, the way in which individuals often sacrifice themselves for the sake of the next generation.

I have now qualified the first two claims on which the one-sex representation of Aristotle rests – that the female always appears as a privation of maleness and that the female is in every sense non-telic. I now turn to the third claim, which is the idea that a type or kind of animal is defined solely in terms of the male body/form/function. A very general point tells against this idea. In both the *Metaphysics* and the *GA* Aristotle makes clear that males and females are of one kind (*eidos*).[63]

> In all animals which can move about, female and male are separate; one animal is male and another female, though they are identical in species (or form, *eidos*), just as men and women are both human beings, and stallion and mare are both horses (*GA* 730b33–731a1).

Those who assign the one-sex model to Aristotle appear to ignore the fact that maleness in his ontology is not part of the essence of each kind and so cannot be included in the form.[64] This is implied by their insistence that in each instance of generation the *telos* (the goal of the kind) is the production of a male, which requires Aristotle to think that the form of each animal kind is a male animal of that kind. Yet, this is contradictory from the start, since it implies that a non-essential feature is part of the essence of a kind.

As well as implying this unappealing contradiction, the one-sex model also finds little support from Aristotle's zoology, where evidence abounds of his interest in features of both males and females within a kind. Indeed, in many instances in the biological treatises what it is to be a certain type of animal is closely connected to the female. For example, the way a whole species of animal moves depends, in some cases, on where the legs are best placed to assist female reproductive functions, such as protecting young, nursing and sitting on eggs (e.g. *IA* 711b29–33). A similar principle applies to the classification of animal types at *GA* 732a26–732b14, where these are demarcated primarily on the basis of the female activities of gestation and parturition.

[63] The distinction between the sexes is material and bodily (*Metaph*. I.9). Plato also recognized sex difference as *per accidens*. *Rep*. 454c.

[64] Some philosophical commentators insist that maleness is part of the father's form. See Balme (1987b), Morsink (1982), Furth (1988) and Cooper (1988) and the discussion of the male *dunamis* in the next section on hereditary resemblance.

Meanwhile, when detailing the characteristics of types of animals in the *HA*, Aristotle often discusses their mothering roles.[65] One could argue that, since the survival of kinds depends on the continued existence of both females and males, the form of the animal is not proper only to one or other of the sexes.[66] Indeed it seems plausible to suppose that both descriptions would be required in order to fully understand the kind, even though in essence it is sexless.

There is no use denying that the one-sex model is neater than the idea that Aristotle thought of the female as simultaneously a success and a failure. However, it is an inaccurate characterization of his views in light of *GA* IV.1–2, where the female principle can be opposed to the male principle in a robust way – as a complementary function. For Aristotle, there is a *logos*, an account, of male and female and each has an *ergon*, function or manner of operating (*GA* 716a18–19, 729a25–6, 731b18–20). Normally, entities which share the same *logos* (*Top.* 101b37–102a5, *A. Po.* 73a34–9, *Metaph.* 1030a6) and *ergon* (*EN* 1097b24) will share the same essence; so differences in these point to distinct essences for the items in question. In the case of sex difference, however, males and females (within the same kind) cannot differ in essence (*Metaph.* I.9). It would be neater, therefore, if the female were like the slave (*Pol.* 1260a12–14)[67] or the bestial human (*EN* 1145a29–33), members of the kind which are marked out only by a failure to reach the relevant *ergon*. But Aristotle does not only characterize the female in that manner; instead being female is for him in this odd position: it is both disabled and differently abled.

65 'Among horses, when one mare has died the mares that graze together rear each others' foals. And in general the horse kind seems to have by nature a parental affection' (*HA* 611a11–14). See also *HA* 611a15–16, where deer are deemed intelligent (*phronimos*) because the female gives birth alongside the road.

66 See Tress (1997) 71.

67 For an alternative interpretation of Aristotle on slaves, see Schofield (1990), who thinks that they can be simultaneously human and living tools (15).

9

ARISTOTLE ON HEREDITY

9.1 The dilemma concerning Aristotle on heredity

After detailing sexual differentiation, Aristotle moves on to give an account of how offspring come to resemble parents and ancestors on both sides.[1] This section of the *GA* has attracted much interest in the past, and continues to be a popular focus of scholarly attention. The analysis often centres on the problem of how, if the male contributes form and the female matter, they can both share the capacity to generate offspring similar to themselves or their families. In this section I will detail how Aristotle thinks that hereditary resemblance comes about and will clarify some of the technical jargon he uses in this context. Next, I will set out the problems many have seen with reconciling this account with Aristotelian metaphysics on the one hand and the theory of generation found in *GA* I and II on the other. In the past, responses to these perceived difficulties tended to undermine or ignore the more positive characterizations of the female role throughout the *GA* as a whole.[2] More recent work, however, remedies these oversights and attempts to find other ways to bypass what is seen by some as the hylomorphic problem.[3] Rather than finding new ways to reconcile seeming difficulties, I will argue that careful analysis reveals there to be no real problems to resolve. The theory of inheritance which Aristotle sets out in *GA* IV.3 is effective and fits with what has gone before. Indeed, rather than contradicting his earlier stance on the contributions of male and female, this section elaborates and expands on it. The presence of shaping capacities within the female contribution, which is part of the

[1] Good birth may come from either the male or female side (*Rh.* 1361a8–9).

[2] Here I focus on the views of Balme (1987b) and Cooper (1988).

[3] In the main, I will discuss Henry (2006a, 2006b) and Gelber (2010). Gelber argues that there is no contradiction or tension; I am in agreement with her on this score.

account here given of inherited resemblances, is not anomalous to Aristotle's theory of generation but is one of its fundamental tenets.

9.2 Aristotle's account of hereditary resemblance

The section on heredity begins with the thought that it occurs due to the state of the mixture (*summetria*) of the two generative residues, as with sexual differentiation (*GA* 767a30).[4] The state of generative residues always varies (*GA* 725b26–30) and so no two offspring are produced exactly alike in every way, even when they have the same parents.[5] Each will take after their parents and ancestors in some features and so will differ from other individuals within their kind. Although Aristotle's detailed account of hereditary resemblance uses new terminology, the account is clearly continuous with the theory of generative residues put forward in contradistinction to the pangenetic hypothesis in *GA* I and II.

Aristotle says in *GA* I that his theory that semens from both parents are the residues of nutrition will explain hereditary resemblance better than rival theories (722a2–11). The advocates of pangenesis claim that their theory is best able to explain resemblance (*GA* 721b21–2); after all, if semen comes directly from the parents' parts, then the offspring will resemble them in those parts. However, the theory fails to help when addressing resemblance to grandparents, whose parts no longer exist (*GA* 722a8–9, 769a23–5). Take, for example, the light-skinned woman from Elea who had a dark skinned partner, with whom she had a daughter. The daughter was pale, but her child (a boy) was dark-skinned (*GA* 722a10–11). Nothing came directly from the bodily

4 Although hereditary resemblance happens 'due to the same causes' as sexual differentiation, Aristotle's analysis here is much more intricate. *GA* IV.1–2 made use of 'mastery' (*kratein*) (767b22) and 'opposition' (*antikeimenos*; 768a5). *GA* IV.3 also employs *dunamis* in a new way (767b23) and adds the concepts of 'departure from type' (*existatai*, 768a2), *kinêseis*, relapse (*luontai*, 768a16), the *kinêseis* being present in potentiality and in actuality (768b6) and being 'nearest', 'nearer', 'further away' or 'furthest away' (*eggus, egguteron, progonôn, porrôteron*, 767b27) and something being 'next in order' (*echomenon*, 768a10). Many of these terms will be further clarified in what follows.

5 Phenotypic differences exist between genetically identical twins. Given this, coupled with the fact that fraternal twins can look very similar to one another, it is understandable that Aristotle and his contemporaries did not distinguish between the two types of twin.

parts of the daughter which made her son dark-skinned because her skin (a part) was not dark. The dark skin of the grandfather, meanwhile, cannot exist in his daughter's seed and so could not have been contributed on this account. Pangenesis will also struggle to explain resemblance to parental features not manifest in their bodies at the point of conception (e.g. greying hair or beard, *GA* 722a7–8), offspring of one sex resembling the parent of the other sex (*GA* 764b25–8)[6] and offspring with a combination of different parts resembling those of both parents' families (*GA* 768b2–5).

Aristotle claims a heuristic advantage over pangenesis through the idea that influences resulting in resemblance may exist in an unactualized state as, what he terms, *dunameis*.[7] *Metaph*. Delta (V) offers a definition of the term *dunamis* which seems apt in this context.

> *Dunamis* is the source, in general, of *kinêsis* in another thing or in the same thing *qua* other, and also the source of a thing's being moved or changed by another thing or by itself *qua* other (1019a18–20).

Dunameis ought to be taken generally to indicate sources of change, the sorts of transformations that can shape the offspring in its parts and as a whole.[8] These *dunameis* fall into three types: those relating to human (or generic kind of animal), those relating to sex difference and those relating to the individual (*GA* 767b24–6, 767a2–3). The latter sort include *dunameis* for parts of the body (*GA* 768b2–4) and potential *dunameis* of the grandparents and more distant relatives (767b37). A change must occur in order for the offspring to become like Coriscus or a male, and the *dunameis* in the male parent's contribution are the sources of those changes.[9]

6 This is particularly difficult to account for on the pangenetic thesis, says Aristotle, since there is no way to separate the sexual parts from the rest of the parts of the body – parts from the female parent will be female parts and parts from the male, male parts (*GA* 764b25–8, 769a16–23).

7 This sketch of Aristotle's account is similar to that of Henry (2006b). As he does, I will leave this technical term untranslated.

8 One might choose to view them as sets of information, so to speak – somewhat like our concept of genes. And they would appear to exist somehow in the parents themselves, rather than in the generative residues; however, it is difficult to see precisely how this works. *Ibid*. (2006b).

9 See Preus's discussion of *dunamis* (1975) 18–20 and Morsink (1982) 132–40, *idem* (1979) 102. Although Henry seeks to show that *dunameis* comprise an organism's genetic nature, he also admits that 'essentially what an Aristotelian dunamis is, [is] a source of change' (2006b) 436, n.28.

Along with the *dunameis* Aristotle uses the mechanism of mastery (*kratein* or *epikratein*) to explain sexual differentiation and resemblance to parents and parts of parents.

If it [the male *kinêsis*] masters, it will make a male and not a female, and like the generator but not like the mother; but if it should fail to gain mastery, according to whichever power (*dunamin*) it does not master, it makes it deficient in that area (*GA* 767b21–4).[10]

If the *dunameis* 'human' gets mastered then the offspring will not end up as a human (or a particular kind) but will be merely an animal (if it lives). Aristotle will discuss this phenomenon in the next section on deformity. He has already sketched out the causes of sexual differentiation and here offers an even more detailed analysis of what occurs. The male can fail to gain mastery with respect to the sex of the offspring (*GA* 766a19–22), which is a failure of the *dunamis* relating to maleness in general.[11] Heredity involves the third type of *dunameis*, that relating to the individual parent. If this final *dunameis* gets mastered, then the offspring ends up resembling the other parent.

In generation there is a particular and a general, but the particular is more important. For it is a substance. And the offspring comes to be some sort of animal, but at the same time it comes to be a particular thing and a substance. Because of this, the *kineseis* from the *dunameis* of all such characteristics are present in the semens (*en tois spermasi*), and potentially those of ancestors, the most influential ones are always the nearer and the particular. By 'particular' I mean Coriscus and Socrates. Since everything deviates (*existatai*) not into any chance thing (*to tuchon*), but into its opposite (*ton antikeimenon*), and that which does not master in generation must deviate and become the opposite according to whichever *dunamis* the parent and changer did not master. So, if it was according to male, a female is generated and if according to Coriscus or Socrates, then it comes to be not like the father but like the mother, since, just as 'mother' is the opposite of 'father' as a general term, so also the individual mother is the opposite of the individual father (*GA* 767b33–768a9).[12]

10 ὥστε κρατοῦσα [Peck: κρατοῦσα vulg.: κρατούσης] μὲν ἄρρεν τε ποιήσει καὶ οὐ θῆλυ, καὶ ἐοικὸς τῷ γεννῶντι ἀλλ' οὐ τῇ μητρί· μὴ κρατήσασα [Peck: κρατήσασα vulg.: κρατῆσαν] δέ, καθ' ὁποίαν ἂν μὴ κρατήσῃ δύναμιν, τὴν ἔλλειψιν ποιεῖ κατ' αὐτήν.

11 This turn of phrase is possibly prefigured at *GA* 739a17: 'the *dunamis* of the male' (ἡ τοῦ ἄρρενος δύναμις).

12 My translation. *γεννᾷ δὲ καὶ τὸ καθ' ἕκαστον καὶ τὸ γένος, ἀλλὰ μᾶλλον τὸ καθ' ἕκαστον·* τοῦτο γὰρ ἡ οὐσία· καὶ τὸ [Peck: καὶ τὸ vulg.: καὶ γὰρ τὸ] *γιγνόμενον γίνεται μὲν καὶ*

This account is a significant adjunct to Aristotle's theory of generation, providing a more detailed picture of the content of seminal residues: all characteristics of the future offspring are derived from *dunameis* according to which parent gains mastery. When the male does not gain mastery in some *dunamis* it is mastered – *krateisthai* (768a23), *kratêthentos* (*GA* 768a34).[13] Although the text does not explicitly state that the female has to have its own *dunameis*, commentators have long understood this to be an obvious implication. The medieval philosopher Michael of Ephesus, for example, in his commentary on the *GA*, emphasizes the necessity of counterpoised female *dunameis*.

> According to whichever *dunamis*[14] does not gain mastery, it necessarily changes over and becomes the opposite. For Socrates can master according to male and he can be mastered according to 'Socrates', clearly [the *dunamis*] of 'Xanthippe' according to female has been mastered, and [the *dunamis* of her] according to 'Xanthippe' masters, necessarily a male will come to be like its mother . . . and if 'Socrates' is mastered, the offspring would come to be like Xanthippe (Hayduck 1903, 181: 8–15).[15]

As Michael understands it, if the male principle is mastered, then it gets mastered *by the female principle* with respect to the *dunameis* of the individual mother. If the male contribution fails to gain mastery, the situation switches from the *dunamis* 'Socrates' to the *dunamis* 'Xanthippe'.[16] When Aristotle notes that *dunameis* are

ποιόν τι, ἀμα δὲ [Rackman: ἀμα δὲ vulg.: ἀλλὰ] τόδε τι, καὶ τοῦθ' ἡ οὐσία. διόπερ ἀπὸ τῶν δυνάμεων ὑπάρχουσιν αἱ κινήσεις ἐν τοῖς σπέρμασι πάντων τῶν τοιούτων, δυνάμει δὲ καὶ τῶν προγόνων, μᾶλλον δὲ τοῦ ἐγγύτερον ἀεὶ τῶν καθ' ἕκαστόν τινος· λέγω δὲ καθ' ἕκαστον τὸν Κορίσκον καὶ τὸν Σωκράτην. ἐπεὶ δ' ἐξίσταται πᾶν οὐκ εἰς τὸ τυχὸν ἀλλ' εἰς τὸ ἀντικείμενον, καὶ τὸ ἐν τῇ γενέσει μὴ κρατούμενον ἀναγκαῖον ἐξίστασθαι καὶ γίνεσθαι τὸ ἀντικείμενον καθ' ἣν δύναμιν οὐκ ἐκράτησε τὸ γεννῶν καὶ κινοῦν. ἐὰν μὲν οὖν ᾗ ἄρρεν θῆλυ γίνεται, ἐὰν δὲ ᾗ Κορίσκος ἢ Σωκράτης, οὐ τῷ πατρὶ ἐοικὸς ἀλλὰ τῇ μητρὶ γίνεται· ἀντίκειται γὰρ ὥσπερ τῷ ὅλως [ὅλως: PZ_2, totaliter Σ vulg.: ὅλω] πατρὶ μήτηρ, καὶ τῷ καθ' ἕκαστον γεννῶντι ἡ καθ' ἕκαστον γεννῶσα.

13 See also the male being completely mastered (*kratêthentos holôs*) at *GA* 772b32–4.

14 Michael writes that the male has four *dunameis*: 'male', 'Socrates', 'human' and 'animal'. The most particular of these (*idion*) is the male. He does not think that *idion* in the passage refers to individual form (1903, 181).

15 καθ'ὁποίαν ἂν μὴ κρατήσῃ δύναμιν, ἀναγκαῖον ἐξίστασθαι καὶ γίνεσθαι τὸ ἀντικείμενον. ἂν μὲν γὰρ ὁ Σωκράτης κρατήσῃ καθὸ ἄρρεν, καθὸ δὲ Σωκράτης κρατηθῇ, τῆς Ξανθίππης δηλονότι καθὸ θῆλυ κρατηθείσης, καθὸ δὲ Ξανθίππη κρατησάσης, ἀνάγκη ἄρρεν γίνεσθαι ἐοικὸς τῇ μητρί . . . ᾗ δὲ Σωκράτης ἐκρατήθη, γεγονέναι ὅμοιον τῇ Ξανθίππῃ.

16 Xanthippe was Socrates' wife and the mother of his children. The female must also have *dunameis* for parts of the body and for producing offspring like her ancestors.

present in the semens (plural) (*GA* 767b37) it makes clear sense to understand this to mean both male and female seminal contributions.[17] The female, then, has *dunameis* and the female can master with respect to these when the male principle fails.[18]

The next part of Aristotle's theory concerns resemblance to ancestors. Here he focuses on the *kinêseis*, which are derived from the *dunameis* (*GA* 768b12–36). *Kinêsis* means, for Aristotle, change or transformation from one state to another – this can be in quality, quantity, location or substance. In this context *kinêsis* does not mean a change in substance, unlike the male role as *hê archê tês kinêseôs* – (*GA* 729b14), the initiator of substantial change.[19] Translating *kinêseis* in the plural as 'movements' suggests that there are local changes, motions, in the generative residues. It seems very unlikely, though, that Aristotle thought these could contain *kinêseis* in location. That he means simple qualitative change also seems implausible. Although blood (and the residue derived from it) is in a dynamic equilibrium of the elemental powers within it and so contains *kinêseis* of hot to cold, cold to hot, dry to wet and wet to dry, it would be too reductive to say that these informing vehicles are merely heatings and coolings. The term is being used in quite a specific way in the context and so it seems best to leave *kinêseis* untranslated before we come to understand more about the theory.[20]

[17] This is yet another instance where Aristotle refers to the female menses under the umbrella term *sperma* – it is a sort of semen. See Chapter 3.3.

[18] Preus (1975) writes that the male semen 'always "tries" to make it into a male; sometimes however, it fails, due to the recalcitrance of the material ... The same general principles are applied to the degrees of lack of resemblance ... ' (102). However, it is not due to 'recalcitrance' of matter that offspring resemble mothers but due to positive powers therein.

[19] Many commentators take the male action in generation, generally speaking, the way in which he brings form into matter to make the offspring viable as an animal, to happen through his *kinêseis* in the plural – but this analysis rests on weak textual support and is interpretatively problematic. See Chapter 5.1.

[20] See also Henry (2006b) 440, who urges that we ought not to think of *kinêseis* merely in terms of local motions, vibrations or waves since 'Aristotle's theory of inheritance is supposed to be more abstract than this.' In contrast, Coles (1995) 74 offers a mechanistic explanation and analogy in attempting to elucidate *kinêseis*. The meaning and role of *kinêseis* in Aristotle's account will be explored in more detail in what follows, where comparisons to the *kinêseis* involved in conveying sensory information from the sense organs to the heart are also helpful. The way in which *kinêseis* are understood as 'potential' and 'actual' will be discussed later on in this section.

According to Aristotle, in generation *dunameis* are not themselves active but work through *kinêseis*. Resemblance to ancestors occurs when these *kinêseis* undergo *luontai* (loosening or relapse).

The changing is always more to the next (nearest) of the ancestors, both the father's and the mother's. Some of the *kinêseis* are in them 'in actuality', some 'potentially'; 'in actuality', those of the male parent and of general kinds, such as human and animal, and 'potentially', those of female and the ancestors. So, when 'deviating' (*existamenon*), it changes to the opposite, but the *kinêseis* that do the fashioning relapse into the nearer things, e.g. if the *kinêsis* of the male generator relapses, it changes by a very small difference into the *kinêsis* of his father, and secondly, into that of his grandfather. And in the same way also for the female,[21] if the *kinêsis* of the female generator (*hê tês gennôsês*) [relapses, it changes] into the *kinêsis* of her mother, and if not into that, then into the *kinêsis* of her grandmother and likewise for the more remote ancestors (*GA* 768a10–21).[22]

Aristotle thinks that there is a certain order to the way in which animals come to resemble family members in their parts. When a specific part is coming into being (explained after the 'e.g.' in the passage), if the *kinêsis* of the parent relapses, then the next in the bloodline becomes actualized – a father's *kinêsis* will relapse into the paternal grandfather's *dunamis* and the *kinêseis* from it will take over; if these relapse, then the paternal great-grandfather's will be the next in line (*echomenên*) (*GA* 768b9; Cf. 768a17–21). Thus is Aristotle able to explain resemblance that skips a generation, through the elaboration of a point already made earlier on in the work. *GA* I and II argued that the generative residues contain a 'potential' (*dunamis*, 726b10–19) to develop the functional parts of an animal or are potentially able (*dunamei*) to become these (*GA* 737a23–4, 741b8–9). Now, in his more specific study of instances of generation, he clarifies this further. Certain *kinêseis* derived from the *dunameis* shape the parts of offspring and when

21 I follow Lulofs here (Aristotle 1965) in bracketing only the first part of this phrase. Peck brackets it all (Aristotle 1942).

22 My translation. ἀεὶ γὰρ εἰς τὸν ἐχόμενον μεταβαίνει μᾶλλον τῶν προγόνων, καὶ ἐπὶ πατέρων καὶ ἐπὶ [ἐπὶ Peck om.: vulg.] μητέρων. ἔνεισι δ'αἱ μὲν ἐνεργείᾳ τῶν κινήσεων αἱ δὲ δυνάμει, ἐνεργείᾳ μὲν αἱ τοῦ γεννῶντος καὶ τοῦ καθόλω, οἷον ἀνθρώπου καὶ ζῴου, δυνάμει δὲ αἱ τοῦ θήλεος καὶ τῶν προγόνων. μεταβάλλει μὲν οὖν ἐξιστάμενον πρὸς τὰ ἀντικείμενα, λύονται δὲ αἱ κινήσεις αἱ δημιουργοῦσαι εἰς τὰς ἐγγύς, οἷον ἡ τοῦ γεννῶντος ἂν λυθῇ κίνησις, ἐλαχίστῃ διαφορᾷ μεταβαίνει εἰς τὴν τοῦ πατρός, δεύτερον δ' εἰς τὴν τοῦ πάππου· καὶ τοῦτον δὴ τὸν τρόπον, καὶ ἐπὶ τῶν ἀρρένων καὶ ἐπὶ τῶν θηλειῶν, ἡ τῆς γεννώσης εἰς τὴν τῆς μητρός, ἐὰν δὲ μὴ εἰς ταύτην, εἰς τὴν τῆς τήθης· ὁμοίως δὲ καὶ ἐπὶ τῶν ἄνωθεν.

these fail they allow for a falling back onto ancestral *dunameis* and the *kinêseis* derived from these.

Aristotle's theory has one major weakness: it fails to account for resemblance to differently sexed ancestors to the parent. For instance, it seems that he can only explain how a specific maternal *kinêsis* can relapse into that of her mother, the maternal grandmother, even though it is clearly possible for offspring to resemble maternal grandfathers. Indeed this is exactly what occurs in the case he cites as helpful to his analysis: the woman from Elea whose daughter was light-skinned but whose grandson was dark-skinned. The boy's dark skin comes from his mother's side (she is the daughter of the woman who had a dark-skinned lover). But the relapse of this daughter's *kinêseis* should have resulted in her mother's *dunameis* and *kinêseis* taking over development of the skin, the woman of Elea herself, who had light skin. Instead, the *dunameis* and *kinêseis* that take over are from the boy's maternal grandfather; the boy inherits his dark skin. Even given Aristotle's belief that most female offspring take after their female parent and most male ones their male parent (*GA* 767b4), he should have been able to posit that each parent has potential *dunameis* of both their own parents, male and female.[23] The problem seems to be that he has only one mechanism, relapse (*luontai*), to explain how the ancestor's *dunameis*, and the *kinêseis* derived from these, take over to create the parts that resemble them. Say if the father's *kinêsis*, which is endeavouring to make a nose, relapses; Aristotle does not have the means to explain how a nose which fails to end up looking like the paternal grandfather's ends up looking like the paternal grandmother's instead – there is only one relapse. He perhaps needed a relapse and an anti-relapse or a primary and secondary relapse. Since this would have been simple to posit, it seems odd that he did not do so in so far as his own woman-from-Elea example requires it.

[23] 'Despite its brilliance, one of the shortcomings of Aristotle's theory is that it appears to lack any means for explaining how we get resemblance to the father's female ancestors or the mother's male ancestors' Henry (2006b) 447. This difficulty is also discussed by Morsink (1982) 140.

9.3 *GA* IV.3: interpretative issues

In *GA* IV.3 Aristotle does not speak of offspring being deficient (*elleipsin*) in the female's characteristics,[24] nor does he talk of an individual female, such as 'Xanthippe', as a *dunamis*. Instead, he refers to the female parent only by the dismissive 'whoever she may be' (*GA* 768b15). By failing to mention Socrates' wife, Aristotle was not necessarily meaning to dismiss the role of individual women in heredity. It perhaps belies a need for economy.[25] Hereditary resemblance is a minor issue within this work, a much more general statement of the metaphysical requirements for animal generation. If Aristotle had spelled things out more thoroughly for the female side then there would not have been anything anomalous and contradictory in it. He did not need to spend time doing so, noting instead that the processes that result in resemblance are the same on the male and on the female side (*GA* 768a10, 19). However, many would disagree, believing that Aristotle did not say that the female contributes *dunameis* and can master with respect to resemblance because he *could not* make this accord with the theory he espouses earlier on in the text.[26]

The main interpretative dilemma is that, on the surface, this account appears to conflict with his thesis that the male contributes form and the female matter to generation (*GA* 729a9–12, 729b18–9, 732a4–5). Some take it that the only place for the characteristics that are inherited, and then manifest in offspring, in Aristotle's ontology is as material accidents, incidental to essence.[27] For instance, as Aristotle says later on in the *GA*, although it is essential

[24] In contrast, if the male parent fails to gain mastery 'according to whichever *dunameis* it does not master, it makes it deficient in that area' (*GA* 767b23–4).

[25] Etiquette could also have played a part. On the impropriety of naming respectable women in prose at the time, see Schaps (1977) and Pericles' famous exhortation (Thucydides, *History* II.45, 2): 'greatest is the reputation of that woman about whom there is least talk among men, whether in praise or in censure' (1919, 341). Aristotle could have employed a generic nomenclature for the female parent, as he does whenever he uses 'Coriscus' to indicate any male person. For example, *A. Po.* 85a25; *Ph.* 219b21, 227b32; *PA* 644a25; *Mem.* 450b31; *Insomn.* 461b23–5; *Metaph.* 1015b17, 1026b18, 1037a7; *EE* 1240b25. Coriscus was not the name of anyone overtly famous in ancient times (Fraser and Matthews 1987, 1994). However, the name of the spouse of Socrates was so well known that such a strategy would have seemed contrived to his audience.

[26] Lesky (1950), Witt (1985), Föllinger (1996), Bien (1998), Henry (2006b), Boylan (1984) and van der Eijk (2007).

[27] Witt (1985), Sharples (2005) 105 and discussion in Henry (2006a) 275–7.

that certain animals have eyes, and that having an eye with a particular ability to see will be part of their essence, the colour of this eye will be an accident (*GA* 778b13–19). The idea that the male contributes to matter is firmly rejected by Aristotle (*GA* 729b19, 730b11, 737a8–16) and so if the features are entirely accidental they cannot be caused by the male. Many conclude from this that sub-specific inherited features cannot be accidental and must be somehow formal, that is, part of the form that the father contributes. This would make his form an 'individual form', containing everything that makes him look as he does.[28]

The issue of a clash between *GA* IV.3 and reproductive hylomorphism is in fact heightened by this supposed solution, for the female also contributes inherited features, which would seem to imply that her own individual form plays a part.[29] Galen was one of the first to point towards this implication. He complains that in order to account for heredity, such Aristotelians should admit that the female contributes to the form.

> Is it the nature of the power (*dunamis*) [in male semen according to *GA* I] that moulds the foetus to make the eye and nose and eyebrow and each of the other parts, but the nature of the matter to make the nose aquiline or snubbed, the eye blue or black, or does it belong to the same power, would you say, to fashion the nose in the first place and to make it aquiline or snubbed or straight, just as it is not the case, I suppose, that Polyclitus when he moulded the spear-carrier, fashioned the nose and eye himself but left it to the clay to make the nose straight? (*Sem.* II.1.52–54; after de Lacy).

Galen marks this seeming inconsistency as the impetus for his own rejection of reproductive hylomorphism.[30] Thus the idea that *GA* IV.3 and hylomorphism cannot sit together is an ancient one, which persists today. When more modern writers try to solve the supposed conflict by ascribing to Aristotle the view that individual forms are active in generation, two options are available with respect to female inherited resemblances. On the one hand, some think that Aristotle took steps towards the Galenic position and

28 That the form contributed by the father is an individual form is held by Balme (1987b), Furth (1988), Whiting (1992), Cooper (1988) and Henry (2006b).
29 See also Gelber (2010) 185–7 for a précis of these interpretations.
30 Galen accepts other aspects of Aristotle's theory, however, such as the idea that semen is derived from blood, that is, the haematogenous theory. See Chapter 3.

had to modify his hylomorphism in light of his account of hereditary resemblances.[31] Another point of view holds firmly to hylomorphism but rejects the agency of the female in the context of heredity. For this camp, the male contributes individual form and the female no form. This means that all the forming of the offspring must come from the male side, including any resemblances the offspring might bear to the female's side (Cooper 1988). Another option, which I will argue is the best one, is to deny that Aristotle has any individual forms at work in generation and therefore that the vehicles which achieve resemblance, the *kinêseis*, can come from both the male and female side, because these are not formal causes.[32] I will detail each of these interpretations, particularly in relation to the difficult passages within *GA* IV.3 upon which they rely. Every interpretation also depends on a certain understanding of the terminology, which is why it will be crucial to try to explain these in their context. There are, I argue, much more plausible interpretations which make better sense of the meaning of *kinêseis* in particular. Ultimately, a focus on *kinêseis* rather than on form solves the problems in interpretation and translation and fits best with Aristotle's theory of generation, articulated throughout the *GA*. The discussion will end by tying together the earlier descriptions of the production of generative residues in *GA* I and II to Aristotle's account of heredity in *GA* IV.3.

9.4 Individual form interpretations

Balme (1987b, 1990) finds evidence for the idea of individual forms in certain passages of the *Metaphysics*.[33] The incidental

31 Morsink (1982), Dean-Jones (1995) 179 and 194–5, Preus (1970) 13, Peck in Aristotle (1942) x-xv, Witt (1985) 48 and 54 n.20 all suggest that Aristotle has to change his mind about matter and form when faced with explaining resemblance. Furth (1988) 137–41 thinks that Aristotle needed to divide the form in two. Henry (2006a) 286: 'However neat and tidy the traditional interpretation may be, when we trace Aristotle's reproductive hylomorphism through the GA we find that it clearly does not divide the contributions of the mother and father exhaustively into matter and form.'

32 I am in broad agreement with Gelber (2010) and offer some refinements to her view.

33 Indeed, he is attempting to solve a problem that emerges in the *Metaphysics* concerning Aristotelian substances: how can we know or define these if they contain matter (each substance being a combination of form and matter). See Connell (2001) 304–5.

features of parts are contained in these forms. Balme argues that these features are formal rather than material because at any moment, out of the sequence of time, all matter is part of the form.[34]

Socrates considered without regard to past or future consists entirely of informed matter; at any one moment all his matter is determined by form, and it is that form that can be defined (Balme 1990, 52).[35]

If there are individual forms at work in heredity, then a problem emerges of how females can influence heredity. While accepting individual forms in generation, Cooper (1988) offers a solution.

In *GA* I and II, when Aristotle is providing the most abstract analysis of generation, he emphasizes the formal role of the male. Cooper worries that heredity stands as a threat to this general theory and does not want to find that Aristotle 'backed down' on entirely antithetical roles for male and female. Thus, he wants to interpret *GA* IV.3 in a way that he sees as eradicating the 'anomaly' of movement (*kinêsis*) within the female contribution, or 'of' the female parent. On this reading, power (*dunameis*), movement (*kinêsis*) and mastery (*kratein*) never come from the female side. Cooper then finds the key to how Aristotle can still explain resemblance to female parents in a passage occurring at *GA* 768a11–14.

[Peck's translation:] Some of the movements (*kinêseis*) (those of the male parent and those of general kinds, *e.g.*, of human being and animal) are present in <the semen> *in actuality*, others (those of the female and those of the ancestors) are present *potentially*.[36]

Following Peck's translation, Cooper concludes that Aristotle places all generative influences (*kinêseis*), potential and actual, in the male semen. On this view, only 'movements in the form' can contribute to the formation of offspring and these have to come from the male. These male movements, according to him,

34 'Taken out of change, as at a moment, the combined whole is entirely determined in a form which can be grasped and defined. This form necessarily includes all the matter, and is therefore individual...' (Balme 1990, 49).

35 This idea is philosophically problematic. See also Henry (2006a) 275–6, n.15. However, it will be accepted for the sake of argument.

36 ἔνεισι δ' αἱ μὲν ἐνεργείᾳ τῶν κινήσεων, αἱ δὲ δυνάμει, ἐνεργείᾳ μὲν αἱ τοῦ γεννῶντος καὶ τῶν καθόλου, οἷον ἀνθρώπου καὶ ζῴου, δυνάμει δὲ αἱ τοῦ θήλεος καὶ τῶν προγόνων.

are always actual. The female ones are potentially present in the male contribution.

When, therefore, Aristotle says that the male's sperm contains potentially both all the movements of its male forbears in the male line, *and* all the movements of the female it copulates with, including those of all her corresponding forebears, he is saying what he *must* say if he is to explain the phenomena about inherited characteristics while sticking to his basic theory of reproduction, as he has worked that out in *GA* I and II . . . If some of what the offspring's form does is to make it like its mother and her family then the father's sperm, the instrument he uses to move, fashion and shape the matter so as to have that form, simply has to have, in some way or other, those movements potentially in it (30).

Cooper admits that this theory would require a male animal to have potentially within his seed all the particular characteristics of any female animal he might produce offspring with. Interpreted in this manner, Aristotle's theory of heredity sounds very odd.[37] The main strength of the interpretation comes from its ability to make sense of the above passage (*GA* 768a11–14).[38] However, when the reader considers passages further on in his explanation of resemblance to ancestors one is strongly inclined to reject this reading and accept that *dunameis* and *kinêseis* originate in the female contribution to generation.[39] Consider, for example, the following.

The *kinêseis* that do the fashioning relapse into the nearer things, e.g. if the *kinêsis* of the male generator relapses, it changes by a very small difference into the *kinêsis* of his father, and secondly, into that of his grandfather. And in the same way also for the female, if the *kinêsis* of the female generator (*hê tês gennôsês*) [relapses, it changes] into the *kinêsis* of her mother, and if not into that, then into the *kinêsis* of her grandmother (*GA* 768a16–21).

[37] Despite its queerness, Boys-Stones (2007) accepts the implications of Cooper's position: 'Aristotle is clear that the father does the crafting of every characteristic of the child, even those which it shares with its mother.' Reeve (2000) 55 and (1995) 206 also follow Cooper writing 'actual movement . . . ultimately derives from Socrates' form. Of course, this movement has been altered or "deformed" by the menses, but it is still Socrates' movement'. Cooper's position has been rejected by Henry (2006a, 2006b) and Gelber (2010), among others.

[38] Gelber (2010) 203 offers this quite charitable interpretation of Cooper: 'what Aristotle means when he says that maternal *kinêseis* are in the male's semen potentially is that the males *kinêsis* can "elevate" to the level of actuality the female's *kinêseis*, which are present only potentially beforehand'. Although this does not so obviously clash with Aristotle's views, it still appears to be an unnecessarily cumbersome interpretation of his account of the female influences in heredity, as I will detail in what follows.

[39] The *kinêseis* of the ancestors refer to the ancestors of both male and female parents. Föllinger (1996) 177–8 and Henry (2006b).

Here Aristotle states explicitly that there can be a *kinêsis* 'of the female parent' and also a *kinêsis* 'of her mother' and 'of her grandmother'.[40] This passage also suggests that once an embryo is female it immediately inclines to resemble the mother (rather than the father; *GA* 768a20–1) and then to the maternal grandmother, therefore requiring a female starting point for likeness. On the Cooper reading there is no way to account for this as the *kinêseis* which bring about resemblance to the female parent would have had to come from her father not her mother. Only if the *kinêsis* comes from the *dunameis* of the female can a relapse of this *kinêsis* result in resemblance to her ancestors. If the *kinêsis* came from the mother's father, a relapse in it would result in resemblance to his ancestors and never her mother's. Furthermore, even if the male semen could somehow take on instructions for how to form offspring like their partner's mother, it seems incredible that it could also take on instructions for how to form them like all of her ancestors.[41] It is not only unnecessary to lumber Aristotle with such an implausible theory; it is also very difficult to find any evidence for this idea in the text. A straightforward reading of what he says about relapse calls for the reader to accept that the female contributes *kinêseis* relating to herself as an individual and containing the possibility to activate various ancestral *kinêseis*.

Cooper's attempt to accommodate what he takes these three lines of text to say comes at too hefty price if it requires rejecting much more lengthy portions of the chapter.[42] Added to this, one need not interpret *GA* 768a11–14 in the way he chooses. Henry presents two possible ways to read the passage which bypass the Cooper conclusion. In both cases, he reads *eneisi* ('they are in') to refer only to the contributions of males.[43] His first suggestion of how to eliminate

[40] ἐπὶ τῶν θηλειῶν ἡ [κίνησις] τῆς γεννώσης εἰς τὴν τῆς μητρός, ἐὰν δὲ μὴ εἰς ταύτην εἰς τὴν τῆς τήθης· ὁμοίως δὲ καὶ ἐπὶ τῶν ἄνωθεν. *Kinêsis* appears in the singular in this passage, because it a description of a particular part being constructed by a particular *kinêsis* in the blood, of which there are many, both potential and actual.

[41] See also Henry (2006a) 272–3.

[42] Cooper admits that 'the sentence at 768a11–14 is the only explicit indication in his text that Aristotle postulated movements somehow potentially present in the male's fluid capable of imposing on an embryo bodily resemblances to its mother's side of the family' (1988, 27). Contrast Henry (2006a) 273: '*GA* 768a14 is an extremely weak foundation upon which to rest such a controversial interpretation.'

[43] Henry (2006a) 277 n.39: 'I agree with Cooper that 768a11–14 only refers to the movements in the male semen.'

any implication that these contain the female animal's *kinêseis* is to excise the words 'of the female and', which could have been placed there mistakenly by a scribe.[44] Once these words are removed, the passage simply says that the *kinêseis* of the father and of general kinds (e.g. human and animal) are present in activity, while those of ancestors are present potentially. Henry's second suggestion is that even if one were not to agree to the textual emendation, one need not accept the Cooper interpretation. This is because the term used of the female here is *thêlu* – femaleness taken generally. If Aristotle had meant to indicate the individual female parent he would have used *mêtêr*, 'the mother' (e.g. *GA* 768a8, 12, 20, 30, 35, 768b15). The passage would then say that, potentially within it, the male semen contains the capacity to make an offspring female.[45] It seems more plausible to imagine that the male semen is potentially able to form a generic female body than that it contains numerous specific instructions relating to all possible procreative partners. However, in order for this reading to be acceptable, it must accord with Aristotle's descriptions of how sexual differentiation occurs in *GA* IV.1–2. In what follows I argue that this is not possible, and so we cannot accept Henry's second reading of *GA* 768a11–14. I will then offer an alternative interpretation of the passage which bypasses these difficulties.

In *GA* IV.3 Aristotle appears to offer two different accounts of how mastery results in the formation of a female rather than a male embryo.[46] At (1) *GA* 768a3–5 he explains that a female comes about when '*that which is not mastered* departs from type and changes into its opposite'[47], while at (2) *GA* 768b7–8 he states that this happens when '*that which is mastered* departs from type and changes into its opposite'[48] (my emphasis). How can these statements be consistent? The simplest solution is to find that the subject, the 'that' referred to in each statement, differs. It could

44 Henry (2006a) 274 excises αἱ τοῦ θήλεος καὶ.

45 Platt reads the text as Henry does, writing '... if the semen cannot impress its male character upon the embryo, it forms the embryo defectively into a female, and hence the female character may be said to exist *potentially* in the [male] semen' (Aristotle 1910, 768a14 n.2).

46 See Henry (2006b) 447, n.48.

47 μὴ κρατούμενον ἀναγκαῖον ἐξίστασθαι καὶ γίνεσθαι τὸ ἀντικείμενον.

48 κρατούμενον μὲν ἐξίσταται εἰς τὸ ἀντικείμενον.

also be the case that what does the mastering in each instance differs, if we allow (as is clearly implied by his account of sexual differentiation and heredity) that the female can master *qua* female when the male is mastered. Since subject and agent can both vary and there are two statements to consider, there are four possible ways to read each. The subject ('that') is either the male principle or the female one. This subject could then be mastered either by the male *qua* male or by the female *qua* female.[49] Possible meanings of (1) *GA* 768a3–5 and (2) *GA* 768b7–8 are as follows:

(1)^{0} That [the male principle] which does not get mastered [by the male *qua* male] . . . changes into its opposite [from male to female]
(1)^{1} That [the male principle] which does not get mastered [by the female *qua* female] . . . changes into its opposite [from male to female]
(1)^{2} That [the female matter] which does not get mastered [by the male *qua* male] . . . changes into its opposite [from male to female]
(1)^{3} That [the female matter] which does not get mastered [by the female *qua* female] changes into its opposite [from male to female]
(2)^{0} That [the male principle] which gets mastered [by the male *qua* male] . . . changes into its opposite [from male to female]
(2)^{1} That [the male principle] which gets mastered [by the female *qua* female] . . . changes into its opposite [from male to female]
(2)^{2} That [the female matter] which gets mastered [by the male *qua* male] . . . changes into its opposite [from male to female]
(2)^{3} That [the female matter] which gets mastered [by the female *qua* female] . . . changes into its opposite [from male to female].

Before checking for consistency between pairs, it is worth ruling out those options that are nonsensical. Both (1)^{0} and (2)^{0} can be ruled out on the grounds that there is no clear sense in which the male principle would be mastered by the male *qua* male. Thus, it looks like (1)^{1} and (2)^{1} are left to those who think that the male shapes the matter to be female and that *enesi* can refer only to the male contribution.[50] The idea, then, is that it is the male principle itself which must change over into femaleness and then shape the matter to be female. The problem is that this only makes

[49] Although one might think that there are two ways in which something could change into the opposite gender – from male to female or from female to male, both passages concern a change from maleness to femaleness. Furthermore, Aristotle would think it impossible for a female to turn into a male.

[50] For example, Henry (2006b) and Platt in Aristotle (1910).

sense of $(2)^1$. $(2)^1$ posits that if the male principle is mastered (by the matter or female principle) then it changes into its opposite, meaning that the male principle itself will become female or feminized and so will itself become the demiurgic agent of the female animal's femaleness. But if the male principle does not get mastered (as according to $[1]^1$), then it will master by itself *qua* male and so will not change over into its opposite but will remain male. We are left with $(1)^2$, $(1)^3$, $(2)^1$, $(2)^2$ and $(2)^3$, four of which make the female principle the subject.

In order for it to make sense that the female principle is the subject, it must be possible for the female contribution to change from male to female. This need not be problematic. Aristotle thinks that the initial state of the embryo is potentially male. If the semen is well concocted (*eupeptos*), a male will result.[51] Every female is a deviation of a sort resulting from the coldness of the environment or the youth or age of parents (*GA* 766b27–767a2, 767b11–12).[52] The most natural reading of passage (2), then, is to take that which changes to be the female contribution to generation, which becomes the offspring's body, and which, Aristotle imagines, begins or inclines towards maleness, before being thwarted and changing over into femaleness.

Now we must consider whether any other options are to be ruled out. $(1)^3$ cannot make sense since if the female matter is not mastered by the female *qua* female, then it will be mastered by the male *qua* male and will not change from male to female, but will be male. $(2)^2$ is similarly ruled out: if the female matter is mastered by the male *qua* male, then it will become male. Thus, the following are the only remaining combinations that allow for consistency:

[x]

$(1)^2$ That [the female matter] which does not get mastered [by the male *qua* male] . . . changes into its opposite [from male to female]

$(2)^1$ That [the male principle] which gets mastered [by the female *qua* female] . . . changes into its opposite [from male to female]

[51] *GA* 767b16–18. As we now know, he was wrong about this. In fact every (potentially sexed) embryo will become female unless a flood of hormones change the direction of this development in order to produce a male.

[52] I will discuss the meaning and import of 'deviation' in the next section.

[y]
(1)[2] That [the female matter] which does not get mastered [by the male *qua* male] . . . changes into its opposite [from male to female]
(2)[3] That [the female matter] which gets mastered [by the female *qua* female] . . . changes into its opposite [from male to female].

In [x] both subject and agent vary and in [y] the subject is the same and the agent varies. Both possibilities require that the female be capable of mastery *qua* female. This means that the idea that the male contribution always makes a female become female cannot accord with any of the options that remain. Furthermore, although [x] allows for the consistency of the passages, they become much more cumbersome. Accepting [x] requires thinking that sexual differentiation occurs either when the male principle changes over into *its* opposite (and makes the female offspring female) or when the female matter changes over into *its* opposite (becoming female rather than male). Taking the subject as invariant, as only [y] does, is more straightforward. It also makes better sense. The idea that the male principle must change over into female and make the embryo female is not required because, according to Aristotle, the female principle has its own agency to shape the offspring into a female. Since the female blood, from which the embryo is derived, already maintains the female parts of the female's adult body, it is much better to posit that the female contribution itself plays some part in developing female parts. And so the female contributes *dunameis* for femaleness, which activate the *kinêseis* needed to produce a female animal's body. [y] also fits with a theme common to Aristotle's discussions of sexual differentiation and heredity in *GA* IV that male masters *qua* male and female *qua* female. So what varies is not the subject but the agent, that is, sometimes the male masters and sometimes the female does.

[y] also fits with the context of both passages where the idea of dual mastery is very important. (1) occurs when Aristotle is explaining that there are three sets of *dunameis* relating to the individual: sex difference (male or female) and the kind and that the generative residues of parents can master according to these. Both here and in *GA* IV.1–2, when the male does not gain mastery (767b10–11, 23, 768a4–6), it is mastered, and the offspring

changes into its opposite: female (766a14–19, 768a4–6). Sexual differentiation is due to the mastering of the male *qua* male or the mastering of female *qua* female. According to (1)^2 we can read (1) to say that that which does not get mastered is the new body of the offspring (the matter contributed by the female), which must, rather than become a male, change over into its opposite and become female. (2) is a more general summary of some of the terms of the previous explanation. Aristotle is gearing up here to explain that resemblance to ancestors occurs when relapse happens on both maternal and paternal sides. Having been told of the sets of *dunameis* in each contribution to generation, we now see that the female can also master *qua* female and individual mother; this is crucial for understanding how resemblance to female ancestors happens, since the relapse (*luontai*) depends on their being a *kinêseis* there in the first place. Neither passage indicates that the male contribution will make the embryo female, but the cause might be thought to lie in the male contribution in the sense that it is due to its failure to master *qua* male that femaleness occurs.[53] The male must master in general in order to make the embryo a viable animal of a certain kind. However, the female contribution can now master *qua* female and *qua* individual, if circumstances allow.[54] When the male has failed to gain mastery *qua* male, a functioning female anatomy results. Clearly *thelu* in the passage means to indicate femaleness rather than the individual female parent, as Henry argues. However, it is not plausible in light of Aristotle's explanation of sexual differentiation, for the male to be in control of making the embryo female and so we must reject Henry's interpretation.

I will now suggest another way to avoid the Cooper interpretation of *GA* 768a11–14 while respecting the current Greek text. Cooper and Henry assume that the subject of *eneisi*, 'they are in',

[53] It is plausible that once an embryo established by the male efficient cause has its own heart, it must develop its own set of *dunameis* and that these would be established by the plural *kinêseis*, which come from both parents. Henry (2006b) 443, 449 remarks that the embryo must undergo some kind of 'formation of [its] own genetic nature'. However, Aristotle never says anything explicit along these lines.

[54] Although the female has reciprocal *dunameis* for femaleness, and the individual, it lacks the ability to master according to the *dunamis* of general kind, for example, human being.

is only male semen, which would mean that the *kinêseis* of femaleness would have to be in the male contribution. If, however, the subject of *eneisi* is the generative residues of both parents, then femaleness does not have to be from the male. This reading is supported by Aristotle's other use of *eneisi* in *GA* IV.3, when he explains how the parts of the offspring resemble parents' parts.

> The same thing happens with the parts of the body. For often some parts are like the father, some like the mother and some like those of ancestors. For some actual and some potential *kinêseis* of the parts are in them (*eneisi*), as has been said many times (768b2–5).

The passage makes no claim about where the *kinêseis* it mentions exactly reside.[55] The implication is that they are present in both parents' residues or the initial combination of the two. This seems equally to be the case for *GA* 768a11–14. Nothing in the Greek text forces the reader to conclude that the *kinêseis* listed all have to be in the male contribution. Indeed, a less strained reading is better here: the *kinêseis* of femaleness and those related to the female parent's ancestors are present in the female's contribution to generation.[56]

One more terminological intricacy needs to be explored before returning to interpret the passage. Here and elsewhere in *GA* IV, Aristotle distinguishes between *energeia kinêseis* [EK] and *dunamei kinêseis* [DK] (768b5–7). *Energeia* is usually translated as 'actual' and *dunamei* as 'potential', but these translations do not fit the context. The *energeia kinêseis* in the generative residues are not actually changing the semen into an animal; all *kinêseis* in this context remain potential until embryonic development begins.[57] Thus one ought not to think of EK as 'actual' *kinêseis* but rather as the 'current potential' *kinêseis* – the *kinêseis* that will shape the

55 Peck translates *eneisi* as 'in the seminal substance' Aristotle (1942).

56 Thomas Taylor's translation of the passage indicates that he thought *kinêseis* were present in both parents rather than only the male in Aristotle (1808) 388: 'And the motions, indeed, which are inherent in energy, are those of the generator and of universals, as of man [i.e. human] and animal. But the motions which are inherent in capacity are those of the female and of ancestors, both in fathers and mothers.' See also Gelber (2010):.'These *kinêseis* could be potentially present in the menstrual fluid or in the embryo' (203).

57 See also Henry (2006b), who notes that the *kinêseis* are 'present in a kind of de-activated state' until embryological development begins (448).

offspring in due course if the *summetria* is not altered in some way. A *dunamei*, or 'possible potential', *kinêseis*, on the other hand, lies dormant until such time as current ones are mastered or relapse and become ineffective, and allow the possible ones to become current. I would like to suggest, therefore, that EK be translated 'current potential changes' and DK as 'possible potential changes'.

This way of translating the EK and DK gains support from the use of similar terms by Aristotle in other contexts.[58] One interesting example occurs in the *Parva Naturalia* treatise *On Dreams*. Here he is attempting to explain why certain dream images occur in the sleeper, and uses *kinêseis* to refer to perceptual possibilities in the soul.

The sleeper owing to his sleep, and the stimulation of the sense organs, and the other factors which are connected with sensation, is liable to be deceived so that an image which has only a remote resemblance seems to be the object itself. For when a man is asleep, as most of the blood sinks to its source, the internal *kinêseis*, some *dunamei* and some *energeia*, travel with it. They are so conditioned that at a given disturbance one will detach itself and rise to the surface, and if this is destroyed another will do so. Their relation to each other is similar to that of artificial frogs, which rise to the surface of the water as the salt is dissolved. In a similar way *kinêseis* reside in us *dunamei*, but are made *energeia* when the preventing cause is removed; and as they are freed (*luomenai*), in the little blood which remains in the sense organs they begin to change (*kinountai*), bearing a resemblance, as cloud-shapes do, which in their rapid changes we liken now to men and now to centaurs. Each of them . . . is a residue of the actual impression; though the actual (*energeia*) impression has gone, this still inheres, and it is true to say that it is (for instance) like Coriscus, though it is not Coriscus (*Insomn*. 461b8–25).

Aristotle is attempting to account for how we can perceive things in dreams that are not actually there. The images come from *kinêseis* in the blood, left over from actual instances of perception, for example, that of the man Coriscus.[59] Although in dreams one is never actually perceiving, the *kinêseis* can be *energeia* because they are active in the blood and so ready to be thought to be

[58] Henry (2006b, 440–2) finds it useful to compare the discussion of *kinêseis* here with that to do with sight in *GA* V.

[59] The example attests to the complexity and specificity of the term *kinêseis* in this context; they are not merely sense data but something more robust such as the Aristotelian idea of *phantasia* – a joined-up idol.

perceived; other *kinêseis* are possible because they remain inactive until such time as they are released, as Aristotle's metaphor suggests. The parallels with *kinêseis* in generation are not exact, but there is striking similarity both in terms of the difference between the DK and EK and in terms of the thought that traces exist in the blood that give a resemblance to something that is real. In the case of dream perception, the previous perception which left these traces was of an actually existing entity, Coriscus the man. In generation, the blood and residues contain traces of likenesses to the parts of the parents and ancestors – to real individuals, although these *kinêseis* are not really existent in any individual until a particular embryo starts to come into being and develop its parts.[60] The *kinêseis* in the residues at conception are malleable at first, not being actual changes yet. At this point the state of the mixture, that is, the condition of the materials, determines which 'rise to the surface'.

Returning, then, to the passage, my translation reads thus:

> Some of the *kinêseis* in them [i.e. in the male and the female generative residues] are current potential changes, some are possible potential changes, the former are those of the male parent and of general kinds, such as human and animal; the latter are those of femaleness and the ancestors [on both sides] (*GA* 768a11–14).

Aristotle's statement is condensed; it does not give a comprehensive list of all the *kinêseis* contained in the residues. The following *kinêseis* can be found in each generative residue, including those that have been left out in the round brackets.

In the male generative residue, K of:	animal [EK]
	human [EK]
	(maleness [EK])
	individual male parent [EK]
	the male's ancestors [DK]
In the female generative residue, K of:	(maleness [EK])
	femaleness [DK]
	(individual female parent [DK])
	the female's ancestors [DK]

60 There is a certain order to these images – a nearest and more distant image in the soul (*Insomn.* 463b1–2). So also there is a certain order going from nearest to most distant relation in heredity (*GA* 768a10). See also van der Eijk (2007) 420–1 for this connection.

If the *summetria* is ideal (i.e. 'well concocted'), then the current potential changes will remain and the offspring will be a human (or other generic) male and it will resemble the male parent (*GA* 767b16–18). These *kinêseis*, then, are present primarily at first in the conceptus. Only when either female mastery or any relapse occurs, do *kinêseis* in the female relating to femaleness and the individual female and *kinêseis* in the male relating to male ancestors, take over. The possible potential changes from ancestors on his side reside in the male residue, while those coming from the female side reside in hers.

It is worth taking a moment to consider the import of this line of thought to our understanding of Aristotle's male bias. Although the female generative residue does contain the capacity to shape parts like her own under certain conditions, Aristotle thinks that if the seminal residue is 'well concocted' a male like the male parent will result. His explanation of hereditary resemblances relies on the idea of deviation from the smoothest instance. Many commentators believe this means that when offspring resemble anyone but the male parent a failure or impediment to development occurs.[61] Viewing offspring that take after a female parent or other relative as straightforward failures, or deformations, is potentially very misleading, however. If the male contribution does not master according to the *dunameis* relating to male and to him as an individual, Aristotle writes that 'it makes it deficient in that area' (*GA* 767b23).[62] What he cannot mean by this is that this particular instance of generation is, as a whole, a failure. Being deficient in the *dunamis* relating to a particular father does not mean that no *kinêseis* will shape that part; it will not lack the part. So, for instance, when the male principle fails to gain mastery in this way, the embryo might fail to resemble the father in nose shape, but the embryo will still get a nose. The deficiency is not general, but related to a specific individual person: the offspring will be

[61] See, for example, Morsink (1982) 136: 'Aristotle clearly viewed as the ideal case an instance where Coriscus, a male, produces a boy that looks like himself in all respects.' See also Henry (2006a) 278–9. This idea is taken up especially by those who assume that Aristotle espouses a one-sex model, for example Lange (1983).

[62] The term for deficiency here, *elleipsin*, can have connotations of inferiority and failure but it is not a term used elsewhere in the *GA* to indicate the inferiority of the female.

deficient in resemblance to Socrates, for example.[63] The resultant offspring is not literally deformed; it is not deficient in any functional part necessary for its essence.

Aristotle starts his explanation of heredity by saying that he intends to elucidate resemblance to both parents.

Some offspring are like their parents and some are not like them, and some are like the father and some the mother, either according to the whole body or according to each of the parts, and they are more like these [i.e. parents] than ancestors, and more like their ancestors than random people, and males are more like the father, females the mother (*GA* 767a36–767b4).

Shortly after, he notes that 'anyone who does not take after their parent is, in a way, a monstrosity' (767b5–6)[64] and this occurs because 'it is possible for the male sometimes not to gain mastery' (767b10).[65] Offspring resembling females are defective if the 'parents' referred to at 767b7 are only male parents. It is possible, however, to take 'parents' in the passage to refer to both males and females. This is the way the term is employed in the beginning of the passage (at 767a36: 'some offspring are like their parents . . . some are like the father and some the mother'). The idea expressed would then be that the male principle must gain mastery in general, that is, a successful instance of generation has to occur, in order for offspring to resemble either parent. The best result in the process of generation, as was made plain in *GA* IV.1–2, is the production of a male animal. If this male animal resembles a mother or ancestor on either side, then, it could not really be a failure.[66] If Socrates Junior's eyes are blue (like Xanthippe's) rather than brown (like Socrates') this will not affect his ability to concoct male semen once he reaches maturity and so cannot make his birth anything less than a complete success.

In the next chapter, when the *summetria* at conception is disrupted somewhat, although the resultant offspring is not literally deformed, Aristotle sometimes thinks of this situation as a sort of

63 Only by being deficient in resemblance to human would it end up generally deficient, deformed and thereby a failure of generation (*GA* 767b6).

64 ὁ μὴ ἐοικὼς τοῖς γονεῦσιν ἤδη τρόπον τινὰ τέρας ἐστίν.

65 ἐνδεχομένου δὲ μὴ κρατεῖν ποτὲ τοῦ ἄρρεν [τοῦ ἄρρεν: Rackman vulg.: τοῦ ἄρρενος].

66 Aristotle takes care to separate sexual differentiation and heredity in terms of different *dunameis* in the semens.

deformity in an extended sense of that word. But offspring that resemble family members other than the male parent are not deformed in this extended sense. What has happened is that 'possible potentials' have become 'current' ones, after the lapse of those that initially were current. Let's consider again the parallel use of *energeia kinêseis* [EK] and *dunamei kinêseis* [DK] in Aristotle's explanation of dream images (*Insomn.* 461b8–21). Each set of dream images results from the particular circumstances in the sense organs where some images make their way to being perceived while others do not. Aristotle would not wish to say in this context that one dream is superior to another. There is no reason why we need think that just because the individual's *kinêseis* are currently potential, they are superior to possible potential *kinêseis*. Indeed, in the *Insomn.* passage, the DK are poised to become EK if an impediment is removed, rather than this situation being due to an impediment. Whether there are any true impediments involved seems to depend, then, on which perspective one is viewing the process from. If one were to think from the perspective of the male, then the offspring will be like the male parent if nothing impedes it, that is, if the male masters with respect to the *dunameis* for individual male parent (and if the *kinêseis* derived from this *dunameis* do not subsequently relapse). However, when the male contribution is mastered, it is mastered by something and from the perspective of that something an account can be given in a different manner. If the offspring ends up like the female parent, then nothing has impeded the forces that would shape it so. Or, we could say that whatever it was that impedes those particular shaping powers has now been removed; the male's influence has been eradicated so the female's DK can become current. From the perspective of the female contribution to generation, it has mastered with respect to individual female parent and its *dunameis* have been established.

The idea that male offspring resembling other relatives are not inferior to those that resemble the male parent fits with one reading of *summetria*. As explained earlier, the generative residues of the parents differ at different times in their lives and due to external conditions, particularly in different seasons of the year, which explains why offspring are never identical in every feature. But although not all combinations of seminal residues are alike, this does not imply that there is only one successful combination. One may meet the mark in a variety of different ways.[67] Just as health is not the same in every individual because each person's constitution differs (*EN* 1173a24–5), so also one successful instance of generation varies from another.

The fact that Aristotle assumes that if *summetria* is optimal the offspring will resemble the male is partly due to the structure of his theory of hereditary resemblance. The theory requires there to be an optimal *summetria* in which no slight change occurs. His choosing that this situation results in an offspring resembling only the male parent is arbitrary. He could have chosen for this to be an offspring resembling the female parent, or even one that resembles the male in half of its parts, and the female in the other half. It is not the case that his general theory dictates that it would have to be the male.[68] He could find that if nothing happens then the offspring would be female and resemble the female parent. After all, why shouldn't the female matter which becomes the body of the offspring already have tendencies in it to replicate the body of the female parent, being a residue of her blood? External bodily resemblance, as Aristotle noted in *GA* II on the appearance of hybrids, very often follows the pattern of the materials out of which offspring are made and these incline to shape the body of the female parent (738b26–36).[69] Aristotle's sexism is not thereby just an adjunct he was forced into conceding in order for his theory to work – it is an integral part of this theory and indeed, on general terms, can be seen to be in tension with a more

67 In contrast to this view, van der Eijk wishes to maintain that an absence of *summetria* is what results in resemblances to ancestors (2007) 420. But if *summetria* were to be absent, the offspring would be literally deformed, which is not the case in these instances.

68 Resemblance to male parents often preoccupies Aristotle, for instance, when he praises the mare of Pharsalis, a horse that had the tendency to produce offspring resembling the male parent with the epithet 'Honest Wife' (*Dikaia: HA* 586a12–15; *Pol.* 1262a22–4). Resemblance to father might serve as tangible proof of paternity (and in the ancient context, perhaps the only proof available), if this were to be in question. As he makes clear elsewhere, a mother always knows better that the children are hers (*EN* 1168a25–7; cf. *Rh.* 1398a32-b4).

69 See also *HA* 577a9–11.

coherent stance that he could have taken, that the embryo inclines to femaleness.[70]

9.5 Alternative solutions: understanding *kinêsis*

Cooper's solution did not work because the entirety of the *GA* IV.3 account of heredity tells against the thought that the *kinêseis* coming from the female side reside in the male residue. In fact this supposed solution is unnecessary because it assumes two problems that are not in fact difficulties for Aristotle's theory – that of how the male contributes to hereditary resemblance without contributing matter and that of how the female does so without contributing form. The purported difficulties of reconciling *GA* IV.3 with reproductive hylomorphism emerge in part because of a failure to pay due attention to the specificity of the discussion.[71] *GA* I and II contain a general account, focused on substantial change, which must occur in any case of generation. By the time the reader reaches *GA* IV that focus has long receded into the distance; here Aristotle is concerned with describing what happens in particular instances where no two offspring are ever exactly alike in all features. In this context, he does not talk of matter and form but of *kinêseis* derived from *dunameis*, which exist in both parents' generative residues. What is required to fully dispel worries about hylomorphism is for the male to determine how some inessential features of parts come about without contributing any matter and for the female to do the same without contributing form. The solution to this worry lies in understanding that *kinêseis* are neither strictly formal nor strictly material.

Hylomorphism is a problem if *kinêseis* are expressions of form because then the female contribution cannot be associated with them.[72] We met a similar problem with interpretations of the male

70 Aristotle clearly thinks that the female contribution inclines to become a male if nothing happens. See my interpretation of *GA* 768a3–5 and 768b7–8 above. For the idea that his sexism is an adjunct to his theory see Chapter 2.5.

71 See Chapter 2.

72 The *kinêseis* at work in inheritance are thought to be 'information-bearing vehicles through which the parts of Socrates' form are physically transmitted to his offspring in the act of reproduction' (Henry, 2006b, 431). If this were the case, these could only originate in the male. Some feel that Aristotle cannot adhere to strict reproductive

role in Chapter 6.3, where it was thought of as the exclusive conveyer of something like 'genetic information', likened variously to computer programmes or blueprints. Such metaphors presuppose that matter is empty and completely malleable with no specifications of its own and certainly no room for the reciprocal powers found in Aristotle's account of heredity.[73] Part of the difficulty here seems to be a cross-cultural linguistic confusion. The *kinêseis* that 'form' the offspring to be a particular animal are best referred to as 'shaping' it since, although *kinêseis* 'form' the parts of the offspring, Aristotle would not say this was due to *eidos*, form in his technical sense.[74] *Kinêseis* which shape parts of the body are not formal in the sense of the formal cause of generation, linked to the efficient cause, the *archê kineseôs*. This is the male animal's exclusive role.[75] In particular instances, these *kinêseis* are *energeia kinêseis*, current potential changes, and *dunamei kinêseis*, possible potential changes, which eventually, in some combination, depending on the final *summetria*, shape particular parts of the offspring's body. That the female contribution has such *kinêseis* is not anomalous. There is nothing stopping materials in Aristotelian terms from having such specifications. Only if we anachronistically suppose bare matter is there any difficulty. In *GA* IV we learn that the female contribution contains *dunamei kinêseis* which can shape the offspring (768a11–14). This does not contradict but rather elaborates what has already been established in *GA* II, that is, that the matter which the female contributes is poised to become an individual animal of a specific sort.

hylomorphism in *GA* IV.3. For Morsink (1982) 141, he qualifies it; for Furth (1988), he abandons it; and for Henry (2006a) and van der Eijk (1999), he modifies it so that form becomes sentient soul (295).

73 The idea that only male 'form' could bring about any change in offspring is already present in the medieval commentator, Giles of Rome. This understanding of the exclusively dichotomous nature of matter and form is heavily influenced by Platonism, where matter was thought of as entirely passive and resistant. For Giles, the male has to do everything, penetrating into menstrual fluid to imbue it with the very qualities of bone and flesh since matter couldn't have any capacities (Hewson 1975, 72–3, 80, 183 and 194). This is not an accurate representation of Aristotle's position in the *GA*, as I have shown.

74 As Gelber (2010) notes: 'the *kinêseis* the female provides can ... "fashion and shape" the embryo' (208). There seems no need to find this to be remarkable ('remarquable'), as van der Eijk does (2007) 422.

75 Gelber (2010) 167.

As for the male role, that it can bring about resemblance without contributing matter is not problematic when it is considered as the efficient cause. Just as any individual carpenter will make a bed in a particular manner (e.g. by making round instead of square bed-posts) so also any individual male will shape offspring in a particular manner.[76] The carpenter does not have to contribute anything material to bring about the instantiation of certain particular features. So we can see how male and female influences on the shaping of particular parts of the offspring's body accord with their generative role as material and efficient causes respectively. But more needs to be said about how specifications in the maternal contribution and effects of the male efficient cause can both bring about inessential inherited features.[77] When an animal does not resemble one parent it does not change into any chance thing (*tuchon*) (*GA* 768a3). Instead, there is something within the semens of the parents that brings about its formation in a particular way, in accordance with the features of other family members. As inherited features occur in this orderly fashion according to *GA* IV.3, it is very unlikely that they are due to randomness.[78] Ontologically, inherited features are accidental to form and essence. But causally, they are not accidental because they come about due to *kinêseis*, determinate potentials to change offspring in particular ways, within both male and female residues. The features are not part of the form; they are sub-specific. However, they are not left to chance but are somehow caused by the parent as an individual generator.[79]

[76] Gelber (2010) terms this phenomenon the effect of 'last agents', which is due to particular tools and techniques (206).

[77] These features are not part of the form as the individual form view claimed. The generic form view is more attractive in so far as it fits better with Aristotle's general metaphysics, where form cannot be individual and contain material specifications. Furthermore, Aristotle does not indicate anywhere in the *GA* that the form contributed by the male is individual, which seems a very curious lacuna if such a theory was meant to be his explanation of inherited resemblances.

[78] See Henry (2008) 66–7: 'heritable variations occupy an interesting place in Aristotle's ontology: family resemblances are neither accidental by-products (since they are heritable) nor are they present for the sake of anything (since they are sub-specific)'.

[79] Gelber (2010) contrasts accidental causes with *per se* causes, and posits that *per se* causes are so in virtue of something being the right kind of thing to bring about a given effect (193). Thus a doctor is a *per se* cause of healing since this is what he potentially brings about *qua* doctor. Similarly, then, a parent is a *per se* cause of inherited features in so far as this is what he or she brings about *qua* (specific) generator. 'Inherited features

The *kinêseis* operating in heredity are neither form nor matter and therefore can come from both male and female parents. No doubt the fact that form does not properly explain these *kinêseis* led to Aristotle abandoning the use of the term *eidos* in the context of heredity.[80] They are the potentials within the generative residues to bring about sub-specific features of parts of offspring. They can come from the male because he is the cause of an animal coming to be this type with all the relevant parts, and each part will also be a particular part with inessential features – and so will be shaped in a particular manner. The male does not have to contribute matter for this to be the case because EK and DK can be transferred from one parcel of matter to another. Although the male's *kinêseis* cannot exist without any matter at all, these influences are transferred from the male semen before it disperses to the menstrual fluid, and reside in it (just as also happens in the white of the egg).[81] In the case of the female, her contribution contains specifications which are neither formal causes nor vehicles of form, but can come to the fore to shape parts of the offspring when circumstances allow.

Looking more closely at his account of embryological development, one can discover a plausible Aristotelian story of how *kinêseis* bring about resemblance.[82] *Kinêseis* are all potential and can assert themselves after they have left the agent – so in perception, the *kinêseis* which convey to the heart the fact of red are no longer in contact with the eye. In the generative process, in the case of dead ancestors, the original *dunameis* from which the *kinêseis* are derived have entirely ceased to exist. In parents and living relatives, the *kinêseis* present in the conceptus and the embryo are no longer in contact with their source.[83] Now recall that in embryology actual nutritive soul comes before the acquisition of

are the *per se* results of certain *kinêseis*, which are said to be drawn from potentials that the generator has non-accidentally, *qua* generator . . . ' (191).

80 The word for form (*eidos*) does not appear at all in *GA* IV.3. This is also noted by Gelber (2010) 196 in order to make a slightly different point.

81 See Chapter 7.3.

82 Gelber (2010) suggests a brief sketch along the same lines (208).

83 Aristotle elsewhere likens the effect of an agent on a medium (here blood) to the movement of a wave or impulse in a fluid medium: 'when anything has stirred water or air, the part moved moves another, and when the first impulse has ceased, a similar *kinêsis* still continues up to a point, where the agent is not present' (*Insomn.* 464a6–9).

sentient soul, which is achieved once the sense organs are differentiated.[84]

(1) actual nutritive soul brought about by male	(2) actual nutritive soul; actual sentient soul (*hexis*)	(3) birth: actual nutritive and sentient souls (*energeia*)
heart	*flesh, eyes, nose, ears, and so on*	*actualization of the senses*

When the embryo first acquires actual nutritive soul, it will not yet possess the organs and parts it needs for the actualization of sentient soul (3). These parts must develop out of the blood provided by the mother. It is at the point when the embryo begins to develop organs and parts that male and female hereditary influences do battle.[85] Changes (*kinêseis*) have to take place in order to differentiate the sense organs, which, once formed, will resemble the organs of someone or other.[86] At stage (1), the male's sense organs must be potentially (*dunamei*) present, since his own nutritive soul produces specifications for all parts of his body and for the sensitivity of his blood. At this point, the female parent's sense organs are also potentially present in her residue, which is the material making up the embryo's body.

When the male principle actualizes nutritive soul, it also starts the embryo on its way to sentient soul, as the initiator of the automaton is responsible for things that happen as a consequence of that first motion (*GA* 734b10). The male somehow directs the material specifications of parts (i.e. the sense organs) like his own so as to actualize sentient soul. At this point, if the female principle masters according to her own *dunameis* (*qua* individual mother) the *kinêseis* derived from these will take over and form the embryo's parts to be like the mother's parts (*GA*

84 See also Chapter 5.3.

85 I agree with Gelber (2010) 208 that '[o]nce the rudimentary heart is formed by the initial action of the semen upon the *katamênia*, the *kinêseis* of both parents are available for use as tools in the generation of the new organism'.

86 Of course, only the external sense organs will reveal inherited resemblances, for example, the eyes or ears. However, subtle aspects of the central sense organ might show up in certain behaviours, for example gracefulness or musicality. Aristotle mentions resemblances evident in the way in which someone moves at *GA* 722a6.

768a7–8). Xanthippe's menstrual contribution will form eyes like Xanthippe's eyes.[87] Aristotle's theory allows, then, that although actual nutritive soul is first established in the embryo by the father, the task of shaping certain parts of the body can be given over to the potentialities present in that blood, that is, present in the female residue itself.[88] This does not mean that Aristotle backs down from an earlier theory. From the beginning of the *GA* he makes plain that both spermatic residues have qualities which make them able to become parts of an adult body, not in the sense of passive material, but in terms of their functional living condition.[89] Generative residues are useful residues; and the most useful of these is that which is produced later and is that from which each of the parts of the body is directly formed (*GA* 725a11–13). Offspring can resemble either parent because the nourishment in the bodies of each is potentially like all the parts of each individual's body (*GA* 726b10–19).

Galen agrees that the *GA* account of hereditary resemblance and of menses as the residue of blood in Aristotle's physiology implies that the female assists in making parts like her own. However, in his concern to show the superiority of his own theory, he attempts to paint this as inconsistent with Aristotle's general theory. Thus he accuses Aristotle and his followers of failing 'to notice that they give to the matter as many powers (*dunameis*) as they earlier gave to the artisan' (*Sem.* II.1.38–9). This is an unfounded complaint. The male as initiator of substantial change (as 'the artisan') can do what the female matter cannot. But female matter, unlike the materials used in craft productions, contains *kinêseis* derived

87 Preus (1975) 103–4 writes, '"dark eyes" pass over to "light eyes". This is what Aristotle means by going over to the opposite.' This is incorrect; by opposite Aristotle means the opposition of male and female parents, rather than the opposition of the characteristics themselves: green eyes could be opposed to either blue or brown eyes, depending on which parents you were dealing with.

88 This consolidation or combination of the powers that reside in both semens (*spermata*) brings to mind descriptions of generation found in the Hippocratic treatise *Vict* I: 'If the secretion from the man be male and that of the woman female, should the male gain the mastery (*epikratêsê*), the weaker soul combines with the stronger . . . *united they master* (*krateous*) the available matter' (*Vict* 1.28; L VI 502–3). It is understandable why some think that Aristotle adopted medical motifs in his embryology. See for instance van der Eijk (2007).

89 '[N]either a hand nor any other part of the body whatsoever is a hand or any part of the body if it lacks soul or some other *dunamis*' (*GA* 726b23–4).

from *dunameis* (*GA* 767b36).[90] In the case of a single instance of generation, it is perfectly appropriate in Aristotelian terms for the matter involved to contain these *kinêseis* which direct development.[91]

[90] We can note how far this is from the craft analogy; the marble could not be said to be responsible for forming a particular shape or colour of the eye of a statue. Although the materials used in crafts obviously have material potentials or powers, these must be more heavily manipulated by craftsmen.

[91] Although I agree with Gelber (2010) that a metaphysical space is available for non-accidental sub-specific features which both parents contribute, I part ways with her on the fact that she does not differentiate between the plural and singular of the term *kinêsis*. For Gelber the *kinêsis* (singular) which Aristotle attributes to the male contribution in *GA* I and II is the same as the *kinêseis* (plural) which form the offspring's particular parts and make them resemble the parents and/or ancestors. This is problematic in so far as the male *hê archê kineseôs* – the start of the substantial change in *GA* I and II – is related to the form (*eidos*) in generation and the *kinêseis* that come from the mother are not. As I argued earlier, the *kinêseis* Aristotle brings to the fore in *GA* IV are not the same as the singular male *kinêsis* which helps to explain why the female contribution contains *kinêseis* but cannot be *hê archê kinêseos*.

10

TELEOLOGY AND NECESSITY IN THE *GENERATION OF ANIMALS*

10.0 Introduction

This final chapter completes an analysis of key passages from all five Books of the treatise on the *Generation of Animals. GA* III details generation in lower animals, *GA* IV and V the situations that arise in particular instances of generation. I have already discussed two sets of explanations of the second sort, that of sexual differentiation and that of hereditary resemblance. This chapter covers three other topics from *GA* IV–V: infertility or sterility, deformity and incidental differences in animals' parts. The content of this portion of the treatise provides important connections to Aristotle's account of the female animal. It continues the more specific investigations, employing many of the same tools of explanation and analysis. This section further shows how open Aristotle is to considering the effects of surrounding elemental powers on living beings, quite beyond the scope of straightforward teleological explanations. The reader is expected by this stage in the work to understand that generative outcomes in specific instances depend on the *summetria* of male and female and that both generative residues can contribute *kinêseis*.

This chapter will begin with an overview of the debate concerning the role of material necessity in Aristotle's biology. It will then divide into two parts. The first will detail Aristotle's general explanation of infertility and deformity in terms of the materials. Both parental contributions contain *kinêseis*, which will shape the offspring depending on the state of the *summetria*. Aristotle strengthens and elaborates on this point as the most comprehensive and useful explanation, given the content of the *GA* as a whole. This also means that the causes of infertility and deformity cannot be assigned solely to the female principle and that it misrepresents Aristotle's theory to suppose this to be the case.

The second part will explore how materially necessary explanations become increasingly detached from teleological ones in this section of the work. A pattern of explanation which allows for regularities in materials apart from teleology emerges in certain instances. The role of material necessity in Aristotle's biology will be reassessed in this light.

This section concerns occurrences that are not good for the animal, such as deformity. Given that such happenings are out of line with anticipated natural regularities, one might think that Aristotle could say no more than that they are random. This is not the case; instead he takes some time to try to explain how and why they occur.[1] For Aristotle, even though monstrous births are, in one sense, against nature (*para phusin*), they are, in another sense, in accordance with nature (*kata phusin*).

A monstrosity, of course, belongs to that which is against nature (*para phusin*), although it is not against the nature of the whole (*to pan*) but only to nature in the generality of cases. So far as concerns the nature which is always and is by necessity, nothing occurs contrary to that – unnatural occurrences are found only among those things which occur as they do in the generality of cases, but which may occur otherwise . . . what occurs is against a particular order, certainly, but it never happens in a merely random fashion . . . even that which is against nature, is, in a way, in accordance with nature (*kata phusin*) (i.e. whenever the formal nature has not gained control over the material nature) (*GA* 770b9–18).

Material necessity is part of nature. In the whole cosmos (*to pan*), it is natural for hot, cold, wet and dry exchange in a certain regular manner. Living beings are also part of nature, but the regularities found in sublunary substances are only ever 'for the most part' (*epi to polu*), because of their corruptible constitution. It is natural for animals and plants to be born and die, and so it is natural for them to contain within themselves some source of irregularity, some possibility that things will not always be the way they are now because they 'may occur otherwise'. Since sublunary existence is transient, corruptible and prone to disorder, this is part of its

[1] Johnson (2005) acknowledges that Aristotle had a teratology, that is, an account of *terata* – monsters – since they have 'their own regularity'. 'Notice that both teleological and necessitating factors must be invoked in the explanation of the freak, since it is only with reference to ends, and the normal development of the organism, that the differences in the developmental processes, and how they went wrong, can be understood' (199).

nature. Thus, animals are uniform only up to a certain point. They also differ from other individuals in their kind and, although in the majority of cases they will have all the features of the kind, there will be exceptions and so these exceptions are part of what it is to be a natural sublunary entity. Aristotle is clearly interested here in the way in which the elemental powers are natural and play some role in explaining monsters. Thus, he will turn to accounts in terms of material necessity (MN) apart from teleology. Deformity is detrimental to functioning but one can account for it by looking to patterns in the natural materials, and need not simply attribute it to inexplicable chance.

In his explanations of certain sorts of untoward results in generation, Aristotle does not always move straight to providing explanations in terms of MN apart from teleology. Often tangential reference to teleology remains as in his discussion of certain sorts of animal that are prone to disorder or deformity (see Chapter 10.4). Another technique Aristotle employs is to locate a material that exists naturally in an animal, meeting certain requirements for its life, but which has other non-telic effects. Some animals are more prone to such deformities and differences because of certain facts about them. These facts don't necessarily fit any expectations of the relative normative order of kinds – bees are low down but intelligent due to special facts about them (*GA* 760a5–6). Mules are in the middle of the scale but are 'cold' and 'infertile' (*GA* 748a25); humans are the paradigm but suffer from baldness and greying to a greater extent than any other animal (*GA* 785a7–14). Eventually all connections to teleological modes of explanation fail and only the elemental powers are left to explain such occurrences; and so natural regularities occur in the materials themselves. The fact that MN explanations without any direct connections to teleology are prominent in the latter half of the *GA* forces a reassessment of the post-Enlightenment criticism that Aristotle was not interested in material happening without constant reference to forms in nature.[2] All these sorts of explanation show how important it is to read through the entirety of the text in order to

[2] See Theophrastus *Metaph.* 11a1–3.

build up a sense of his explanations in this context and how they relate to teleology.

10.1 Explanations in terms of material necessity in Aristotle's biology

Aristotle's explanations of sexual differentiation, hereditary resemblances, infertility and sterility, deformity and the development of parts incidental to essence all lean heavily on material necessity. The relative neglect of these topics may owe something to a certain discomfort with finding this to be the case.[3] Commentators can be resistant to the idea that material necessity ever provides explanations independent of teleology for Aristotle.[4] The early books of the popular work, the *Physics*, state that all causes and explanations, indeed any regularity in nature, must be due to final causes (*Ph.* II.8). Furthermore, Aristotle's *Analytics* gives us reason to believe that a cause which is not linked to essence (i.e. goals of a kind) will not be properly explanatory. In light of these passages, it is crucial, before considering this portion of the *GA* in more detail, to show how it is that systematic explanations not based on *telei* can fit with Aristotle's philosophy.

There are at least three broad senses of the term 'necessity' in Aristotle's works.[5]

AN. Absolute necessity (in terms of the motion of the heavenly bodies) (*Met.* 1015a24; *PA* 639b23–4; *GA* 778b1; *Ph.* 199b34–5, 202b33; *PA* 639b24; *GC* 337b35).[6]

HN. Hypothetical necessity (relative to some goal) (*PA* 642a13, 639b25–31, *Ph.* II.9).

[3] The latter Books of the *GA*, particularly those portions not concerned with hereditary resemblance, have tended in the past to be ignored or dismissed as irrelevant. Needham (1959) 24: '[T]he fourth book contains a collection of minor items Aristotle has not been able to speak of before ... the work as a whole trails off in a rather unsatisfactory manner'. Morsink (1982) 148 dismisses *GA* V as 'more or less an appendix', while Barthélemy Saint-Hilaire in Aristotle (1887) vol. I, cclix argued for its removal from the *GA*. Gotthelf and Leunissen (2010), in contrast, argue that *GA* V should not be treated as 'a mere appendix' (326).

[4] See, for instance, Scharle (2008). On Aristotle's teleology, see Chapter 7.1.

[5] Interpretations of key passages are all much disputed. See Gotthelf (1976), Cooper (1987) and Johnson (2005) ch. 2.

[6] Some suggest this also subsumes MN below, for example, Cooper (1987) 260.

MN. Material (or pre-Socratic) necessity (i.e. elemental powers and their interactions without reference to goals) (*PA* 642a3–7; *Ph.* 198b12–14; *GA* 789a8-b8; *A. Po.* 94b27–33).[7]

If, as some would have it, Aristotle subsumes all necessary interactions of elemental powers under teleology then materials within animals would always be hypothetically necessary. MN would play no role in his explanations but instead only represent the stance of thinkers with whom he disagrees.[8] This interpretation can emphasize that Aristotle often presents MN as a bad explanation, exhorting his readers to refrain from employing it. In *GA* V, for instance, we can find him contrasting his own type of explanation with that of Democritus.

If animals are to take in nourishment after they have been weaned, it is necessary that they have instruments (*organa*) for working on the food – and so if this happened, as he [Democritus] says, about the time of maturity, nature would have left incomplete something it could have achieved, and the function of nature would happen contrary to nature. For 'by force' is contrary to nature and he says that the genesis of teeth happens by force (788b23–9).

This complaint against Democritus suggests to some that there is nothing for Aristotle that fills the role of MN or the elemental powers acting on their own.[9] In so far as teeth are part of the essence of a kind, they are not caused merely 'by force' or MN. However, Aristotle's further explanation of the flaw in Democritus' approach does not indicate that this is the case.

Democritus, however, omitted to give an account in terms of the final cause and so he refers all things that nature employs to necessity. It is of course true that they are determined by necessity, but at the same time they are for the sake of some purpose, and for the sake of that which is better in each case. And so there is nothing to prevent the teeth being formed and being shed in the way he says; but it is not on that account that it happens, but on account of the final cause (*GA* 789b3–7).

'There is nothing to prevent the teeth being formed and being shed in the way [Democritus] says', that is, due to MN separate from the

7 Leunissen's three types of necessity correspond to these listed in the following way. (1) necessity simpliciter = MN, (2) conditional necessity = HN and (3) unqualified necessity = AN. See Leunissen (2010) 103.

8 Balme in Aristotle (1972), Kullmann (1974) 333 and Preus (1975) ch.4.

9 Sedley (2007) 183 '[Aristotle] considers *natural* purpose to be omnipresent in that world's structure all the way down . . . .'

goals achieved. Indeed, 'it is of course true,' he states, 'that they are determined by necessity'. Aristotle thus leaves it open that MN is a sort of cause or explanation of the formation of teeth.[10]

Another defence of the idea that Aristotle himself never employs MN in the *GA* or elsewhere requires finding that all reference to necessity are references to HN, and thus underpinned by teleology. HN is necessity that refers to and depends on a given goal one wishes to achieve (*Ph.* II.9; *PA* 639b21–640a10, 642a1–13, 31-b4). So, if one wishes to make an axe that splits wood well, then one has to use something hard such as iron. The iron is the hypothetically necessary material.[11] There is some dispute about what constitutes a hypothetically necessary material in nature as opposed to the craft example given. Some hold that organs, such as the heart or eye, are themselves hypothetically necessary.[12] Clearly, it is the case that in order to see, an animal needs an eye, but this does not seem to be a HN. Rather, it is part of the essence of the animal and necessary, in a fourth sense; of the necessity of one thing following from another in a definition.[13] The necessity for an eye flows directly from the human essence (*PA* 640a33-b1).[14] HN emerges when it comes to nature selecting (so to speak) what to make the eye from. Like hardness in the case of the saw, fluidity is required in order for the eye to serve the

[10] In this case, these results are (or happen to be?) also 'for the sake of something'. See Chapter 10.4.

[11] *PA* 642a9–13: 'And this is, as it were, hypothetical necessary; for just as, since the axe must split, it is a necessity that it be hard, and if hard, then made of bronze or iron, so too since the body is an instrument (for each of the parts is for the sake of something, and likewise also the whole), it is therefore a necessity that it be of such a character and constituted from such things, if that is to be.'

[12] For a summary of the controversy see Johnson (2005) 191–2.

[13] There is a fifth sense of necessity as 'forced' against a thing's natural tendency, for example, throwing a stone upwards (*A. Po.* 95a1–3; *Metaph.* 1015a26–33; *EN* 1110a1–3). Leunissen (2010) terms this 'enforced necessity' (103). This sense is not relevant in biology (*PA* 642a4–6).

[14] Cooper (1987) 254. Another example might be that the point at which parturition occurs is connected to the overall goals of the kind. Aristotle would reject contemporary explanations based on contingent factors. One such theory posits that offspring are born when nourishment in the womb becomes insufficient (*Nat. Puer.* 30, L VII 530–41). The fault with this view is that it cannot explain the regularity with which the length of pregnancy occurs in certain types of creature. What it is to be that animal – its essential nature – includes the gestation period (777a33–4, Cf. 769b25). However, Aristotle does note an exception to this general rule: the human kind whose gestation period varies (*GA* 772b8–10).

function of seeing (*Sens.* 438b7–8, 438a13–16; *GA* 779b23–5; *PA* 656b3). Most often in nature, HN is the elemental composition best suited to serve the ends in question.

Within the *GA* HN plays a significant role in Aristotle's account of body parts and their development, although he does not explicitly label it as such. Prevalent in the *GA* and other biological works are parts that develop due to MN, parts whose coming to be does not flow directly from the essence of the kind. These parts are not, then, hypothetically necessary, but they could be connected to final causes in some way. Consider those parts and events which Aristotle says occur both due to 'necessity' *and* due to final causes. Examples include respiration (*PA* 642a31-b2), the formation of cephalopod ink (679a25–30), deer horns (663b12–14), the epiglottis (665a6–10), hair on the human head (658b2–8), feet (694b5–9), lard about kidneys (672a12–15), female births (*GA* 767b8–15), mammaries (776b31–2) and the drawing up of semen by the uterus (739b9–15). Examples abound in the *GA*, in the earlier Books as well as the later ones.[15] Let us look more closely at a specific passage in which Aristotle describes the formation of flesh and bone in the embryo.

> Heat and cooling are both employed by nature. Each has the *dunamis*, grounded in necessity, of making one thing into this and another thing into that; but in the case of the forming of the embryo it is for a purpose that their power of heating and cooling is exerted and that each of the parts is formed, flesh made soft – as heating and cooling make it such, partly owing to necessity, partly for a purpose – sinew solid and elastic, bone solid and brittle. Skin is formed as the flesh solidifies, just as scum forms on boiled liquids. Its formation is due not merely to its being on the outside, but also to the fact that glutinous substance remains on the surface because it cannot evaporate. In blooded animals the glutinous substance is more fatty than in bloodless in which it is dry, and on this account the outer parts of the latter are testaceous or crustaceous ....all these things are formed partly as a result of necessity, partly also not of necessity but for a purpose (*GA* 743a36-b18).

One might worry that this description is incoherent. How could it be that Aristotle is entitled to say that each of MN and teleology

[15] In the case of some oviparous kinds (e.g. fish who lay an imperfect egg), he explains that eggs grow after they exit the womb due to necessity – because of the presence of *pneuma*, which expands and for the best, because there is no room inside the uterus for them to grow (*GA* 755a31–3). See also Aristotle's description of lactation (*GA* 776b29–35).

only partially make up one complete explanation? After all, the two explanations are often presented by him as exclusive. Giving them both then over-determines the cause. Respiration, for instance, ought to either happen because it aims to aid an animal's survival or due to the hot and wet conditions of the materials (*PA* 642a31-b2). Some argue that, in fact, Aristotle's material account in these instances constitutes a complete and comprehensive explanation, which then means the teleological account merely represents an attractive way to present research.[16] Others urge that although MN does give an account of the processes involved, it is not adequately explanatory in the Aristotelian sense because it does not explain why they happen.[17] On this view X is both necessary and for the best, as the passages proclaim, in so far as this is yet another instance in which soul uses MN as a tool, only this time to produce a part that isn't strictly speaking essential, that is, for survival, but that is a subsidiary or luxury one. Aristotle describes such parts in his *PA*.

> And it must be supposed that these variations either have some reference to the activities of the creatures and to their essential nature, or else bring them some advantage or disadvantage. For example, the eyes of some creatures are hard in substance, of others, fluid; some have eyelids, others do not. In both cases the difference is for the sake of greater accuracy of vision (*PA* 648a15–19).

In certain cases, a hypothetically necessary feature might require further additional protection. So for instance, since eyes are fluid, if the animal in question is terrestrial, then it will need eyelids (*PA* 657a31–6). The presence of eyelids is not, strictly speaking, part of the essence of an animal with eyes. Instead, eyelids are due to necessity in the sense that fluid has a certain nature which makes it vulnerable to long exposure to the air

[16] For this rather modernizing view of Aristotle see, especially, Wieland (1962) and Nussbaum (1978) 59–99. Many associate this reading with Kant's 'as if' teleology. See Johnson (2005) 182–4.

[17] A full defence of either position requires much more time and space than is available here. See, for example, Cooper (1987) 264: 'Democritean necessity does indeed exist and has a role to play in the formation of living things, but … where it does make a contribution it only does so because the materials whose necessary action is in question are themselves necessitated hypothetically. On this interpretation … Aristotle "subsumes" Democritean necessity under hypothetical necessity, in the explanation of living things.'

(which surrounds an animal's eye). This air would dry the eye out if an eyelid did not develop to protect it. So we might say that this sort of thing is indirectly connected to the animal's essence. The embryology at *GA* II describes how the best nourishment goes to form the essential parts, that is, those required by essence. The leftovers from that stage of development are used by the animal's nature for non-necessary parts (such as eyelids), which develop later (*GA* 744b12–26).[18] In these cases, then, explanations in terms of MN and teleology are neither incompatible nor over-determined. The MN explanation does stand apart; the materials interact without being initially goal-directed. The results of their material interactions can be linked to teleology because they end up being used by the animal's nature or soul to bring the creature some advantage. This means that, rather than being non-teleological, explanations in terms of MN are here indirectly teleological; they form a 'secondary' teleology.[19]

Explicit mention is made of materially necessary explanations in *GA* V.

> In some cases [what we will discuss] has nothing to do with the account (*logos*) of the animal's being; instead of that, we are to assume that these things come to be by necessity, and so their causes must be referred back to the matter and to the source which initiated their *kinêsis*. Remember what was said at the beginning, at the outset of our discussion.[20] ... [These] characteristics are formed during the process which do not fall under the heading just given, and the cause of them is to be looked for in the *kinesis*, i.e., the process of formation – we must assume that they acquire their differences within the actual process of construction (*GA* 778a34-b16).

A recent interpretation of the above text proposes that whereas strong teleological modes of explanation were most prominent in *GA* I-IV, non-teleological ones come to the forefront in *GA* V.[21] On

[18] Leunissen (2010) 89–95 divides these into 'subsidiary' and 'luxury' parts. Nature is like a 'good housekeeper' (*GA* 744b15–16) in so far as it makes use of leftovers to make a beneficial part, not strictly required by form. These parts which come to be last in the animal (*GA* 734a16–30) are hypothetically necessary in a more restricted and qualified way. See also Gotthelf and Leunisen (2010), 341–5.

[19] A phrase coined by Leunissen (2010).

[20] The reference here is to *PA* 640a10–12, establishing both another key link between the two treatises and also that *GA* V is no mere adjunct but a part of the whole project of providing an account of all or most aspects of animal generation.

[21] See Gotthelf and Leunissen (2010) 335.

this reading, Aristotle is seen to use non-teleological modes of explanation to detail inessential differences in parts of the body which develop after birth due to aging, disease, environment and food consumed.[22] These parts are taken to be connected to teleology in the 'secondary' or indirect manner described above.

The idea that Aristotle's explanations in terms of MN in the *GA* constitute a 'secondary' teleology has many advantages as an interpretation, first in terms of placing Aristotle's interests within a teleological framework, and thus providing interpretative continuity with his *oeuvre* more generally. It also usefully links the discussion in *GA* V to Aristotle's description of embryological development in *GA* II, and the differentiation he makes there between essential and subsidiary parts (744b11–28). However, these insights come from restrictively focusing on only one portion of *GA* V. Rather than non-teleological modes being exclusively employed in *GA* V, they already take over as the favoured mode of explanation in *GA* IV, particularly when it comes to accounting for deformity.[23] Also, indirect links to teleological explanation do not always follow the above pattern whereby nature uses external matter to develop subsidiary and, especially, luxury parts after birth; in fact, they are broader. Close readings of these texts reveal other types of explanations of processes and events surrounding generation in *GA* IV and V. Even in the passages in which necessity is coupled with the 'for the sake of which', subsidiary parts are not always indicated. Consider, for example, the occurrence of menstruation in women.

> Thus the production of this residue by females is, on the one hand, the result of necessity, and the reasons have been given: the female system cannot effect concoction, and therefore of necessity residue must be formed not only from the useless nourishment, but also in the blood-vessels, and when there is a full complement of it in those very fine blood-vessels, it must overflow. On the other hand, in order to serve the better purpose, the end, nature diverts it to this place and employs it there for the sake of generation, in order that it may become another creature of the same kind as it would have become, since even as it is, it is potentially the same in character as the body whose secretion it is (*GA* 738a33-b4).

[22] *Ibid.*, 338. [23] This will be detailed in sections 10.4–5.

Here menstrual blood comes from a MN process which is not goal-directed. It does not get used by the individual animal's nature for its benefit like a subsidiary part, but is instead employed to benefit the kind.[24] Furthermore, this passage shows that beneficial results are not always development of body parts, but include the production of residues.[25]

In *GA* IV and V Aristotle is also interested in cases in which it is natural for body parts to form in particular ways, although the results do not aim at the ends of the kind and cannot be linked to goals. As Aristotle states in his *Parts of Animals*,

> True, nature sometimes uses even products to advantage, but this does not justify our seeking a final cause in all – but while some things exist for the sake of an end, many other things necessarily come about too because of them (677a17–19).

There is space for explanations in terms of MN to come apart from teleology. But one must be careful about terminology in setting these out, since it is not appropriate to characterize such explanations as 'efficient causes'. Efficient causation is a technical way to characterize initiating forces within a teleological framework – marking them as intrinsic or natural. For Aristotle, strictly speaking, there is no 'efficient' cause when motive and material factors are separate from goals. An 'accidental' efficient cause, for instance, the confectioner who makes a healthy snack (*Metaph.* 1027a2–7), is a cause in a sense but is not properly the efficient cause.[26] This explains how some events and processes look like they have goals, that is, they appear similar to those that are the result of teleology, but in actual fact, do not. For instance, crops that get wet due to unseasonal rain are not ruined for a reason, that is, due to a God's malicious intention (Zeus' wrath) (*Ph.* II.8). Similarly, crops grow in the summertime not because inanimate elemental powers aim for this to occur, as if there were a benevolent intention in the natural world, but rather because, in this

[24] There is no benefit to the individual, which the account in terms of 'secondary teleology' requires. Leunissen (2010) speculates that '[e]ven though the presence of female menses is necessary for reproduction, Aristotle treats these menses as subsidiary parts, perhaps because they play no *active* role in reproduction ... ' (91). Leunissen considers the menses to be a subsidiary part but I would contest this.

[25] Semen, and other residues, are not parts (strictly speaking) of the body (*GA* 724b29).

[26] See Chapter 6.5.

instance, humans use the elemental powers for their own purposes. Here, the efficient cause is not located in the elements themselves apart from the bigger picture of how they are used by intentional agents. Without the backdrop of human purposes, the summer rain is merely 'accidentally' an efficient cause, and really does not count as one at all. One can understand the same pattern holding for forms and natures in Aristotle which use elemental powers to aid their ends.[27] But in *GA* IV and V this positive feature of natural elemental forces is not always present. Instead, accidental efficient causes (which aren't really efficient causes at all) can clash with the animal's aims, like crops that spoil on the threshing floor (*Ph.* 198b22). In this way, the elemental powers and their interactions can at times stand on their own outside the framework of teleology.

Even in these instances, Aristotle usually attempts to link the two together. If the animate goals of (i) survival, (ii) flourishing and (iii) reproduction[28] are impeded, then an animal can:

(i)¹ fail to survive (i.e. early or untimely death)
(ii)¹ fail to flourish (i.e. impairment of the senses)
(iii)¹ fail to generate another like itself (i.e. sterility, failure to complete the process of generation, and the production of deformed offspring).

One can characterize the effects of non-telic material impulses as either:

(1) Beneficial
(2) Detrimental or
(3) Neutral (with respect to these goals).

Cases of (1) have already been discussed; these are parts or residues that can either benefit the individual or the kind. Other differences are detrimental (2) to animate goals – the main instances being those of infertility, sterility [(iii)¹] and deformity

[27] This account of *Ph.* II.8 follows Leunissen (2010) 31: '[T]he coming to be of crops is due to the goal-directed actions of human beings who make use of the natural material potentials of regular winter rain (and seeds) and impose the art of agriculture on what is naturally available, in order to secure the regular coming to be of beneficial outcomes . . . Serving an anthropocentric goal is something that is imposed on natural processes from the outside, not something that is intrinsic to their internal natures.'
[28] See Table 7.1.

[(i)' and (ii)'].[29] These phenomena cannot be explained directly or indirectly as teleological. Instead, Aristotle accounts for them in terms of a combination of features that are essential, which interact and produce results due to MN and an animal's material nature. Aristotle's normative ordering of animal kinds, which he introduced in *GA* I and II, and further elaborated in *GA* III, forms an important part of the account of why and how such impediments come about. The hierarchy places some animals at a disadvantage in terms of their material make-up. Those animals that are weaker and colder are generally less able to use and control the material potentials in their bodies. However, the normal order is not always important; in many circumstances peculiar facts about the animal explain the phenomenon. Neutral results (3) include the variations in parts described in *GA* V which are a necessary part of the birth of any particular individual animal. Although these features are not part of the essence of an animal, in so far as they are accidental, they can be explained following a similar pattern to (2). Neither pattern of explanation covers all instances, some of which will be much more tangentially connected to teleology, while others will be due to some facts about materials existing within certain animals' bodies.

10.2 Explanations of infertility

Many of Aristotle's explanations of infertility rely on his normative ordering of kinds; better, hotter, stronger animals are more fertile. This extended sense does not necessarily indicate that anything unnatural or untoward is involved and so these explanations link together with teleological ones. Sterility, on the other hand, does not fit into that way of looking at matters; explanations of this phenomenon rely much more on MN apart from any teleological framework. I will detail each strategy in turn in order to illustrate this contrast.

[29] Aristotle is also interested in certain negative effects on (ii) flourishing in *GA* V. For instance, he discusses loss of hearing (781a34-b2) and impairment of sight (780a17–22, 780b1–3). A combination of conditions within the animal and influences from the climate help to account for these.

The extended sense of infertile is similar to the extended sense of deformed Aristotle uses elsewhere. Any fully functioning living being can be 'deformed' in an extended sense if it does not measure up to the best animal – the human male.[30] This sense of deformity can even be used to characterize whole species of animal. For instance, the lobster is 'deformed' in that it is a matter of chance which claw is bigger, and all quadrupeds are defective because they do not stand upright (*PA* 686a31-b6). Meanwhile fish are 'even more stunted' than birds for they have neither legs nor hands nor wings (*PA* 695b2–6).[31] Another example of this extended sense is Aristotle's characterization of the female, which he notoriously claims is 'as it were a deformed male'.[32] Females and non-human animals are deformed in the sense that they fail to achieve the highest end. One living goal which particularly interests Aristotle in the *GA* is that of generating to type (iii); those that cannot achieve this are, in a sense, deformed. Thus, infertility counts as a sort of deformity. This applies to animals that produce incomplete or unfinished offspring or offspring that are unlike parents.[33] They suffer from a minor sort of infertility, because they do achieve a complete copy eventually.[34] A normative hierarchy of kinds can be ordered in terms of generative ability (see Table 10.1).

Sterility is the failure to produce another living animal. Infertility, on the other hand, allows for the production of living offspring which are in some manner defective, or unlike the parent(s). Such imperfections can in a sense be natural, since they occur in all instances in certain types of animal, for example,

30 See Chapter 7.1 and Lloyd (1983) 26–43.

31 See also seals, *HA* 498a32-b4 and *IA* 714b10–12. Gotthelf and Leunissen (2010) 340.

32 *GA* 737a25–30, a passage cited often by those feminist scholars who emphasize Aristotle's role in propagating a 'one-sex' model. Irigaray (1985a), du Bois (1988), Allen (1985) and Saïd (1983) 102. The term he uses for 'deformed' here comes from the word *pêros* – which indicates an infirmity of the body, translated generally as 'disabled in limb, maimed'. This does not mean, crudely speaking, that the female lacks a penis. The presence of a penis is a sign not a cause of this heating capacity, as is the presence of a lung (*GA* 732b32). The cause is the heart. See Chapter 8.4.

33 '[T]he offspring that does not resemble its parents is already, in a way, a monstrosity' (767b7).

34 'Nature's rule is that the perfect offspring shall be produced by the more perfect sort of parent' (*GA* 733a2–3). See also *de An.* 415a26-b7; *Metaph.* 1033b32–1034a2. The constant production of females counts, in one way, as this type of infertility since the female offspring is unlike its male parent (*GA* 766b27–767a34, *Pol.* 1335a12–24).

Table 10.1 *Hierarchy of generative success according to Aristotle*

Human male
Human female
Other large live-bearing animals (cloven-hoofed) [bear a single offspring]
Medium sized live-bearing animals (solid-hoofed) [bear few offspring]
Small live-bearing animals (many-toed) [bear many offspring]
Selachians: internally egg-layers, externally live-bearers
Egg-layers with hard shelled, double-coloured eggs
Those that do not need to sit on eggs (i.e. quadrupeds)
Those that need to sit on eggs (i.e. birds)
Egg-layers with soft shelled, single coloured eggs
Larvae-producers
——— (below this line, animals are literally deformed)
Fertile hybrids
Infertile hybrids
Spontaneously generated animals

those that produce larvae. There is little to differentiate larvae producers and fertile hybrids in terms of fertility since both produce offspring but neither generate another like themselves. The larvae-producers are superior, however, because their offspring eventually become like them, after the animal emerges from the pupae. Fertile hybrids almost always produce offspring that are part one and part another animal kind, and so will always be unlike either parent. Those animals below fertile hybrids are unable to produce any offspring whatsoever, like or unlike, and so are sterile. Aristotle explains sterility in the same way he explains particular instances of generative failure, which can affect any animal at any time – in terms of MN.

In a particular instance, in order for generation to be achieved there must be *summetria*.

Male and female . . . must stand in a right proportional relationship (*summetrias*) to one another, since everything that is formed either by art or by nature exists in virtue of some due proportion. Now if the hot is too powerful it dries up fluid things; if it is very deficient it fails to make them set; what it must have in relation to the object which is being fashioned, is the mean proportional (*GA* 767a16–20, Cf. 772a10–21, 743a27–36).

When attempted conception fails, the correct *summetria* may not happen for a variety of reasons: (1) deformed generative parts (e.g. blocked passages *GA* 747a12), (2) old age (which is a dry and cool condition) (3) youth (presumably too hot and wet), (4) fatness and (5) being diseased (746b20–32). Alternatively, it could be due to the heat or cold in the surrounding environs, which disrupts the balance of the initial mixture (*GA* 767a15–35).[35] None of these explanations are connected in any systematic manner to teleology. Animals that are congenitally sterile are not necessarily low down on the scale of kinds.[36] What unites them, along with those that are infertile due to old age, youth, disease or injury, is a lack of the necessary material preconditions within their bodies: usually they are too fluid and/or cold. What happens in the actual process of construction of a particular animal will be crucial, highlighting yet again the centrality of *summetria* to Aristotle's theory. Similar considerations apply to accounts of literal deformities in offspring, as the next two sections show.

10.3 General explanations of deformity

MN is particularly important when it comes to Aristotelian explanations of monstrous births.[37]

As for monsters, they are not necessary so far as the purposive or final cause is concerned, yet *per accidens* they are necessary, since we must take it that their origin at any rate is located here (*GA* 767b13–15).[38]

35 *Summetria* includes ideal seasonal factors (*GA* 777b25). One example of this effect is that when it is cold and the wind blows from the east, female offspring are engendered (*HA* 573b32–574a1). Because of this, Aristotle seriously considers the idea that animals ought to copulate facing a certain direction (*GA* 767a9–14). See also *Politics* VII.16, where we are encouraged to study medical manuals in order to determine the best time of year for conception.

36 Excepting testacea, which are peculiar in other ways. See Chapter 7.4.

37 Aristotle here criticizes those who believe that deformed people are part animal (769b14–18). Such views persisted well into the early modern period. See Bates (2005). A hybrid is impossible, says Aristotle, if the animals in question have different gestational periods (746a32–3).

38 In other words, the material necessity. See *GA* 770a31: 'the cause of such things (monsters) is in the material'.

By noting that embryos can become damaged *in utero* Aristotle accepts some role for a cruder account of birth defects, similar to the Hippocratic constriction hypothesis. When the womb is too narrow, the child cannot grow properly, just like a cucumber that is growing in a confined place or a tree that must twist around a stone.[39] If a strong child is produced by strong parents, this is said to be fitting or appropriate (*epieikes*), whereas, if a weak child is constricted in the womb (of the strong parent), this is said to be merely 'necessary' (*anagkê, Genit.* 9.2, L VII 482–3).[40] Aristotle agrees that 'necessity', that is, MN, explains deformity in such cases. If he is now in the business of providing non-telic explanations, one might wonder how his approach differs from that of his usual opponents.[41] Aristotle emphasizes the continuity of his account. The account of deformity comes directly after that of hereditary resemblance (*GA* 769b3–7), which had followed on from sexual differentiation.[42] He was no doubt aware of other works which connected heredity to deformity, most of which merely claim that deformity is inherited. He agrees that in some instances, acquired characteristics, including deformities, can be passed on to the next generation (*GA* 721b28, *HA* 585b28–34).[43] However, he does not regard this explanation as comprehensive (*GA* 724a4–6; *Aër* 14). In fact, most instances of deformity are not passed down the generations. Instead there are intricate MN causes that lie behind them. There is a sense, however, in which the material processes that lead to resemblances to relatives are important here.

Normally, offspring resemble some relative; when this does not occur, deformity has begun to take hold.

39 *Genit.* 9–11; See Hippocrates (1981) 139–46.

40 If the womb is too open, nourishment will escape (*Mul.* I.13, L VIII 50–3); this is also an 'unnatural' condition (*para phusin; Mul.* I.25, L VIII 64–9).

41 Aristotle agrees in principle with the Pre-Socratic project of demystifying the occurrence of monstrous or deformed animals and humans, something that was treated as portenteous throughout early history. See in particular his discussion of Democritean theory at *GA* 769b31–4.

42 'And indeed this is what comes next to be treated after what we have already dealt with [i.e. heredity] – the causes of monsters (*terata*) ... ' (*GA* 769b11–12). All these particularities are closely allied because they can all be related back to the state of the mixture of male and female residues and the circumstances in which conception and early development occur.

43 This would be termed Lamarkian inheritance in modern biology.

> Some [offspring] are like none of their family, but nonetheless are like some human; some do not resemble a human as regards the way they look but already resemble a monster. For, the offspring that does not resemble its parents is already, in a way, a monstrosity, since in these cases nature has in a way strayed (*parekbebêke*) from the kind (*genos*). The first stage of this is when a female comes to be rather than a male (*GA* 767b6–9).

In the passage, Aristotle indicates that there are gradual steps away from a smooth instance of generation which will eventually lead to deformity. The smoothest instance is when the offspring is male and shaped like the father (the mixture is *eupeptos* – i.e. well concocted [*GA* 767b16]). The steps towards deformity would appear to look like this:

1. If the offspring is not going to be male, then it will be female (*GA* 766b15–16)
2. If the offspring is not going to be like the father, then it will probably be like the mother (768a35)
3. If the offspring is not going to be like the father or the mother, then it will probably be like some of the ancestors (768a28–768b10)
4. If the offspring is not going to be like any ancestor, then it will probably be like other humans (768b10–13, 767b4–5)
5. If the offspring is not going to be like other humans, then it is a monster and a failure of generation (767b5–7).

Several issues of relevance to the female in Aristotle's theory occur to one immediately. First of all, the reader might think that the first stage towards deformity is for offspring to be female. The second thought might be that it is due to the female matter and its recalcitrance that all failures of generation, including deformities, occur. Both of these ideas are misguided, as I will explain.

According to Aristotle's theory, Table 10.2 represents the possible variations that can occur in any viable instance of generation within a kind. At the very limits of the first scale are those males that only resemble someone in the kind, unrelated to them. Nearest to these are those offspring that resemble maternal great grandparents, then paternal great grandparents, then maternal grandparents, then paternal grandparents and finally the female parent (in between the asterisks). The ideal is at the centre of the two uneven extremes: a male resembling the male parent. At the very limits of the second scale are females who merely resemble some other animal in their kind. Deformed offspring, properly

Table 10.2 *Possible generative outcomes in terms of hereditary resemblance*

Key: R = resembling; a = male (*arren*); t = female (*thelu*); p = pater/nal; m = mater/nal; g = grandparent; gg = great-grandparent; k = any animal in the kind.

aRk – aRmgg – aRmg – aRm – *aRp* – aRpg – aRpgg – aRk

\ /

V

tRk – tRpgg – tRpg – tRp – Female offspring – – tRm – tRmg – tRmgg – tRk

speaking those that are nonfunctional and defective, fall off the ends not of the small scale, but of either scale – a male *or* a female that does not come to be a member of its kind, and so resembles only an animal. Thus it is misguided to imagine that, despite the awkward expression at *GA* 767b9 ('the first stage is when a female comes to be') (1) is a necessary step before (2)–(5). After all, males can, of course, fail to resemble parents and relatives and so reach (4). An offspring is deformed 'in a way' when it does not resemble any of its family members (*GA* 767b4–7) (4).[44] Furthermore, males can be deformed in a literal sense and reach (5). In fact, being male predisposes an animal to some types of deformity since male embryos move around more *in utero* (*GA* 775a9).[45]

The key to Aristotle's account of deformity is not found in sexual differentiation, and the initial birth of a female, but rather in the processes that result in hereditary resemblance (*GA* 769b11–18). Aristotle takes pains to separate sexual differentiation and heredity. In each individual there is a *dunamis* for their sex which is separate from the *dunamis* for them as an individual parent, which determines whether offspring resemble them

44 Possibly, this would occur if the *kinêsis* went any further back than three generations. It is indeed unusual to find that someone resembles their great great grand-parents, and if they did, in an era before photographic technology, there would have been no way to verify this.

45 Aristotle may have heard from medical authorities that stillborns are more likely to be male than female. This has subsequently been proven to be the case. See Lowe and Record (1951) 34.

(*GA* 767b23–6, 768a6–7).[46] When the *kinêseis* of the parents relapse or loosen, those of grandparents and great-grandparents can take over and complete the job of shaping the parts (in their own particular manner). At a certain point, though, no effective *kinêseis* will be left, and offspring will be undifferentiated and misshapen (*GA* 768b33–5). When Aristotle states that the first stage of this process is when a female comes to be, he cannot be suggesting that femaleness leads to deformity. Rather, he is reminding the reader of the ways in which generation in particular instances can vary. The occurrence of the variation of femaleness reminds us that the less extreme variations in parts (which result in resemblances) can go to extremes.

Other interpretations do not recognize that deformity is confined to cases that fall off either side of the above. Instead, it is assumed that, for Aristotle, if offspring are not male and like the father, something untoward has occurred, perhaps even counting as a degeneration or aberration.[47] The viability of this reading depends on how we understand the terms *parekbebêke* (*GA* 767b7, past perfect of *parekbaino*) and *existatai* (768a2, 15, 768b8, from *existathai*), which Aristotle uses of the processes whereby a female comes to be, when offspring resemble someone in the family other than the male parent and when deformity occurs (as in *GA* 767b8 above). Whether the former two types of alteration count as deviations or deformities as well as the case of deformity depends on whether the terms necessarily indicate that something disastrous has occurred.

Some of the meanings of the term *existathai* listed in LJS' *Greek Lexicon* include the following:

(1) displace, change
(2) drive out of one's senses
(3) digress (in an argument)
(4) cease from, abandon
(5) depart from type, degenerate from one's own nature (as part of a phrase).

[46] This is how a female like a male parent can occur; the *dunamis qua* male fails, but that *qua* individual parent succeeds (*GA* 768a28–31). '[C]hildren often resemble parents of the opposite sex' (*GA* 769b5).

[47] See Chapter 8.4.

The fifth meaning is most often assumed in Aristotle's description of sexual differentiation and heredity in the *GA*.[48] This meaning is clearly applicable in one instance in the text where semen cannot be a 'colliquescence' (*suntêgma*) because it is unnatural. Aristotle remarks there that 'everything that undergoes colliquescence gets destroyed and departs from (*existatai*) its proper nature (*tês phuseos*)' (*GA* 725a28). It is not necessarily the case that this meaning applies in other instances, however. Various meanings are apparent in the corpus. The fifth one appears at *Rh.* 1390b28–31, which explains that the well-born (*eugenes*) are those that have not 'degenerated from the family type (*genos*)'. The highly gifted can thus degenerate (*existatai*) into maniacs.[49] *Parekbaino* can have similar connotations when it means a violation or flaw, a departure from excellence (*GA* 771a12, *EN* 1160b20, *Pol.* 1275b1, 1282b12).[50] However, in other places the first meaning, that is, change, is more viable. This meaning does not carry the negative connotations of departing from type and being unnatural or bad. Take, for example, the occurrence of *parekbainô* in the *EN*, where it indicates that small differences do not render an action or feeling vicious. Virtue is hitting a mean (*mesos*) between two extremes, or rather, between excess and deficiency in feeling and action. For example, generosity involves acting and feeling generous, which lies between acting and feeling parsimonious and acting and feeling profligate (*EN* 1121a10–16).[51] However, it is not the case that only one single feeling or action would count; instead there are a series of these that exist within the bounds of virtue. Some will be nearer to parsimony and some nearer to profligacy, but all count as virtuous and therefore good. '[W]e do not blame one who diverges

48 For Peck (in Aristotle, 1942), the term means '[l]oses or alters its character; degenerates'. He writes 'The force of *existasthai* can be seen from the phrase *existêsi kai phtheirei tên phusin* (*Eth. Nic.* 1119a23); cf. *G. & C.* 323b28, *Phys.* 261a20 (*tês phuseôs, tês ousias, existasthai*), and 725a28 above.'

49 See also *HA* 488b19 (thorough-breeds do not degenerate from their type), Theophratus *CP* VI.7.5 on souring of wine and Plato *Rep.* 380d.

50 It can also mean 'deviant' in the moral sense of rule-breaking. See *Pol.* 1325b6.

51 Aristotle's account of this virtue is, in fact, much more complicated because 'the generous person will aim at what is fine in his giving and will give correctly; for he will give to the right people, the right amounts, at the right time, and all the other things that are implied by correct giving' (*EN* 1120a23–6). Virtue as a mean seems to be a starting point rather than the only way to elucidate the virtues. See especially Hursthouse (2006).

(*parekbainôn*) a little from the right course, whether on the side of the too much or the too little' (*EN* 1109b18–20; see also 1126a35). In the ethical context, it is only outside a certain range that actions and feelings become wrong and blameworthy. In particular instances of generation, it could well be that there are a number of results within the limits of the normal and natural which count as successful, including both male and female animals that resemble others in their kind. Within given limits, the results can be deemed good.

A passage from the *Politics* suggests this same meaning for *parekbainô* and its correlates. Here, when discussing extremes in democracies and oligarchies, Aristotle explains that, in parallel to them, 'a nose which deviates (*parekbebêke*) from perfect straightness by being either hooked or snub is still a fine nose and looks good (*kalos*)'. Only when the shape is taken to either extreme will the product not be good and in fact, cease to be a nose at all. This principle then applies to other parts of the body (*Pol.* 1309b23–30). Although Aristotle is here referring to the process by which an artist tries to recreate a body part, the description is strikingly similar to the way in which nature forms the parts of the embryo in the particular ways that make it come to resemble someone. Within limits, snub or hook noses are not only serviceable and thus successful noses, but also aesthetically pleasing. Only when the shape is far out of proportion does the process properly deviate and deformity result such that it would not be like a nose at all.

Good outcomes are the result of a balance of sorts, as is emphasized in the *GA* with regard to *summetria* and also in the ethics, where virtue is imagined to be a measured response.[52] There will, however, be variation within the normal range. When the terms *existathai* and *parekbainô* are applied to these variations one need not rush to translate as 'deviation', that is, a failure or fault. The variations can, however, be normatively ordered to a lesser degree in certain contexts. Aristotle's ontology allows that within the range of what counts as natural or fully functional members of

52 See Hursthouse (2006) on connection between what she terms biomedical *krasis* and virtue as mean (97–9, 109–11).

the kind, variations can be deemed better or worse, as with political constitutions, where oligarchy and democracy function fine but are deemed worse than the best constitution (*Pol.* 1309b20–35, *EN* 1160b20). In order to see which connotations apply to the terms when they are used of the birth of females and of offspring that do not resemble male parents, we must first consider the difference between literal and real deviation in the process of generation. In real deviations, a part such as a nose will not be a (functioning) nose, or parts will be swapped around (*GA* 773a8–9). If the deviation is just over the limits of normality, then the animal can still live but is surely monstrous.[53] This is a more extreme instance than the female condition, since females are functioning members of the kind.[54] However, the idea of the female as sub-functional is viable also in so far as Aristotle thinks that the female heart is, in a sense, defective because it cannot concoct male semen (*GA* 765b35, 766b16–17). Thus the female would be like the living deviant animal; not so impaired that its heart does not function but impaired enough to count as deformed ('in a way'). Thus, on Aristotle's view, the female can be considered as either normal or deviant.[55]

When *existathai* means to depart from type – or to 'degenerate from its own nature' this indicates unnatural failure. If the term is translated as 'departs from type'[56] then the connotations are negative, making it seem that something unnatural is occurring. Consider, for example, the usual translation of a passage where when Aristotle explains how a female occurs.

> Everything when it departs from type (*existatai*) passes not into any casual thing but into its own opposite. Thus, that which does not get mastered must of necessity depart from type (*existasthai*) and become the opposite (*GA* 768a2–5).

This way of rendering the passage may be misleadingly negative. Consider that the pattern Aristotle imagines for the forces that

[53] *GA* 771a12: 'Those which only deviate (*parekbainonta*) a little from nature commonly live; not so those which depart further . . . '

[54] In a sense, the female can be 'perfect' or complete as a functioning animal (*GA* 715a21, 737b11).

[55] Earlier on I concluded that it occupies the unusual position of being in one sense a success and in another sense a failure of generation, of being both disabled and differently abled. See Chapter 8.5.

[56] By, for instance, Peck in Aristotle (1942).

shape offspring to be male or female in generation (and to resemble family members) is that of alteration. In this type of change, one contrary exchanges for another. Male and female act as contraries; when one is diminished, the other takes over – the same substance becomes female rather than male (*GA* 766a21–5). When the proto-embryo deviates it is not into any chance thing but into its opposite, femaleness (*GA* 768a3–4). As Aristotle explains in his work *On Coming to Be and Passing Away*:

> [U]nless two things are contraries or made up of contraries, one cannot displace (*existêsi*) the other from its natural condition. But since only such things as possess contrariety or are themselves actual contraries – and not only chance things – are naturally adapted to be acted upon and to act, but agent and patient must be alike and identical in broader kind, but unlike and contrary in narrower kind (*GC* 323b28–34).[57]

Although development has been changed and diverted from maleness and then proceeds towards femaleness, the result is still natural; it is unnatural to males, but natural to females, or rather, natural to humankind or dogkind, the broader natural grouping to which both contraries belong. Thus, it would be legitimate to translate the above passage as:

> Everything when changed (*existatai*) passes not into any casual thing but into its own opposite. Thus, that which does not get mastered must of necessity change (*existasthai*) and become the opposite (*GA* 768a2–5).

When the male nature is diverted, that course of development is abandoned, but an offspring comes to be and a contrary power makes it so. This reasoning also applies when offspring do not end up resembling the male animal.

> Usually the natural course of events is that when that of the male gains mastery – and when *it is mastered* – it will do so both *qua* male and *qua* individual father, since the difference between the two is a small one ... (my emphasis) (*GA* 768a22–4, my emphasis).

57 It is best not to translate *genos* and *eidos* as genus and species as this wrongly imposes modern classificatory objectives onto Aristotle. In biology, the terms refer to various orderings of animals relative to the view and point of the inquiry at hand. See Pellegrin (1987) and Balme (1962b).

The most natural thing is not just when a male offspring resembles its male parent, when the male masters *qua* male and *qua* individual male parent, but also when *it is mastered* in both ways. When it is mastered, it is mastered by the *dunamis* of female *qua* female and most often simultaneously by the *dunamis* of the individual female parent, which is the opposite of the individual male parent (*GA* 768a8, 26). Although they are separate, there is a stickiness about these two propensities such that 'males usually resemble their fathers and females their mothers' (*GA* 768a25–6, Cf. 767b4). Thus, it is also most natural when a female comes to resemble the female parent.

In so far as maleness and femaleness are contraries, this does not mean that they are equally good outcomes. Contraries can be hierarchized – one being better than the other within a set (e.g. hot and cold). Despite this, it is still more accurate on Aristotle's theory not to take the meaning of *existasthai* to be 'deviation' in the sense that an offspring which is not male and like the father is deformed or unnatural. In generation, it is only outside the first set of absolute limits on normality that the result is a literal abomination (*GA* 771a12). There is no reason, then, to load these terms with negative connotations. Instead we ought to think of femaleness and resemblance to relatives other than the male as 'changes' from the straightest path, which still reach the destination or goal. The path is not optimal, but it still counts as natural and, even, as good.[58]

The preceding discussion has helped to clarify what counts as a *bona fide* deformity. Literal deformities occur when processes which normally bring about resemblance to parents and ancestors cross over a line. If we find out why these changes occur then we are closer to understanding how a natural occurrence can become unnatural at its extremes. Many regard the cause of all failures and deformities to be the female contribution. A polemical passage where Aristotle argues with Democritus constitutes the main source for this idea (*GA* 769b32-770a8). For Democritus, deformity occurs when two sets of (male) semen, disrupt each other – one arriving later than the other. Aristotle counters that this

[58] Peck thinks the word *existasthai* indicates degradation because of its use in a passage in the *EN* 1119a23. This passage uses the term, however, in a very specific way to mean 'going out of one's mind' – pain makes us go out of our mind (meaning 2).

explanation is too complicated. Rather than add this implausible situation with respect to the male semen, it is better to locate the cause in the materials.

Now if we are really obliged to refer the cause to the semen that comes from the male, then, I suppose these are the lines on which we should make our explanation; but from every point of view we ought preferably to hold that the seat of the cause is the material and in the foetations as they take shape (*GA* 770a4–8).

Reference to 'the material' that is the cause of deformity is unspecified here – it could indicate either the matter supplied by the female or the bodies of the new animals, a combination of the two generative residues and the environmental conditions.[59] To assume the former relies on the idea that the female material is recalcitrant and acts as an impediment to embryological development. But this does not accord with the full description of the female contribution we have encountered throughout the *GA* and also overlooks Aristotle's more detailed account of the causes of deformity.

The earlier chapters of *GA* IV have already made plain that both sexes' generative residues manifest *dunameis* relative to sex and resemblance to individual parts of parents and ancestors through *kinêseis* (768a10–21). As the generative residues interact, some of these *dunameis* are overpowered by others resulting in an offspring that is either male or female and resembles some members of his or her family in their parts. All of these *kinêseis* ensure that the parts form correctly and end up functional, once the process of development is complete. As Aristotle explains with reference to heredity, *kinêsis* in the parents' residues can loosen or relapse (*luomenôn*) resulting in the latent *kinêseis* of the ancestors coming to the surface. Ancestral *kinêseis* then take over the task of normal embryonic development. This relapse to ancestors occurs because the agent gets acted on by that upon which it acts (*GA* 768b16–21). In the case of deformity of the whole animal or in its parts, Aristotle posits that the ancestral *kinêseis* themselves are weakened. He explains that 'the cause why sometimes the offspring is human but resembles none of them [relatives] and why at

[59] Peck in Aristotle (1942) takes the material to be that 'supplied by the female' in a footnote (421).

other times it proceeds thus but the final offspring seems not to be human but only some animal, which is called a monstrosity' (*GA* 769b7–10) is as follows.

In the end, when the *kinêseis* relapse (*luomenôn*), and the matter has not been mastered (*ou kratoumenês*), the most general thing remains and this is animal (*GA* 769b11–13).

Deformity occurs when no more *dunameis* relating to any ancestor (and thus any member of this animal's kind) remain, making it impossible for the part or parts to be formed properly. Clearly, then, deformity is the result of the failure of both parents' powers (*dunameis*) along with those of their ancestors.

Peck's influential translation of *GA* 769a11–13 incorporates the assumption that deformity is caused by the recalcitrance of the female contribution, which is how he interprets *GA* 770a4–8 above. The *kinêseis* in this passage are thereby taken to be entirely from the male.[60] And so he adds the following brackets: 'in the end, when the *kinêseis* <that came from the male> relapse and the material <that comes from the female> does not get mastered, what remains is that which is most "general"' (*GA* 769b11–13).[61] The soundness of assuming that these *kinêseis* come solely from the male ought to be challenged by the fact that Aristotle appeals to the same explanatory principles here as in his account of heredity and sexual differentiation, which requires *kinêseis* coming from both the male and the female sides.[62] Although the idea that formal nature fails to control material nature in instances of deformity (*GA* 770b16–17) may tempt us to understand the female matter as resisting the active, formative male principle, this is not a viable interpretation.[63] In a successful instance of animal generation, the female matter does not resist formation of the

[60] *GA* 770a31: 'cause of things [i.e. deformity] is in the material.' Peck in Aristotle (1942): 'not in the [male] semen'.

[61] Peck in Aristotle (1942) 417. Some Platonists interpret this passage as saying that matter overpowers (*kratein*) form. See Giles of Rome in Hewson (1975).

[62] See Chapter 9.5.

[63] Explanations of congenital deformities in the medieval period very often assigned blame to the female parent. This cannot properly be traced to Aristotle, as some suggest (see for instance Miller 2010) who are perhaps lead to believe this due to Peck's note and other inaccurate translations of this portion of the *GA*.

new animal like itself – it assists.[64] In order that matter be undifferentiated to the extent of deformity, even the *kinêseis* in the female must be obscured.

Aristotle makes clear that in most instances, deformities occur due to untoward material factors. A resistance to being formed to resemble any member of the kind is due to the 'bulk and coldness of that which is being concocted and differentiated' – some combination of the matter comprising both male and female residues and the external weather conditions (*GA* 768b26–8).[65] As with female birth and hereditary resemblance, this phenomenon is to be explained by the state of the mixture or *summetria*. Some examples of deformities that are the result of factors completely apart from teleology include the malformation of bodily parts and the misplacement or exchange of parts. For example, when the liver and spleen exchange places, *kinêseis* are diverted thus changing the position of the material (*GA* 773a7–8). In these cases, the combined generative residues have been unable to establish the appropriate *kinêseis* – the *kinêseis* of both parents and all ancestors have been thwarted. In other types of case, Aristotle's explanations of deformity can be connected to teleology, but not in a way one might expect, as I will explain in the next sub-section.

10.4 Explanations of proneness to deformity

Towards the end of the *GA* Aristotle incorporates explanations in terms of material necessity in many different ways, allowing for some autonomy for MN. Aristotle's discussion of types of animal that are prone to produce deformed offspring (*GA* IV) and parallel

64 The resistance given by the female contribution in the case of heredity amounts merely to a resistance to being formed in a particular way: to resemble the father.

65 One might think that deformity can only be due to the failure of the male principle, since deformity is the absence of an essential feature and only the male could determine any formal features. Although the male principle is the formal cause, one cannot infer from this that the female and ancestors play no role in developing essential parts. Most parts, for example, nose or arm, are functional and required for the form, as well as being accidentally snub or long or whatever else. '[The offspring] is formed so as to possess the generic characteristics and at the same time comes to be a particular individual' (*GA* 767b35–6). '[A]ll particular individuals are accompanied by this [generic] characteristic.' (*GA* 768b13–14). These parts are formed by the *kinêseis* that come from parents and ancestors and are functional but also particular due, in some sense, to these shaping *kinêseis*.

points concerning inessential features (*GA* V), reveals another set of the explanations operative in specific instances of generation with emphasis yet again on the *summetria*. Here I will first consider proneness to deformity. Not only birds (*GA* 770a14–16), but also some quadrupeds, suffer from the conjoining of offspring. Some of these deformities are tied to features that only certain animals possess. Aristotle explains thus:

Deformity occurs often in the many-toed because they are multiparous and their offspring are incomplete, like the dog. For most of these are born blind . . . The way is prepared by nature *(proôdopoiêtai tē phusei)* for the generation of monstrosities as regards the offspring not being like the parents because of imperfection. And a monstrosity is in the class of things that are unlike [the parents] (*GA* 770a38–770b6).

The phrase 'the way is prepared by nature' *(proôdopoiêtai tê phusei)* is puzzling. Since deformity is unnatural or against nature, why and how could nature prepare for it? The comment takes us back to the initial characterization of monsters as in some sense according to nature, as well as being against nature (*GA* 770b9–13). All animals, as sublunary living beings, incorporate the possibility of irregularity in their constitution as part of what they are. Furthermore, some are better at maintaining living goals than others, very generally those with compromised material constitutions. As we know, for Aristotle, animals come in degrees of perfection. One way for him to explain why lower animals are worse is that their material natures restrain the general goals sought by all living things (survival, generation and flourishing), that is, 'their offspring are incomplete' and 'the offspring' are not 'like the parents because of imperfection'. Some lower animals, for example, birds, and some middle of the scale animals, for example, dogs, are prone to a certain degree of infertility in so far as their offspring are always incomplete, in an extended sense. Birds produce eggs and the many-toed produce blind offspring with undifferentiated digits (*GA* 742a7–10). However, this relative infertility does not cause their proneness to certain birth defects – or at least not directly and so is not neatly tied to teleology. Instead 'deformity occurs . . . because they are multiparous'.[66] The

[66] '[T]his type of monstrosity hardly happens at all in monoparous animals, but rather in the multiparous and especially in birds . . . they lay many twin eggs, for the foetations grow together because of close proximity' (*GA* 770a7–16).

Aristotelian explanation of certain types of deformity, then, relies on a propensity of certain kinds to produce many offspring at once. In order to find out why it is that nature has 'prepared the way' for deformity in these kinds we need to investigate what sort of feature multiparity is.

Multiparity, or rather a distinction between multiparity (*polutokos*) and monoparity (*monotokos*), is first mentioned in *GA* II with reference to infertile hybrids (748a17). A general discussion of the phenomenon also occurs in *GA* III, covering fish, reptiles and insects (749b26–750a35, 753a27–35, 755a31–36).[67] The account in *GA* IV is more detailed and describes how a cluster of features tend to result in animals typically producing one or more offspring per birth.[68] These features include:

(1) The size of the animal's body.

> The reason for producing few and many offspring is the largeness or smallness of their bodies and not the fact that the type (*to genos*) is cloven-hoofed, solid-hoofed or many-toed (*GA* 771b6–9).

(2) The amount of generative fluid that the animal typically produces.

> Large animals are not multiparous, because nutriment is used up on the size of their bodies, but in smaller animals nature takes away from the size and adds the excess to the spermatic residue (*GA* 771a28–31).

(3) The degree of heat and *dunamis* in the generative fluid (*GA* 772a12).

The earlier account in *GA* I also connected body size to monoparity and multiparity. The generative products are residues of the blood that go to make up all the parts of the body. In certain small animals these residues are not used to construct its body or other possible parts of its anatomy, such as feathers or horns (*GA* 718b15, 725a32–4). In the case of small or 'naked' animals such as these, a lot of blood and residue can become generative and,

67 *GA* 749a14: the same statement will serve for them all; 753a1: 'we don't need to study reptile eggs, because they are just like bird eggs'; *GA* 761a2: wasps and hornets reproduce in the same way as bees, so we do not need to discuss them separately.

68 Initially, Aristotle postpones an investigation into multiparity writing 'the cause why this occurs and the cause why they are multiparous must be stated later' (*GA* 770b3–4). When he picks up the question again, it is with a view to finding out if there is any relevant connection between multiparity and deformities that involve redundant parts and monoparity and those that involve deficient parts (*GA* 771a14–16).

consequently, the animal will be multiparous. We might wonder: is an animal small because it is multiparous or multiparous because it is small? Sometimes Aristotle expresses it as if the former were the case (*GA* 749b26, 750a6, 771a28–31), which would make multiparity the cause of body size. However, this seems to allow a contingency to body size that sits ill with its being an essential feature of a kind. Size being such a feature is reinforced by the fact that a larger amount of generative residue does not result in a larger animal (*GA* 771b23–6).[69] Aristotle explains:[70]

> As animals are being completed there is a certain size of each – a limit of bigger and smaller. They would not become bigger or smaller than the size they have; as to excess or deficiency in size, it is in between maximum and minimum limits for them. So a human being or any other animal comes to be big or small, thus, also the spermatic matter from which it comes to be is not indeterminate but has a limit of bigger and smaller, so that the offspring cannot come to be from any amount of it whatsoever (*GA* 771b33–772a4).[71]

Since body size is essential, the abundance of residue is a kind of material necessity following on from essence. Nothing about the formal essence of the creature requires it to produce a great amount of generative residue (2). Incidentally, it turns out that if the generative residues of both males and females are more abundant, then the combined seed can be divided up into more offspring, particularly since these, like their parents, will be smallish in size.[72]

[69] See also *GA* 745a5–9: 'The bones continue growing only up to a certain point, for as all animals have a limit to their size, this involves a limit to the growth of the bones ... because the bones set the limit for an animal's size' and *de An.* 416a17: There is 'a limit and formula of both size and growth'.

[70] This constitutes a reply to the question at *GA* 771b23–26: 'what is the cause why it [greater amount of semen] does not make one large animal as fig juice acts with regard to the watery part of milk?' The fact that body size, along with gestational period, is part of essence helps to account for why hybrids are only possible if body size and gestational periods are very close (*GA* 746a33). Aristotle previously described generation as like the coagulation of milk into cheese at *GA* 729a12–14, 737a14–15 and 739b22–3.

[71] ἀλλ' ὥσπερ καὶ τελειομένων τῶν ζῴων ἔστιν ἑκάστου τι μέγεθος καὶ ἐπὶ τὸ μεῖζον καὶ ἐπὶ τὸ ἔλαττον, ὧν οὔτ' ἂν μεῖζον γένοιτο οὔτ' ἔλαττον, ἀλλ' ἐν τῷ μεταξὺ διαστήματι τοῦ μεγέθους λαμβάνουσι πρὸς ἄλληλα τὴν ὑπεροχὴν καὶ τὴν ἔλλειψιν, καὶ γίνεται μείζων ὁ δ' ἐλάττων ἄνθρωπος καὶ τῶν ἄλλων ζῴων ὁτιοῦν, οὕτω καὶ ἐξ ἧς γίνεται ὕλης σπερματικῆς, οὐκ ἔστιν ἀόριστος οὔτ' ἐπὶ τὸ πλεῖον οὔτ' ἐπὶ τὸ ἔλαττον, ὥστ' ἐξ ὁποσησοῦν γίνεσθαι τῷ πλήθει.

[72] Note that the excessive residues are not in this case used to make up 'subsidiary' or 'luxury' parts, although the pattern is in other ways similar to Leunissen's (2010) secondary teleology. The animal's nature uses material necessity to further generate to type, one of the goals of its kind.

In trying to discover connections between an animal type and multiparity, (1) body size and (2) abundance of generative residue are not sufficient. On its own, body size is neither a necessary nor a sufficient condition for multiparity. It is not sufficient because the copious amounts of generative residue that small animals are prone to produce (2) must have the correct qualities to be effectively generative (3). The residue is not merely abundant but must also have more *dunamis* in it (*GA* 772a8–10). It is not necessary because, as we find out later on, some large animals can be multiparous, although this is 'unnatural' (*GA* 772a35-772b1). Let's now consider feature (3), the degree of heat and *dunamis*, which seems so crucial to determining whether being multiparous is natural to a kind. In the more general account we have been told that the female contribution is bulky whereas the male one more dynamic (*GA* 726b31–3).[73] *GA* IV provides an account in somewhat similar terms.

> Since it seems there must be some proportional relationship (*summetria*) to one another – the residue of the female and that of the male (in those where males emit semen), in multiparous animals, the male immediately emits that which, when divided up, sets many offspring, and the female contributes an appropriate amount such that being set, many offspring are produced (772a17–22).

From these passages, one might suppose that a combination of abundant female residue and extra-dynamic male residue is required for multiparity. Males copulate often, while females emit a lot of matter (*hulê*) (*GA* 749b16). Unlike the craft analogy, however, the male contribution cannot simply shape whatever amount of matter it chances upon. A large amount of undynamic female residue will arrest development and result in its failure altogether (*GA* 766b37–767a2). Therefore, despite there having to be more matter to construct the bodies of numerous offspring in these cases, it is misleading to imagine that the necessary dynamism required for multiparity is restricted to the male contribution. Multiparity depends instead on the male and female residues sharing the same character:

[73] For males, 'the final product resulting from all that amount of nourishment is quite small' (*GA* 765b34). 'Though only a very small quantity of semen be emitted, the exhaustion which follows is quite conspicuous' (*GA* 725b7).

A foetation is not formed out of any casual quantity: if there is too little or too much, none will be formed, because there is a definite limit set both to the *dunamis* of the material which is acted upon and to that of the heat which acts upon it (*GA* 772a26–30).

The *dunamis*, potential, in the female contribution, must match that in the male contribution.[74] So far we can see that, more often than not, the property of producing many offspring at one time requires (1) a certain body size (although this is not necessary), which results in (2) an abundance of generative residues, but (1) and (2) give rise to multiparity only when there is also (3) the correct amount of *dunamis* in these residues. Multiparity occurs, then, when generative residues of both parents are dynamic enough together to establish *summetria* and so to begin lots of new individual lives.

Whether an animal type is prone to deformity or not, multiparity can be natural to it. In the many-toed and the pig, as well as numerous varieties of fish and other egg-layers, multiparity is natural. This is because, although multiparity is not an essential feature, it is the result of a combination or coincidence of other essential features. If animals have (1) body size and (3) heat of generative fluid, then they will almost always produce many offspring at once – and this is then a consequence of a fixed nature. If only heat (3) is present without a small determinate size (1), then multiparity does not flow from the essence of a kind, despite there being some regularity to it under certain conditions. In these cases the phenomenon is much further removed from natural features in animals and so we have to locate its key causes in the external conditions. This occurs in human beings, as Aristotle explains at the end of the section on multiparity in *GA* IV.[75] Human twins

[74] Aristotle's view that the female contribution is a sort of semen with living potentials predominates in this section. So, for example, after using *sperma* to signify the residue which is overabundant in prolific animals, Aristotle moves straight away to an example of female birds. He writes that some birds die after having laid too many eggs, because their own nutriment is expended on the generative residue (*GA* 750a28–9). The female contribution, the egg, is referred to as 'semen' here. There is also the case of the lioness, who produces less 'semen' after using it all up on numerous offspring in her lifetime (750a31–750b1). See also 751a15–19, where the release of eggs is likened to male ejaculation.

[75] This follows the pattern of placing accounts of unusual or exceptional cases towards the end of his explanation of the phenomenon in question. See Chapter 2. Another case of twins as a deformity occurs in mares (*HA* 575b34–576a2).

tend to occur in hotter countries, such as Egypt, where the state of external elemental powers results in more abundant and dynamic generative residues (*HA* 584b31–2).[76] The presence of more potent generative residues mean that, even though humans are large animals, there is enough *dunameis* in the combination of male and female to establish two offspring at the outset. The occurrence of twins in human beings can partly be explained by particular human characteristics. Their superior heating capacity, for instance, allows the female to complete the offspring once there are two or more. Humans' variable gestational period also aids this outcome (*GA* 772b7–12).[77] However, even if all human beings lived in hot places like Egypt, resulting in the regular production of twins, this method of generation would still not be linked to the human essence in any straightforward way and so counts as a deformity.[78]

Multiparity on its own is not sufficient to create a proneness to conjoined offspring. Although these deformities are more common in these animals because they are multiparous, multiparity on its own can be unproblematic. So, for example, in the pig, an abundance of generative residues (2), resulting from their size (3) and relative nakedness, combine with a stronger or hotter given nature (2) which then results in the subsequent completion of all offspring.

> [The pig] brings the nourishment [of its many offspring] to completion because of its fine physical condition: it is like a rich soil which can provide plants with sufficient and indeed abundant nourishment (*GA* 774b25–7).

[76] Twin lambs can occur because of good pasturage (*HA* 573b31). Good nutrition is now thought to be a factor in the increase in natural twin births in humans (Steinman 2006).

[77] It is not clear how a variable gestational period helps to explain how twins are possible. It is also a very curious feature since gestational periods are fixed in every other kind of animal. Aristotle appears to simply accept this fact, conveyed to him by women or midwives, without further justification (*HA* 584a35–584b1). See also *GA* 776a22.

[78] '[Twins] seem rather to be monsters, because their formation is contrary to the general rule and to what is usual' (772a36–772b1). Aristotle disagrees with the Hippocratic notion that twinning is natural because of the symmetrical structure of the uterus. *Vict.* I.30, 1 (L VI 504). We can contrast the superstitious attitude of some peoples that twins are animals and not human and so not to be raised by a human community. See Meek (1937) 291. Unlike other deformities, bearing twins does not seem to be detrimental to the living goal of generating to type (iii). However, before modern obstetrics, the risks to mother and babies of multiple pregnancy would have been considerable. It might have seemed to Aristotle, then, that the human female is not designed for generative success in the case of twins.

Multiparity may coincidentally be beneficial in that it helps to ensure the continuation of the kind (iii). However, whether natural or not, it can also have effects that are detrimental to an animal's goals, that is, the proneness to deformity which preoccupies Aristotle in *GA* IV. Animals with colder material natures tend to be less able to maintain their own bodies (survive) (ii) and produce another just like themselves (reproduce) (iii). Although this is their particular way to exist and so it is natural to them, it also allows more influence to come from material necessity, which acts counter to living goals. The effect of external seasonal variation can influence the outcomes of generation in any animal.[79] However, these tendencies are much more pronounced in lower animals. The way is prepared for literal failure to generate another like itself because the animal's material nature results in the combination of multiparity and relative coldness. The female parent in these cases is not able to adequately complete or nourish the large number of offspring they produce at once.[80]

Those many-toed animals which produce their offspring in an imperfect state, all produce numerous offspring, and on that account while the foetations are quite young they are able to nourish them, but once they have advanced in growth and have attained some size their bodies are unable to bring the nourishing of them to completion, and so discharge them just as the larvae-producing animals do, for indeed their young, like the larvae, are practically unarticulated when born, e.g. those of the fox, the bear, the lion, and similarly with some of the others; moreover, practically all of them are blind, e.g. the ones just mentioned and in addition those of the dog, the wolf and the jackal (*GA* 774b10–17).

They are unable to complete offspring and so even small differences in external conditions result in offspring that are literally incomplete.

Monstrosities occur more frequently in goats and sheep, because they are more multiparous; and still more frequently in the many-toed, because animals of this sort are multiparous and the offspring are not perfected when born (e.g. the dog) . . . the way to the production of monsters has been already prepared for nature by the fact that they generate offspring which, owing to its imperfect state,

79 Even in the highest, see *Pol.* 1335b1–2.

80 Complete failure to generate, that is, sterility, can also be due to a failure to adequately nourish the foetus; see *GA* 748b29–30 on mules.

is unlike its parents, for monstrosities come under the class of offspring which is unlike its parents (*GA* 770a37–770b6).

Several interesting points emerge from this discussion. The first is that accounts of some phenomena require reference to features that are inessential but natural and explicable – here multiparity.[81] The second is that when the same features turn out to be unnatural, explanations further detached from teleology in terms of the material 'facts' of the case are presented. Multiparity can be either natural or unnatural depending on the causes that lie behind it. If those causes originate in an animal's nature, even its material nature, then it will be natural. And if they do not, then it will be unnatural. Natural multiparity can be detrimental to living goals. When combined with natural coldness, it can result in a proneness to deformity. We may express this more generally: when a feature occurs due to a coincidence of features necessary for essence, then it is 'natural' even though not essential, and can lead to defects or deformities in combination with other features of a kind or with the material situation an animal finds itself in. This is one way in which explanations in terms of MN can be connected to, and are more or less dependent on, teleological accounts in the latter portion of the *GA*. And this is very similar to the sorts of explanations Aristotle employs in *GA* V, as the following sub-section illustrates.[82]

If features are 'unnatural' then they cannot be, in this way, connected to teleological explanations. The usual clues will be missing. Aristotle's expectations about how various factors degrade through his normative scale of living beings won't be as useful (*GA* 722a34-733b2). It would appear to be a generalization that in hotter, better animals sex separation is the norm (c) generation occurs more efficiently (a) and less deformity, metaphorical or literal, is apparent (b).[83] These patterns only help to explain cases

[81] Aristotle finishes the section by explaining the connection he set out to discuss. Redundancy of parts is like unnatural twins (*GA* 772b13–26) and both are due to excessive generative residues or excessive *dunamis* in them. Deficiency of parts is like abortions, that is, failed pregnancies (*GA* 772b35–773a2), and is due to deficient generative residues or deficient *dunamis* in them, as is sterility (e.g. in mules *GA* 748a32–3, 748b13–14).

[82] None of these are explanations in terms of 'secondary teleology' (Leunissen 2010).

[83] See Chapter 7.1–2.

in which proneness to certain deformities is quite regular. Most other problem births are less tied to whether the animal is high up the scale or not but have very specific material causes, relating to particular facts about the kind. One example is the fusing together of offspring, which is not due to multiparity but to the shape of a particular animal's uterus.[84] Another case is that of the birth of human twins, discussed earlier. Both are to be explained in terms of material facts, largely separate from a teleological framework.

10.5 Explanations of proneness to environmental effects in subsidiary parts

GA V explains features of animals that are neither beneficial nor detrimental and are incidental to essence, in a similar way to his account of proneness to some deformities. These parts differ in the way they do due to a combination of the material results of certain essential features and weaknesses in an animal's constitution when faced with environment contingencies. At the beginning of *GA* V he sets out the project of giving an account of such features.

> We must now study the *pathêmata* by which the parts of animals differ. I mean such *pathêmata* of the parts as light and dark eyes, and high and deep voice and colour difference of hair or feathers. Some of the variations happen to be present to an entire class of animals, some not; in humans, for example, this [the latter] happens most of all. Further, changes that come with ages of life are present similarly in all animals, others occur in an opposite way, as those concerning the voice and hair colour. For some animals do not go noticeably grey toward old age, but humans suffer this most of all other animals. And some [*pathêmata*] follow immediately on birth, and others come to be clear as time advances and the animal becomes aged (*GA* 778a16–29).

Gotthelf and Leunissen (2010) posit that *GA* V continues the project of *GA* I–IV of explaining a new animal's development.[85] In *GA* I and II, the emphasis was on internal development before birth, whereas *GA* V covers the differentiation and formation of

[84] The snake (*GA* 770a26–7).

[85] Aristotle makes it clear in *GA* V that some differences between animal parts have no final causes and are the result of material and efficient causes acting apart (778a29–778b1). Strictly speaking there are no efficient causes without formal and final causes. However, there is an efficient cause in an accidental sense. See Sauvé Meyer (1992) and Chapter 6.4.

parts after birth, most of which are external. Differences between such parts happen predominantly because materials initially unconnected to form and soul are then used by the animal's nature to aid its ends. This reading provides a rationale for the content of *GA* V; Aristotle is completing his account of the ends, his teleological account, by adding those parts that do not occur because of a direct or primary teleology, but due to this secondary sort. Of course Gotthelf and Leunissen are aware of instances where no advantage accrues, but as their emphasis is on the continuity of the teleological account in the *GA*, they focus on cases of materials left over from the developmental process *in utero*, which are then used by the individual to develop subsidiary or 'luxury' parts. But *GA* V also covers the occurrence of other features that do not fit that pattern. Inessential features are similar to multiparity: they can be uniform and natural or divergent within a kind and thus unnatural. The former can be affected by the external conditions an animal finds itself in but are much less prone to be so than the latter. Explanations, connected and unconnected to teleology, will be set out more clearly in my exposition of some of the content of *GA* V.

Aristotle begins by noting that if a variation is 'proper to the particular class of animals', then it is part of their *logos* and a teleological account is appropriate (*GA* 778b12–13). In general, the type of hair, voice and eyes an animal has are all very important for the particular life it leads.[86] For instance, animals need a particular kind of voice in order to scare away other animals or to call mates closer.[87] Also, the eyes have two kinds of keenness of vision – long-distance and distinctness – both of which are needed for different purposes, depending on whether the animal seeks prey that are far away or searches for foodstuffs close to it.[88] These purposive variations are then contrasted to parts which differ within one given type of animal, and are not essential. In humans, for instance, the differences that occur between individuals in their hair, voice and eyes, are merely accidental and not related directly to their essence.

[86] External coverings and voice are most affected by the environment (*GA* 778a16–20).
[87] The horse's voice (*HA* 609b16); mating calls (*HA* 536a11–15).
[88] For example, eagles need keen vision, *HA* 620a1–2.

> Each thing, then, is 'for the sake of something,' while as regards their process of formation, all those characteristics which are contained in its *logos*, or are subservient to some end, or are an end in themselves – these come to be formed on account of this cause as well as the remaining causes. Other characteristics, however, are formed during the process which do not fall under the headings just given, and the cause of them is to be looked for in the *kinêseis*, i.e., the process of formation – we must assume that they acquire their differences within the actual process of construction. Thus (to take an example) X will of necessity possess an eye (because that characteristic is included in the essence of the animal as posited), and it will – also of necessity – possess a particular sort of eye, but the latter is a different mode of necessity from the former, and is derived from the fact that it is naturally constituted to act and to be acted upon in this or that way (*GA* 778b12–19).

The contrast made here between different types of necessity distinguishes the necessity for parts included in the essence of the creature (a sort of logical necessity: 'X will of necessity possess an eye'), the hypothetical necessity (HN) that certain material makes up this eye ('all those characteristics which are ... subservient to some end') and finally a Pre-Socratic material necessity (MN) ('it will of necessity possess a particular sort of eye'). Eye colour depends on the fluid in the eye: light eyes have a small amount of fluid, while dark eyes have a lot of fluid (*GA* 779b28–30). Although eyes must be fluid, and certain types of eye will need to have a set amount of fluid in order to carry out their function, the colour that manifests because of the amount of fluid in an eye is incidental.[89] We might think of it as a sign or symptom rather than a cause.

As in proneness to deformity, where multiparity was seen to emerge from an essential feature, or combination of essential features, so also the character of the parts of the body, although inessential, can flow directly from essence. Let's take the case of eye colour that does not vary within a kind. This appears to occur when the amount of fluid present in the eye is hypothetically necessary – thus following directly from the animal's essence. In the owl, for example, the exact amount of fluid is crucial to the keenness of its night vision, which all members of the kind require

89 Incidental features are discussed at *Ph.* 196b24–29 where Aristotle notes that house-building is an essential feature of the artisan but that paleness and musicality are incidental.

(*HA* 619b13–17).[90] Since the amount of fluid will not vary, their eye colour won't either (*GA* 780b13). This is why eye colour, when uniform as in owls, can give us some clue about the functional requirements of the eye involved – just as the presence of bile, although itself the accidental consequence of material interactions, can tell us about essential features of animals (i.e. that the animal must remain alive through the process of nourishing itself, *PA* 677a13–19). When eye colour differs within a kind such links to goals cannot be justifiably made; these differences are then not signs of any underlying more important difference, indicative of essence.

Aristotle's favoured examples of differences of this latter sort occur in humans.[91]

> The eyes of human beings show in practice a multiplicity of colour; some are blue, some greyish-blue, some dark, some yellow. In the case of other animals . . . the individuals of any class do not differ from each other (*GA* 779a34–779b1).[92]

Formally, human lives require the type of keen vision which distinguishes objects (*GA* 781b19–20), since such acute perception enables participation in ethical and scientific inquiry. This hypothetically necessary requirement is met by the thinness of human skin and the fact that human sense organs are pure and less earthy (*GA* 780b27, 781b21–2). However, there is no hypothetical necessity for a set *amount* of eye fluid. Thus, this will differ and so also will the attendant colour.

In addition to eye colour, Aristotle is also interested in hair, which constantly grows and is replaced and so, he reasons, is constructed from the least digested materials in an animal's body (*GA* 725a14).[93] This is the case both initially, in the mixture of

90 Long-range senses give advance warning of predators and prey (*de An.* 434b27; *Sens.* 436b18–22). See *de An.* III.12–13.

91 As he remarks several times in the first paragraph of the *GA* V (778a22, 25–26). Gotthelf and Leunissen (2010) note that this is the case but do not attempt to explain why.

92 See also *HA* 492a4–7.

93 'Hair and similar things . . . continue growing so long as they are there at all, and they grow more during diseases, and when old age advances, and when the body is wasting. This is because old age and disease mean that less nourishment is expended on the supreme parts of the body and therefore more residue is left over' (*GA* 745a11–16). The female is less hairy because she releases less concocted stuff along with the menses which would have gone to make up the hair (*GA* 727a17). See Boylan (1982) on the order of production of the useful residues.

nutritive residues that make up the embryo's body and eventually result in the formation of hair just before or shortly after birth, and throughout subsequent life.[94] Even in animals that require the same sort of body coverings, being luxury and not essential, these parts are not directly related to essence and so not as strongly controlled by the animal's soul. Because hair is made from less digested matter it is not very different from external food and more prone to external influences than other parts. When Aristotle explains the weakness in hair and the subsequent alterations in it, his focus shifts from teleological to non-teleological. For example, the external food consumed can explain variegated coats.

The more omnivorous animals are more variegated, as we should expect . . . for of course if the various sorts of nourishment they take are the causes of the change, we shall expect to find that variegated kinds of nourishment make the *kinêseis* which the nourishment undergoes and the residues which result from it more variegated and it is out of the residues that hair, feathers and skin are formed (*GA* 786a35–786b5).[95]

Next, there is a MN explanation in terms of the shape and quality of the skin. Skin differs due to its relative fluidity, that is, its greasiness or wateriness.[96] Thick hair comes out of loose skin, thin hair out of compact skin (*GA* 782a32–782b3). Finally, the decline in hair quality in instances of old age, malnutrition or disease happens because of the increasing influence of the external environment on these parts.

The shift to non-teleological explanations is particularly clear in human hair and skin alterations, which Aristotle spends a good deal of time discussing. Human head hair differs greatly

94 When nutriment is well-digested, it is more stable in its constitution. In embryological development, it goes to form the heart and other essential parts (*GA* 741b25–26, 743b25–30).

95 Of course, an animal's coat could serve as a sign of its eating habits and these habits do form part of its essence (*HA* 487a10–13).

96 Thick hair comes from earthy skin which has large pores in it. Also, the contributory cause is that the matter is greasy and therefore does not dry off, so that the hair is bigger and thicker. In the case of thin hair, the pores are narrow and the matter is watery (*GA* 783a4–5). For another example of a material explanation see *GA* 783a25–28 on sea urchins. Although hair is like external food, it still has features that the animal's soul determines. For instance, the general fluidity of an animal's body, its heat, the size of its pores and the thickness of its skin affect the particular way in which the residues ooze out and are set.

between groups (*GA* 782b33), between individuals (*GA* 785b32–3) and between time periods in an individual's life (781b30–782a20), showing how susceptible it is to external influence (782b33–83a2).[97]

One of the peculiarities of human hair is that its colour never follows skin colour, as is normal (*GA* 785b3–5), due to its weakness.

As regards humans, the skin is not a cause [of hair colour]. For some white [i.e. fair] people have very black hair. The reason is that the human has the thinnest skin of all animals, as to its size, and because of this it has no strength to change the hair, but the skin changes the colour itself because of its weakness, and also becomes darker by the agency of sun and wind, while the hair does not change with it (*GA* 785b6–13).

Human skin is so weak it cannot retain its own colour when it meets with the external elements; this means that it also cannot secure the colour of the hair that grows out of it. The explanation for hair going grey is in terms of MN – the hair undergoes a kind of moulding (*GA* 784b4–21).[98] The skin and hair of aging people, which becomes even weaker, succumbs easily to this external effect. The explanations offered here are not connected to teleology in a secondary manner; instead the effects are very loosely tied to teleological explanations. They can be connected only to the HN for thin skin, required for the extra perceptual sensitivity required for intelligence (*GA* 781b19–23).

Aristotle explains balding in terms of a proneness to go bald.

Humans bear summer and winter in different ages of life. Because of this, no one becomes bald before having sexual intercourse. That is when those who are more prone by nature [go bald]. For the head is by nature the coldest part of the body and sexual intercourse cools. For pure natural heat is excreted. So it is reasonable that the head is the first to suffer (*GA* 783b26–32).

The idea of proneness by nature is now familiar from the section on deformity (Chapter 10.4) and a similar pattern of explanation is

97 Although greying and balding occurs in some other animals as they age, it is most prominent in humans. For animals that all have white hair, the whiteness 'happens naturally and not by misfortune (*pathei*)' (*GA* 785b1–2).

98 For similar cases of bodily materials interacting with external matter to produce inessential features see *GA* 782b29–783a1 on the straightness and curliness of hair and 788a16–20 on the voice.

apparent. Only a certain subset of human males are 'prone by nature' to lose the hair on their heads. Again, this proneness can be tangentially connected to aspects of the human essence. Humans need very cold heads in order to counterbalance their extra hot hearts (which ensures perceptive sensitivity, etc. *PA* 653a27-b2, 650b20–6). Also, they have to produce a lot of semen in order to set another hot creature like themselves. These features result in baldness in conjunction with certain climatic conditions.[99] Aristotle's explanations of weakness in parts, or of the external environment impeding goals, sometimes depend on certain animals being weaker and colder in their constitutions than others. These patterns do not adequately explain how it is that humans have such difficulty maintaining the colour and texture of their external parts, however, since humans are the highest and best animals. Instead, these are due to particular material facts about the kind.

The latter portion of the *GA* focuses on the coming to be of animals, rather than their being. Any particular instance of generation will necessarily involve the development of individual parts of the body, most of which display inessential characteristics. Readers can learn from generation in specific instances as detailed in *GA* IV and V (including the attempt to answer such questions as 'Why is this particular animal male?', 'Why did these twins come about?' and 'Why do they have blue eyes?') about the process more generally. Aristotle's account of these features shows his acceptance that there are regularities that occur right back in the parents' seeds which strive to perpetuate themselves without clear purpose or advantage. Male and female contribute residues of blood (or its analogue) containing *kinêseis* (both current potential and possible potential *kinêseis*) derived from *dunameis*. It then depends on the combination of the two sexes' contributions plus climatic conditions what happens next. Some of the features discussed, such as skin, hair and eye colour, are recognized in human beings as those inherited from parents or other family members. They occur from one generation to the next, or skip a generation,

99 Curiously the same principle does not apply to the female, although female humans evacuate more menstrual blood than any other type of animal (*GA* 728b15–6, 776b26–7; *HA* 521a26–7, 572b30, 582b29).

and then recur. Grandma's blue eyes recur in the grandchild. Because inherited characteristics come from definite powers within the parents' generative fluids, these features relate to a propensity within the animal. It is not the eye or hair colour *per se* that is included in any powers that come from parents, but the propensity to create offspring with particular thickness/thinness of skin and fluidity of eye.[100] The colours are consequences of those factors. The features that come to be after birth can also be inherited, such as the propensity to become bald or go grey.[101] Thus we can see how, for Aristotle, each generative residue contains specific *dunameis* and *kineseis* which strive for regularity even when goals are absent.[102]

These findings tell against the idea that Aristotle's teleological perspective limited his ability to engage appropriately in investigations of the material world, thereby inhibiting advances in bio-chemistry (see Chapter 6). Indeed an interest in the material processes which shape the bodies of new animals is very much apparent in *GA* V and depends on the state of the mixture rather than the ends in view. The elemental potentials in the environment will affect how the potentials in the residues of both parents manifest themselves. Sometimes the full formal potentials will be realized, but there will also always be the co-occurrence of inessential features. An eye will form that can see but it will also have a certain constitution that is non-telic – and this material constitution will then result in it being coloured in a particular way. The colour will be due to a 'bio-chemical' process linking it to material regularities in the bodies of parents or ancestors. We learn something, then, about Aristotle's actual practice when accounting

100 Gelber (2010) 188–91 very clearly explains the difference between inherited features and these – these are accidental and due only to environmental contingencies (e.g. thickness of hair).

101 One might object that all human beings have the propensity to go grey; however, it is clearly the case that some people display this earlier than others: at 20 rather than 40.

102 Henry (2008) notes the unusual position of inherited characteristics; although they are inessential, they come from definite powers (*dunameis*) within the parents. 'Heritable variations, such as eye colour and nose shape, occupy an interesting place in Aristotle's ontology: they are not accidental by-products of development, since each one is the per se object of a corresponding *dunamis* (*GA* IV.3); yet they are not present for the sake of anything ... ' (71). Later anti-teleologists claimed that Aristotle did not allow for regularity without purpose; but this seems dubious in light of his position on inherited features. See Sedley (2007) ch. 2 on the Epicureans.

for particular occurrences in more detail in *GA* IV and V. Thus we can see that it is important to sometimes view the biological texts on their own terms and not always to seek patterns that one might expect from Aristotle's works such as the *Posterior Analytics*.

10.6 Teleology, material necessity and the generation of animals

Given the number of explanations in the *GA* that depend on reference to material necessity apart from teleology, the idea that Aristotelian teleology reaches all the way down to the elemental powers ought be rejected.[103] However, we need not go to the other extreme and posit that material necessity opposes the goals of formal and final causes.[104] Although that idea can explain aging and death, it sits ill with Aristotle's claim that animals have a determinate life span, which is part of their essence.[105] His view falls somewhere between these opposing positions and the *GA* offers much help in discovering where.

Aristotle is concerned with environmental effects on animal generation in the later Books of the *GA*, in concert with his more advanced study of particular instances of generation. Accounts in terms of MN cover many different areas. On the one hand, we can see an interaction between the matter internal to the animal and that which exists in the environment external to it. Sometimes these two fall in line and aid living goals; at other times the results of the interactions are neutral with respect to teleology; yet others represent an impairment or impose limitations on

103 Sedley (2007) ch. 6. Balme in Aristotle (1972) 80–2, citing *Mete.* 390a4f., *Metaph.* 1075a19–22f. Those who claim that 'all material and efficient causation in nature depends on formal and final causation' (Scharle 2008, 29) go too far. Aristotle's *Meteorologica* discounts such a statement. See Lennox (2012).

104 Gill (1989a) 204: '[T]he deterioration and destruction of organism is caused by their own constituents, which tend to move off to their proper places, if given a chance . . . composites are always on the verge of annihilation on account of their own lower material properties, and the project of remaining the same and avoiding decay is one that demands considerable exertion.' *Idem* (234): 'Wearying, decline, and natural destruction are the unfortunate lot of generated composites, even the most perfect. This tendency to lapse into a simpler state is the profound problem that the elements, which are worked up into unities, pose to the survival of those unities'.

105 Aging is a natural process linked to essence rather than an attack from foreign elemental powers. See, especially, King (2001); *Resp.* 478b21–27.

survival, flourishing and generative goals. This can happen either on a case by case basis or with some regularity, if the conditions are usually concurrent (e.g. specific instances of failure to conceive as opposed to sterility). Thus, on the basis of the *GA* we can see that when reading Aristotle we must assume certain limits to the 'for the sake of something and towards the best', and should not expect it for all cases without qualification.[106]

The fact that the four elemental powers can and do stand on their own apart from teleology offers some rationale for the scale of achievement and value Aristotle maps onto the variety of living beings in the world.[107] As *GA* II explains, being in existence is better than non-existence and living is better than not living (*GA* 731b30). Since the four elements must exist, life at least permeates as much of them as is possible. This means that some animals will be cold and watery – which is better than no life existing in cold and watery places. However, because that which is cold and watery cannot be as powerful, strong or sensitive, since heat is required for so many soul functions, then there will be types of animal that do not best meet the highest goals that all living beings share. And these lesser animals are going to be more influenced by materials outside their bodies – they will be more easily disrupted and impaired. Natural aims are sometimes limited, according to Aristotle in the *GA*.

> It is heat and cooling in their various manifestations which up to a certain due proportion (*summetrias*) bring about the generation of things, and beyond that point their dissolution; and the limits of these processes, both as regards their beginnings and their end, are controlled by the movements of these heavenly bodies . . . .Nature's aim, then, is to measure the generations and endings of things by the measures of these bodies, but she cannot bring this about exactly on account of the indeterminateness of matter and the existence of a plurality of principles which impede the natural processes of generation and dissolution and are so often the causes of things occurring contrary to nature (777b28–778a9).

[106] Theophrastus *Metaph.* 11a1–3.

[107] Someone might protest that different types of animal are not differentiated by matter but rather by their differing souls. The main types of soul are nutritive, sentient and intellectual. All animals possess sentient souls, so it could not distinguish one from the other. Instead, it seems that differences in perceptive ability rest in bodily differences, such as the presence of certain sense organs. These differences are fundamentally material.

Aristotle's view is not, however, the tidy idea that animals follow exactly a line of degradation with reference to heat and relative perfection. For one thing, all animals are detrimentally affected by the elements to some extent since all require nutrition. In most cases, the food which accords with an animal is subdued and converted into the animate bodily system of that particular type. However, some of the powers of external food remain and these can affect body size (*HA* 606a11–25), gestation, sexual differentiation, fertility, deformity and which inessential features of the body come to be.[108] Indeed, the coldness of food always has a negative effect on cognition, which is the reason why we sleep (*Somn.* III).

Aristotle's concession to the influence of the environment on the successful outcomes of generation, and on health and bodily fitness more generally, is shared with many of his medical contemporaries.[109] His idea that 'the reason why any animal is long-lived really is that its "blend" is about the same in comparison with the air which is around it' (*GA* 777b7–9) echoes the Hippocratic notion that a 'temperate climate' (*he krasis ton horeon*) produces the best, healthiest and largest animals and plants (*Aër.* 12).[110] Indeed, like the Hippocratic doctors, Aristotle understood that a significant part of the everyday existence of most animals involves getting the right food, water and encountering the right weather conditions for adequate self-maintenance.[111] Aristotle thinks that by employing causes 'for the sake of which' in his analysis of animals, he can provide satisfying and useful explanations. But he is not an absolutist in this respect, and remains keenly aware of the potential negative effects of environmental factors on animals.

108 The condition of the generative residues and fertility are influenced by nutrition and the weather (*GA* 738a16–22, 767a28–34). Other examples include the voice, which differs due to cooler or warmer air (788a16–20) and the white discharges in women which are affected by their diet (728a6–9).

109 Van der Eijk attests to Aristotle's interest in contemporary medical theory (1999) 494 and (2007) 410–12.

110 *Pol.* 1327b24–34 goes further by linking climate with character traits in different ethnic groups, again echoing this same Hippocratic author.

111 Disease, illness, premature death and infertility are linked by early doctors to the external environment in most cases, especially the water and food consumed. *Aêr.* 4–7, L II 18–33, notes that barrenness in women is due to the hard cold water of a region and also states that the influence of water on health is very great.

Aristotle has expectations about how animal kinds will degrade with respect to animate goals based on a general hierarchy of kinds, details of which are to be found throughout *GA* I-III. Very generally, coldness renders animals weaker and less able to thrive and gives the animal less purchase on the ability to control its material nature. This sort of weakness is a very important explanatory tool for Aristotle but it does not necessarily correspond to the place an animal sits in the hierarchy. And perhaps because this is the case, he has to investigate more specific facts about the kind. Such methodology is apparent when he notes a natural tendency in some sorts towards abnormalities or peculiarities in generation. Horse and ass, for example, have a natural proneness to sterility; this is part of their materiality. Because both are naturally cold and infertile, what happens when they mate seems explicable – that is, they produce sterile offspring. Other cases of sterility are unnatural, in so far as they are due to very specific faults (such as deformed generative parts) and so cannot be linked to a proneness found in that animal's material nature. There can also be a proneness to certain sorts of birth defect, as is found among the group Aristotle terms 'the many-toed'. Multiparity and the relative inability to complete offspring by the time of birth in this kind frequently result in embryos fusing onto each other, a propensity found in their nature. Again the results are detrimental to living goals. This tendency can be opposed to instances in which birth defects are much more unnatural, that is, due to the effects of untoward external environment on the *summetria*, which suppress the formative *kineseis* in the generative residues. In *GA* V eye and hair variations are most common in human beings, because they are prone to such differences, due to weakness and coldness in the relevant parts. These differences are neutral with respect to living goals. But the parallel with proneness to sterility and proneness to deformity remains since such natural tendencies can be contrasted with very specific instances of, for example, hair loss or greying due to extreme weather conditions or disease. In so far as they are large and warm, human beings fit with the expectation of separation of the sexes (c) and success at generation (a). However, relative completeness of offspring at birth (b) is not their strongest suit (*GA* 779a25).[112] So, although Aristotle's explanations often include the idea that lower animals will be more prone to sterility and deformity, he is not rigid in this. In fact, it seems rather to be a way of setting up research, by providing broad guidelines that are helpful, but not always applicable in particular situations, where peculiar material facts do not fit and cannot be made to do so.

112 Human infants cannot walk and need extensive hand feeding. They cannot even hold up their own heads (*GA* 744a33) and so require that heads be held to the breast rather than finding it themselves like almost all other mammals.

11

CONCLUSION

This book has analysed Aristotle's views on female animals, much evidence for which comes from his zoological writings, particularly those concerned with animal generation. We have learnt how little evidence there actually is for a complete polarizing of the roles of the sexes. If the work entitled *On the Generation of Animals* is not treated selectively, then many positive claims about female nature therein cannot be obscured. In particular, it becomes clear that the female contributes a sort of semen or seed complementary to that of the male. And this key idea is extraordinarily productive of explanations which include:

(1) How it is that one offspring results from the combination of two parents: Male and female residues are jointly potential; the new animal is the locus of substantial change and so is, metaphysically, the single actualization of the parents' joint potentials.
(2) Explanations of embryonic development: The generative residues in both parents are designed to go towards each part of the living body and are potentially the same in function and character as those parts. This explains how the living embryonic body is generated.
(3) The origin of the new animal's soul: The new animal's soul is this joint actualization, so comes from neither parent in isolation from the other. It begins with a collection of dormant potentialities; the body parts that eventually form provide material underpinnings for the actualization of these potentialities (e.g. eyes for seeing, legs for walking). Psychic capacities emerge gradually after conception.
(4) The reasons why copulation in healthy individuals does not always result in generation: There must exist *summetria*, or a correct proportional relationship, between male and female contributions to generation (and also the material conditions in which copulation takes place).
(5) The reasons why generation sometimes occurs but fails to reach the goal of replication: Numerous alternate outcomes of generation are explained in terms of variations in and failures of *summetria*, for example, accidental and inherited features of parts, and female, infertile or deformed offspring.

(6) The reason why infants develop in certain ways after birth: The generative residues of both parents not only shape offspring but shape them in particular ways, for each is a unique individual with current and possible potentials in their blood and residues.
(7) The reasons for sexual differentiation in embryonic development: The female condition is not as fortuitous as the male one but shares in one sense in the ultimate ends of generation -the attempt to keep the species in existence. Aristotle explains that the intricate structure and function of the female body as well as maternal behaviours in the natural world are necessary for the goal of successfully generating.

As sophisticated and effective as parts of his explanation of the generation of animals are, the theory is deeply informed by sexism. With respect to this sexism, it is important to steer a path between two untenable interpretative extremes: vilification on the one hand and exoneration on the other. The former previously resulted in the dismissal of most of the content of the *Generation of Animals* and other zoological works, especially the detailed analyses of lower animals and particular instances and exceptional circumstances, where the female role is more clearly represented. The reasons for Aristotle's sexism are not available to us. Attempts to find such reasons or connect his biology with his politics are often unhelpful and sometimes detrimental to our understanding of his views on gender. Although sexism is integral to his theory of generation, it is unwarranted to view the text as mere ideological ramblings. We cannot dismiss his more complex and interesting works on these grounds.

On the other hand, any attempt at exoneration from a charge of sexism is seriously problematic. Aristotle's main conclusions about the nature of semen and the physiology of generative processes are overlaid by sexism, when they did not have to be. He enters the arena of explanations of animal generation with one key commitment – the centrality of the heart and its production of the internal nutriment, blood. Blood is manufactured by the animal's soul so as to carry out psychic and living capacities and to form the material parts appropriate for these functions, with a specificity fitted to each kind. Thus, it is all important for Aristotle that semen be blood-like in its origin, constitution and function.[1] He believes

[1] 'Both the residue of males and the menstrual residue of females are naturally blood-based' (*GA* 776b11–12).

that the female contributes blood-like stuff to the embryo through the umbilicus throughout pregnancy (in the paradigm blooded live-bearers). The female in higher animals also bleeds from its generative organs and this regular menstruation corresponds with fertility. The female's contribution to generation is blood-like and seminal. The puzzle, then, would appear to be about the male and how his contribution might be equally blood-like, in crucial respects similar to the all important internal nutriment.

The indispensability of the male is the strongest empirical evidence Aristotle brings to bear to show that male and female contributions to generation differ. If they were exactly the same, there would be a problem of the redundancy of the male sex, since the female typically does the work of gestation and foetal nourishment. As it turns out, the male contribution also has to be derived from blood but it differs from the female's obviously bloody contribution, as it is white and emitted at sexual climax. The fact that both male and female generative contributions are blood-like is not at all sexist, neither is the fact that they differ in some respects. One need not hierarchize two different but equally crucial contributions to one goal. Most of the evidence that Aristotle is aware of and documents supports an equal but different hypothesis. One can only guess that this hypothesis was not put forward by him or any of his contemporaries because it was not acceptable, due to deeply ingrained sexism. His statement in *GA* II that it is better that male animals be separate from female animals (732a7–12), and better in some metaphysical sense that the two sexes are separated, rests on the idea that male animals will be superior with respect to flourishing capacities, that is, perception and knowledge (Chapter 8.6). This must be based on the perceived disadvantages of human child-rearing to women. For surely the evidence from the rest of zoology is inconclusive. Not all female animals are encumbered by their reproductive role. In some species, females lay eggs that they leave and do not take any notice of afterwards. Some male animals even take more care over the young than females in their kind (*HA* 621a20-b2). Even in those kinds closest to humans, lactating live-bearers, the female animal's extra reproductive tasks constitute only a minor inconvenience rather than a major impediment to the flourishing capacities in which Aristotle is so interested. This is what makes Plato's comparison of human to canine females so effective. In terms of physical burdens, Plato argues, female humans could treat their reproductive role as a short-term inconvenience which need not prevent them from gaining knowledge or holding political office (*Rep.* 451d-452a, 454d-3, 460d).

If we consider the metaphysical picture, Aristotle's sexism is not integral after all. Indeed, his theories surrounding the generation of animals need not have been sexist in order to operate successfully on a conceptual basis, lending much support to the view noted in Chapter 1.5 that sexism is some kind of adjunct to theory for Aristotle.[2] The fact that there are different and joint potentials coming from male and female parents does not require one of the pair to be superior to the other. A *summetria* can result from the combination of two forces or entities that are equal to begin with or it can come about through the combination of one inferior and one superior entity. Thus Aristotle can fit what he takes to be the appropriate and acceptable characterization of sex difference, that male is superior to female, to the metaphysical framework that explains how two are needed to begin one new living being. The fact that nothing in the empirical evidence forced him to do so is of course what we now believe. And it explains how it is that Aristotle himself must still be held culpable for imposing sexist assumptions onto a set of natural processes that did not require it. He cannot be exonerated.

The fact that Aristotle values male fertility over the more obvious achievements on this front by females follows a long tradition in Greek ideas, especially those which equate the male contribution with plant seeds, agriculture representing technological control over the environment. Concern about controlling fertility can find a place within Aristotle's intellectualism, if one were to speculate that widely. Philosophy begins with the desire to understand and understanding requires taking command of the material at hand and organizing it into a rigid scheme (*A. Po.*). In ethics, it is crucial to be able to control your own state of

[2] Deslauriers (1998) 139. Cf. Lennox (1985), Nussbaum (1998), Mayhew (1999), Wardy (1993) 23 and Mayhew (2004).

happiness, despite the vicissitudes of fortune (*EN* 1100b23–1101a14). As tempting as it might be to make such general inferences, it assumes a much more comprehensive view than we can ever find in Aristotle's works. In both his logic and his ethics, despite seeming efforts at absolutism, Aristotle is hardly ever so rigid. Although the *Posterior Analytics* provides guidelines for achievement in intellectual endeavour, in the practice of zoology in particular, Aristotle relaxes these rules. In ethics, he concedes readily that we cannot entirely (or even mostly) escape our good or bad fortunes. You cannot be happy if you 'look utterly repulsive or are ill-born, solitary or childless' (*EN* 1099b3–4). And for the children, despite advice about how to procreate the best ones in his *Politics* VII.16, Aristotle is very much aware that this is one of those things over which an individual has very little control. Generative success for anyone is a matter of luck; a person cannot ensure that they have children and, even if they do manage it, that these children will be long-lived and healthy. Although Aristotle wanted to emphasize the remarkable powers of forms or natures, which use matter such as food, water and warmth from the environment to make beautifully controlled internal systems, the powers of these are not infallible. Some animals, due to their bodily composition, will be more prone to particular problems and those with typically strong constitutions may be liable to certain weaknesses or be unlucky. Aristotle's outlook, evident in the zoological account of generation, is of a world dominated not only by forms and self-perpetuating kinds but also by the forces of nature that are not so structured: by the necessary and the material.

The study of female animals not only brings to our notice Aristotle's theory of generation, but also expands our understanding of his thought more generally. Indeed, once we take seriously the positive aspects of the female role, great insight into Aristotle's methodology and his philosophy becomes available. There is much to learn here about material necessity, the form/matter relation and the soul. Perhaps most striking of all, though, is what this study of the female animal has shown us about the way in which Aristotle investigates the natural world. Certain portions of *GA* I and II provide broad outlines of the research project he is undertaking, connecting these with other lines of thought in other works.

GA III–V (and some parts of *GA* I and II) contain the results of more specific investigations. Understanding this procedure makes it possible to see the work as a whole as coherent and consistent regarding the part played in generation by the female animal (Chapter 2.6). There is no need, on the score of inconsistency, for us to experience philosophical squeamishness about the content of the latter Books of the work. There is also no good reason to dismiss or ignore portions of *GA* III–V from our study of Aristotle's philosophy when we find that they do not contain any indication of the sorts of axioms, differentiae or demonstrations that were so important in the *Analytics*. Many portions of this work (and others) do not show any straightforward application of more general principles of explanation. So, although Aristotle structures his analysis with reference to expected patterns in nature, as any scientist will, these are not rigidly applied. Aristotle's scientific sensitivity is exemplified by his approach to the study of females in nature, which is just one of the many reasons why it is of such importance.

BIBLIOGRAPHY

Ancient texts, translations and commentaries

Albertus Magnus (1890–99a) *De Natura et Origine Animae*, in *Opera Omnia*, ed. A. Borgnet, trans. M. T. Dougherty, Chapter 4 (Paris: Apud Ludovicum Vives).

(1890-99b) *Quaestiones Super De Animalibus*, in *Opera Omnia*, ed. A. Borgnet, Book XV, Quest. 20 (Paris: Apud Ludovicum Vives).

(1916–20) *De Animalibus*, 2 vols., ed. H. Stadler (Münster: Beiträge zur Geschichte der Philosophie des Mittelalters).

Alexander of Aphrodisias (1887) *On Aristotle's De Anima*, ed. I. Bruns, Supplementum *C.A.G.* 16/11 (Berlin: Georg Reimer).

(1996) *On Aristotle Meteorology* 4, trans., comm. E. Lewis (London: Duckworth).

Aquinas, St Thomas (1952) *On the Power of God* (*Quaestiones Disputae De Potentia Dei*), trans. the English Dominican Fathers (Westminster, MD: Newman Press).

Aristotle (1808) *The Treatises of Aristotle*, vol. 4 (*The Generation of Animals*), trans. T. Taylor (London: Robert Wilks).

(1887) *Traité de la Génération des Animaux d'Aristote*, trans. J. Barthélemy Saint-Hilaire, J., 2 vols. (Paris: A. Durand).

(1906) *De Sensu and De Memoria*, comm. G.R.T. Ross (Cambridge: Cambridge University Press).

(1910) *Aristotle's Generation of Animals*, trans., notes A. Platt (Oxford: Clarendon).

(1912) *De Motu Animalium, De Incessu Animalium*, trans., notes A.S.L. Farquharson, vol. 5 of *The Works of Aristotle Translated into English* eds. J.A. Smith and W.D. Ross (Oxford: Clarendon Press).

(1924) *Aristotle's Metaphysics*, ed., intro., comm. W. D. Ross, 2 vols. (Oxford: Clarendon).

(1933) *The Metaphysics Books I–IX*, trans. H. Tredennick (Cambridge, MA: Harvard University Press).

(1936a) *Aristotle: Minor Works*, ed., trans. W.S. Hett (Cambridge, MA: Harvard University Press).

(1936b) *Aristotle: On the Soul, Parva Naturalia, On Breath*, ed., trans. W.S. Hett (Cambridge, MA: Harvard University Press).

(1937) *Parts of Animals, Movement of Animals, Progression of Animals*, ed., trans. A.L. Peck (Cambridge, MA: Harvard University Press).

(1939) *On the Heavens*, trans. W.K.C. Guthrie (Cambridge, MA: Harvard University Press).
(1942) *The Generation of Animals*, ed., trans. A.L. Peck (Cambridge, MA: Harvard University Press).
(1944) *Aristotle's Chemical Treatise Meteorologica*, trans., comm. I. Düring (Göteborg: Elanders Boktrychken Aktiebolag).
(1952) *Aristotle: Meteorologica*, ed., trans. H.P.D. Lee (Cambridge, MA: Harvard University Press).
(1955a) *Parva Naturalia*, comm. W.D. Ross (Oxford: Oxford University Press).
(1955b) *On Sophistical Refutations, On Coming to Be and Passing Away, On the Cosmos*, ed., trans. D.J. Furley (Cambridge, MA: Harvard University Press).
(1956) *Aristote: Les Parties des Animaux*, ed., trans. P. Louis (Paris: Belles Lettres).
(1957) *Aristotelis Metaphysica*, ed. W. Jaeger (Oxford: Oxford University Press).
(1961) *Aristote: Generation des Animaux*, ed., trans. P. Louis (Paris: Belles Lettres).
(1962) *Aristotle, The Politics*, trans. T.A. Sinclair (London: Guild Publishing).
(1964–9) *Aristote: Histoire des Animaux*, 3 vols., ed., trans. P. Louis (Paris: Belles Lettres).
(1965a) *De Generatione Animalium*, ed. H.J. Drossaart Lulofs (Oxford: Oxford University Press).
(1965b) *Historia Animalium*, 2 vols., ed., trans. A.L. Peck (Cambridge, MA: Harvard University Press).
(1970) *Aristotle's Physics Books I and II*, trans., intro., notes W. Charlton (Oxford: Clarendon).
(1972) *De Partibus Animalium I and De Generatione Animalium I*, trans., comm. D.M. Balme (Oxford: Clarendon).
(1982) *De Generatione et Corruptione*, ed., trans. C.J.F. Williams (Oxford: Clarendon)
(1984) *Complete Works*, 2 vols., ed. J. Barnes (Princeton: Princeton University Press).
(1984) *Aristotle's Politics*, trans. C. Lord (London: MIT Press).
(1991) *The History of Animals Books VII-X*, ed., trans. D.M. Balme (Cambridge, MA: Harvard University Press).
(1992) *De Partibus Animalium I and De Generatione Animalium I (with passages from II. 1–3) with a Report on Recent Work and an Additional Bibliography*, A. Gotthelf (Oxford: Oxford University Press).
(1993) *Aristotle De Anima Books II and III*, trans., comm. D.H. Hamlyn (Oxford: Clarendon).
(1994) *Aristotle Posterior Analytics*, trans., comm. J. Barnes (2nd edn.) (Oxford: Clarendon).

(1998) *Metaphysics of Aristotle*, trans., intro. H. Lawson-Tancred (London: Penguin Books).

(2001) *Aristotle on the Parts of Animals I-IV* ed., intro., notes. J. Lennox (Oxford: Clarendon).

Augustine (1969) 'Tractus in Iohannis Evangelium' in *Oeuvres de Saint Augustin*, vol. 71 (Paris: Desclée De Brouwer).

(1972) 'De Genesis da Litteram', in *Oeuvres de Saint Augustin*, vol. 8. (Paris: Desclée De Brouwer).

The Holy Bible (1987) *Authorized King James Version: Containing the Old and New Testaments* (Cambridge: Cambridge University Press).

(1930) *The New Testament in the Original Greek*, text revised by B.F. Westcott and F.J.A. Hort (London: Macmillan).

Censorinus (1983) *De Die Natali*, ed. N. Sallmann (Leipzig: Teubner).

Diocles of Carystus (2000) *A Collection of Fragments*, trans., comm. by P. van der Eijk (Leiden: Brill).

Diogenes Laertius (1925) *Lives of Eminent Philosophers*, trans. R.D. Hicks, 2 vols. (London: Heinemann).

Galen [Claudius] (1907–1909) *De Usu Partium*, ed. G. Helmreich, 2 vols. (Leipzig: Teubner).

(1968) *On the Usefulness of the Parts of the Body*, ed., trans. M.T. May, 2 vols. (Ithaca: Cornell University Press).

(1971) *De Uteri Dissectione*, ed., trans. D. Nickel (Berlin: Akademie Verlag).

(1992) *On Semen*, ed., trans., comm. P. de Lacy (Berlin: Akademie Verlag).

Hero of Alexandria (1995) *On Automaton Making*, trans., notes in S. Murphy, 'Heron of Alexandria's On Automaton-Making', *History of Technology* 17, 1–44.

Herodotus (1920) *Histories*, trans. A.D. Godley (Cambridge, MA: Harvard University Press).

Herophilus (see H. von Staden 1989).

Hesiod (1943) 'Theogony' 'Works and Days' in *Hesiod, The Homeric Hymns and Homerica*, trans. H.G. Evelyn-White (Cambridge, MA: Harvard University Press).

Hippocrates (1839–61) *Oeuvres Completes d'Hippocrate*, ed., trans., comm. E. Littré, 10 vols. (Paris: Librarie de l'Academie Royale de Medicine).

(1923) *Ancient Medicine, Airs Waters Places, Epidemics I & III, The Oath, Precepts, Nutriment*, ed., trans. W.H.S. Jones (Cambridge, MA: Harvard University Press).

(1931) *Nature of Man, Regimen in Health, Humours, Aphorisms, Regimen I–III, Dreams (Regimen IV)*, trans. W.H.S. Jones (Cambridge, MA: Harvard University Press).

(1970) Hippocrate Tome XI. *De la Génération, De la Nature de l'Enfant, Des Maladies IV, du Foetus de Huit Mois*, ed., trans. R. Joly (Paris: Belles Lettres).

(1981) *The Hippocratic Treatises 'On Generation', 'On the Nature of the Child', 'On Diseases IV'*, ed., trans., comm. I.M. Lonie (Berlin: Akademie Verlag).

(1995) *Places in Man, Glands, Fleshes, Prorrhetic 1–2, Physician, Use of Liquids, Ulcers, Haemorrhoids and Fistulas*, ed., trans. P. Potter (Cambridge, MA: Harvard University Press).

(2006) Hippokrates, *Über die Natur des Kindes (De geniture und De natura pueri)*, ed., trans. F. Giorgianni (Wiesbaden: Dr. Ludwig Reichert Verlag).

(2008) *Nature de la Femme*, ed., trans. F. Bourbon (Paris: Les Belles Lettres).

Michael of Ephesus (1903) *[Ioannis Philoponi] Michaelis Ephesii, In Libros De Generatione Animalium Commentaria*, ed. M. Hayduck (Berlin: Reimer).

Ovid (1916) *Metamorphoses*, trans. F.J. Miller (London: Heinemann).

Philo of Alexandria (1929) *On the Account of the World's Creation Given by Moses, Allegorical Interpretations of Genesis II–III*, trans. F.H. Colson and G.H. Whitaker (London: William Heinemann).

(1937) *On the Decalogue, On the Special Laws*, trans. F.H. Colson (Cambridge, MA: Harvard University Press).

Philoponus, John (1897) *In Aristotelis Libros De Generatione et Corruptione Commentaria*, de. H. Vitelli (Berlin: Reimer).

(1999) *On Aristotle, On Coming-to-Be and Perishing I.*1–5, trans. C.J.F. Williams (London: Duckworth).

Pindar (1997) *Olympian Odes, Pythian Odes*, trans. W.H. Race (Cambridge, MA: Harvard University Press).

Plato (1997) (1937) *Plato's Cosmology: The Timaeus*, trans., comm. F.M. Cornford (London: Routledge and Kegan Paul).

(1929) *Timaeus, Critias, Cleitophon, Menexenus, Epistles*, trans. R.G. Bury (London: Heinemann).

Complete Works, ed. J.M. Cooper (Indianapolis, IN: Hackett).

(2000) *Plato Timaeus*, trans. D.J. Zeyl (Indianapolis, IN: Hackett).

The Pre-Socratics (1903) *Die Fragmente der Vorsokratiker*, ed. H. Diels and W. Kranz, 3 vols. (Berlin: Weidmann).

(1983) *The Pre-Socratic Philosophers: A Critical Edition with a Selection of Texts*, 2nd edn., J.S. Kirk, R.E. Raven and M. Schofield (Cambridge: Cambridge University Press).

(2010) *The Texts of Early Greek Philosophy: The Complete Fragments and Selected Testimonies of the Major Presocratics Pt I.*, ed., trans. D.W. Graham (Cambridge: Cambridge University Press).

Soranus of Ephesus (1927) *Gynaeciorum*, ed. J. Ilberg (Leipzig: Teubner).

(1956) *Soranus' Gynecology*, trans. O. Temkin (Baltimore: Johns Hopkins Press).

Theophrastus (1916) *Enquiry into Plants*, trans. A. Hort (Cambridge, MA: Harvard University Press).

(1923) *On Aristotle's Metaphysics*, trans. A. Hort (Cambridge, MA: Harvard University Press).

(1993) *Metaphysics*, intr., trans., comm. M. van Raalte (Leiden: Brill).
Thucydides (1919) *History of the Peloponnesian War Books I-II*, trans. C. Foster Smith, vol. 1 (London: Heinemann).

Modern Works

Ackrill, J.L. (1973) 'Aristotle's Definition of Psychê', *Proceedings of the Aristotelian Society*, 119–33. Repr. in Barnes *et al.* 1979, vol. 4, pp. 65–75.
(1981) *Aristotle the Philosopher* (Oxford: Oxford University Press).
Adelmann, H.B. (1966) *Marcello Malpighi and the Evolution of Embryology*, vol. 2 (Ithaca, NY: Cornell University Press).
Alexander, S. (1920) *Space, Time, and Deity*, 2 vols. (London: Macmillan).
Allen, G. (2009) 'Rebel with Two Causes: Hans Driesch' in O. Harman and M. Dietrich (eds.), *Rebels, Mavericks, and Heretics in Biology* (New Haven: Yale University Press), pp. 37–64.
Allen, P. (1985) *The Concept of Woman: The Aristotelian Revolution, 750 BC–AD 1250* (Montreal: Eden Press).
Anonymous (1986) *Aristotle's Master-Piece, Aristotle's Compleate Master Piece, Aristotle's Book of Problems, Aristotle's Last Legacy* (New York: Garland Facsimile Series).
Annas, J. (1981) *An Introduction to Plato's Republic* (Oxford: Oxford University Press).
Antony, L.M. and C. Witt (1993) (eds.) *A Mind of One's Own: Feminist Essays on Reason and Objectivity* (Boulder, CO: Westview Press).
Aughterson, K. (1995) (ed.) *Renaissance Woman a Sourcebook: Constructions of Femininity in England* (London: Routledge).
Bachofen, J.J. (1861) *Das Mutterrecht: Eine Untersuchung über die Gynaikokratie der alten Welt nach ihrer religiösen und rechtlichen Natur* (Stuttgart: Krais and Hoffmann).
Balme, D.M. (1962a) 'Development of biology in Aristotle and Theophrastus: Theory of Spontaneous Generation', *Phronesis* 7, 91–104.
(1962b) 'GENOS and EIDOS in Aristotle's biology', *Classical Quarterly* 12, 81–98.
(1985) 'Aristotle *Historia Animalium* Book Ten' in J. Wiesner (ed.) *Aristoteles, Werk und Wirkung* vol. 1 (Berlin: Walter de Gruyter), pp. 191–206.
(1987a) 'Teleology and Necessity' in Gotthelf and Lennox (1987), pp. 275–85.
(1987b) 'Aristotle's Biology Was Not Essentialist' in Gotthelf and Lennox (1987), pp. 291–312 (first appeared in *AGP* 1980, 62, 1–12).
(1987c) 'Aristotle's Use of Division and Differentiae' in Gotthelf and Lennox (1987), pp. 69–89.
(1990) 'Matter in the Definition: A Reply to G.E.R. Lloyd' in Pellegrin and Devereux (1990), pp. 49–54.
Barnes, J., M. Schofield and R. Sorabji (1975–9) (eds.) *Articles on Aristotle*, 4 vols. (London: Duckworth).

Bibliography

Barnes, J. (1981) *Aristotle* (Oxford: Oxford University Press).

(1995a) (ed.) *The Cambridge Companion to Aristotle* (Cambridge: Cambridge University Press).

(1995b) 'Metaphysics' in Barnes (1995a), pp. 66–108.

Bar-On, Bat Ami (1994) (ed.) *Engendering Origins: Critical Feminist Readings in Plato and Aristotle* (Albany: State University of New York Press).

Bates, A.W. (2005) *Emblematic Monsters: Unnatural Conceptions and Deformed Births in Early Modern Europe* (Amsterdam: Rodopi B.V.).

Beare, J.I. (1906) *Greek Theories of Elementary Cognition: From Alcmaeon to Aristotle* (Oxford: Clarendon).

Bechtel, W. and R. Richardson (1998) 'Vitalism' in E. Craig (ed.) *The Routledge Encyclopedia of Philosophy* (retrieved July 09, 2012 from www.rep.routledge.com/article/Q109).

Berman, R. (1989) 'From Aristotle's Dualism to Materialist Dialectics: Feminist Transformation of Science and Society' in Jaggar and Bordo (1989), pp. 224–55.

Berryman, S. (2003) 'Ancient Automata and Mechanical Explanation' *Phronesis* 48, 344–69.

(2007) 'Teleology Without Tears: Aristotle and the Role of Mechanistic Conceptions of Organisms', *Canadian Journal of Philosophy* 37, 351–69.

Bien, C.G. (1998) 'Der "Bruch" in Aristoteles Darstellung des Zeugungsbeitrags von Mann und Frau', *Medizinhistorisches Journal* 33, 3–17.

Birke, L. (1998) 'Biological Sciences' in Jaggar and Young (1998), pp. 194–203.

Bleier, R. (1984) *Science and Gender: A Critique of Biology and Its Theories on Women* (New York: Pergamon Press).

Blundell, S. (1995) *Women in Ancient Greece* (London: British Museum Press).

Bolton, R. (1987), 'Definition and Scientific Method in Aristotle's *Posterior Analytics* and *Generation of Animals*' in Gotthelf and Lennox (1987), pp. 120–66.

Bonitz, H. (1955) *Index Aristotelicus* (Akademische Druck– und Verlagsanstalt: Graz).

Bordo, S. (1987) (ed.) *The Flight to Objectivity: Essays on Cartesianism and Culture* (Albany: State University of New York Press).

(1993) *Unbearable Weight: Feminism, Western Culture, and the Body* (Berkeley: University of California Press).

Bos, A.P. (2003) *The Soul and Its Instrumental Body: A Reinterpretation of Aristotle's Philosophy of Living Nature* (Leiden: E.J. Brill).

(2009) 'Aristotle on Soul and Soul "Parts" in Semen (*GA* 2.1, 735a4-22)', *Mnemosyne* 62, 378–400.

Bowen, A. (ed.) (1991) *Science and Philosophy in Classical Greece* (New York: Garland).

Boylan, M. (1981) 'Mechanism and Teleology in Aristotle's Biology', *Apeiron* 15, 96–102.

(1982) 'The Digestive and "Circulatory" Systems in Aristotle's Biology', *Journal of the History of Biology* 15, 89–118.

(1983) *Method and Practice in Aristotle's Biology* (Washington, DC: University Press of America).

(1984) 'The Galenic and Hippocratic Challenges to Aristotle's Conception Theory', *Journal of the History of Biology* 17, 83–112.

Boys-Stones, G. (2007) 'Physiognomy and Ancient Psychological Theory' in S.R. Swain (ed.) *Seeing the Face, Seeing the Soul* (Oxford: Oxford University Press), pp. 19–124.

Briffault, R. (1927) *The Mothers: A Study in the Origins of Sentiments and Institutions* (London: Allen and Unwin).

Broad, C. (1925) *The Mind and Its Place in Nature* (London: Routledge and Kegan Paul).

Broadie, S. [Waterlow] (1982) *Nature, Change and Agency in Aristotle's Physics* (Oxford: Clarendon).

(1992) 'Aristotle's Perceptual Realism', *Southern Journal of Philosophy* 31, Supple., 137–60.

Bronstein, D. (2006) 'Review of A.P. Bos, *The Soul and Its Instrumental Body*, 2003', *Ancient Philosophy* 26, 422–7.

Burnyeat, M.F. (1992) 'Is an Aristotelian Philosophy of Mind Still Credible?' in Nussbaum and Rorty (1992), pp. 15–26.

(1995) 'How Much Happens When Aristotle Sees Red and Hears Middle C? Remarks on De Anima 2.7-8' in Nussbaum and Rorty (1995), pp. 421–34.

(2002) 'De Anima II 5', *Phronesis* 47, 28–90.

Cadden, J. (1993) *The Meaning of Sexual Difference in the Middle Ages* (Cambridge: Cambridge University Press).

Cameron, R. (2003) 'The Ontology of Aristotle's Final Cause', *Apeiron* 35, 153–79.

(2010) 'Aristotle's Teleology', *Philosophical Compass* 5, 1096–1106.

Cantarella, E. (1987) *Pandora's Daughters*, trans. M.B. Fant (Baltimore: Johns Hopkins University Press). Originally published in Italian as *L'Ambiguo Malanno* (Rome: Deitori Riuniti, 1981).

Caston, V. (1997) 'Epiphenomenalisms, Ancient and Modern', *Philosophical Review* 106, 309–61.

(2005) 'The Spirit and the Letter: Aristotle on Perception' in R. Salles (ed.) *Metaphysics, Soul and Ethics in Ancient Thought* (Oxford: Oxford University Press), pp. 245–320.

(2007) 'Aristotle's Psychology' in M.L. Gill and P. Pellegrin (eds.) *Blackwell Companion to Ancient Philosophy* (Oxford: Blackwell), pp. 316–46

(2009) 'Commentary on Charles', In J.J. Cleary and G.M. Gurtler (eds.), *Proceedings of the Boston Area Colloquium on Ancient Philosophy* 24, 30–47.

Cavarero, A. (1995) *In Spite of Plato: A Feminist Rewriting of Ancient Philosophy*, trans. S. Anderlini-D'Onofrio (Cambridge: Polity). Originally published in Italian as *Nonostante Platone: Figure femminili nella filosofia antica* (Rome: Editori Riuniti, 1990).

Charles, D. (1988) 'Aristotle on Hypothetical Necessity and Irreducibility', *Pacific Philosophical Quarterly* 69, 1–53.

(1997a) 'Method and Argument in the Study of Aristotle: A Critical Notice of *The Cambridge Companion to Aristotle*', *Oxford Studies in Ancient Philosophy* 15, 231–57.

(1997b) 'Aristotle and the Unity and Essence of Biological Kinds' in Kullmann and Föllinger (1997), pp. 26–42.

(2000) *Aristotle on Meaning and Essence* (Oxford University Press).

(2009a) 'Aristotle's Psychological Theory' in J.J. Cleary and G.M. Gurtler (eds.) *Proceedings of the Boston Area Colloquium on Ancient Philosophy* 24, 1–29.

(2009b) 'Aristotle on Desire and Action' in D. Frede and B. Reis (eds.) *Body and Soul in Ancient Philosophy* (Berlin: Walter de Gruyter), pp. 291–307.

Cherniss, H. (1935) *Aristotle's Criticism of Pre-Socratic Philosophy* (Baltimore, MD: Johns Hopkins University Press).

Clack, B. (1999) *Misogyny in the Western Philosophical Tradition: A Reader* (London: Macmillan).

Code, A. (1984) 'The Aporematic Approach to Primary Being in Metaphysics Z', *Canadian Journal of Philosophy*, Supple. 10, 1–20.

(1987) 'Soul as Efficient Cause in Aristotle's Embryology', *Philosophical Topics* 15, 51–9.

(1997) 'The Priority of Final Causes Over Efficient Causes in Aristotle's *PA*' in Kullman and Föllinger (1997), pp. 127–43.

Code, A. and M. Moravcsik (1992) 'Explaining Various Forms of Living' in Nussbaum and Rorty (1992), pp. 129–45.

Cohen, S.M. (1989) 'Aristotle on Heat, Cold and Teleological Explanation', *Ancient Philosophy* 9, 255–70.

Coles, A. (1995) 'Biomedical Models of Reproduction in the Fifth Century BC and Aristotle's *Generation of Animals*', *Phronesis* 60, 48–88.

Connell, S.M. (2000) 'Aristotle and Galen on Sex Difference and Reproduction: A New Approach to an Ancient Rivalry', *Studies in the History and Philosophy of Science* 31, 405–27.

(2001) 'An Integrated Approach to Aristotle as a Biological Philosopher', *Review of Metaphysics* 55, 297–322.

Cook, K.C. (1996) 'Sexual Inequality in Aristotle's Theories of Reproduction and Inheritance' in J.K. Ward, *Feminism and Ancient Philosophy* (London: Routledge), pp. 51–67.

Cooper, J. (1987) 'Hypothetical Necessity and Natural Teleology' in Gotthelf and Lennox (1987), pp. 243–74.

(1988), 'Metaphysics in Aristotle's Embryology', *Proceedings of the Cambridge Philological Society*, 14–41. Repr. in Pellegrin and Devereux (1990), pp. 55–84.

Cornford, F.M. (1912) *From Religion to Philosophy: A Study in the Origins of Western Philosophy* (London: E. Arnold).
Crane, T. (2001) *The Elements of Mind* (Oxford: Oxford University Press).
Darwin, C. (1868) *The Variation of Animals and Plants Under Domestication* (London: John Murray).
Davidson, H.A. (1992) *Alfarabi, Avicenna, and Averroes, On Intellect* (Oxford: Oxford University Press).
Davis, E.G. (1971) *The First Sex* (New York: Penguin).
Dawkins, R. (1976) *The Selfish Gene* (Oxford: Oxford University Press).
Dean-Jones, L.A. (1992) 'The Politics of Pleasure: Female Sexual Appetite in the Hippocratic Corpus' in D. C. Stanton (ed.), *Discourses of Sexuality: From Aristotle to AIDS* (Ann Arbor: University of Michigan Press), pp. 48–77.
(1994) *Women's Bodies in Classical Greek Science* (Oxford: Oxford University Press).
(2012) 'Clinical Gynecology and Aristotle's Biology: The Composition of *HA X*', *Apeiron* 42, 180–200.
De Beauvoir, S. (1953) *The Second Sex*, trans. H.M. Parshley (New York: Knopf). Originally published in French as *Le Deuxieme Sexe* (Paris: Gallimard, 1949).
DeForest, M. (1993) *Women's Power, Man's Game* (Wauconda, IL: Bolchazy-Carducci).
Delbruck, M. (1971) 'Aristotle-Totle-Totle' in J. Monod and E. Borek (eds.) *Of Microbes and Life* (New York: Columbia University Press), pp. 50–5.
De Ley, H. (1978) 'Empedocles' Sexual Theory: A Note on Fragment B63', *Antiquité Classique* 67, 153–62.
(1980) 'Democritean Notes on Aristotle's *GA*', *Hermes* 108, 129–53.
Demaitre, L. and A. Traville (1980) 'Human Embryology and the Development in the Works of Albertus Magnus' in J.A. Weisheipl (ed.), *Albertus Magnus and the Sciences* (Toronto: Pontifical Institute of Mediaeval Studies).
Demont, P. (1978) 'Remarques sur le sense de τρέφω', *Revue des Études Greques* 41, 358–84.
Deslauriers, M. (1998) 'Sex and Essence in Aristotle's Metaphysics and Biology' in Freeland (1998a), pp. 138–67.
Detel, W. (1997) 'Why All Animals Have a Stomach: Demonstration and Axiomatization in Aristotle's Parts of Animals' in Kullmann and Föllinger (1997), pp. 63–84.
Driesch, H. (1908) *The Science and Philosophy of the Organism: The Gifford Lectures Delivered Before the University of Aberdeen in the Years* 1907–8, 2 vols. (London: Adam and Charles Black).
(1914a) *The Problem of Individuality: Lectures Delivered Before the University of London,* 1913 (London: Macmillan).
(1914b) *The History and Theory of Vitalism*, trans. C.K. Ogden (London: Macmillan).

du Bois, P. (1988) *Sowing the Body : Psychoanalysis and Ancient Representations of Women* (Chicago: University of Chicago Press).

Dudley, J. (2012) *Aristotle's Concept of Chance: Accident, Cause, Necessity, and Determinism* (Albany: State University of New York Press).

Duprat, G.L. (1899) 'La Théorie du Pneuma Chez Aristote', *Archiv für Geschichte der Philosophie* 12 (Berlin: Reimer), 305–21.

Düring, I. (1966) *Aristoteles: Darstellung und Interpretation seines Denkens* (Heidelberg: Winter).

Elshtain, J. B. (1981) *Public Man, Private Woman: Women in Social and Political Thought* (Princeton: Princeton University Press).

Feyerabend, P. (1975) *Against Method: Outline of an Anarchism Theory of Knowledge* (London: New Left Books).

Fine, C. (2010) *Delusions of Gender: The Real Science Behind Sex Differences* (New York: W.W. Norton and Company, Inc.)

Flemming, R. (2000) *Medicine and the Making of Roman Women* (Oxford: Oxford University Press).

Fodor, J. (1974) 'Special Sciences', *Synthèse* 28, 97–115.

Föllinger, S. (1996) *Differenz und Gleichheit: Das Geschlechterverhältnis in der Sicht griechischer Philosophen des 4. bis 1. Jahrhunderts v. Chr.* Hermes Einzelschriften 74 (Stuttgart: Franz Steiner).

(2005) 'Dialogische Elemente in der antiken Fachliteratur' in T. Fögen (ed.) *Antike Fachtexte/Ancient Technical Texts* (New York: Walter de Gruyter), pp. 221–34.

Fortenbaugh, W.W. (1977) 'Aristotle on Slaves and Women' in Barnes et al. (1975–9), vol. 2, pp. 135–40.

Fox Keller, E. (1985) *Reflections on Gender and Science* (London: Yale University Press).

(1989) (ed.) *Three Cultures: Fifteen Lectures on the Confrontation of Academic Cultures* (Rotterdam: Universitaire Pers).

(2000) *The Century of the Gene* (Cambridge, MA: Harvard University Press).

Fox Keller, E. and H.E. Longino (eds.) (1996) *Feminism and Science* (Oxford: Oxford University Press).

Fraser, P.M. and E. Matthews (1987–1996) *A Lexicon of Greek Personal Names*, 2 vols. (Oxford: Oxford University Press).

Frazer, J.G. (1922) *The Golden Bough: A Study in Magic and Religion* (New York: Macmillan).

Frede, M. (1987) *Essays in Ancient Philosophy* (Oxford: Oxford University Press).

(1992) 'Aristotle's Conception of the Soul', in Nussbaum and Rorty (1992), pp. 93–107.

Freeland, C. (1986) 'Aristotle on Possibilities and Capacites' *Ancient Philosophy* 6, 69–89.

(1987) 'Aristotle on Bodies, Matter and Potentiality' in Gotthelf and Lennox (1987), pp. 392–407.

(1991) 'Accidental Causes and Real Explanations' in Judson (ed.) (1991), pp. 49–72.

(1994) 'Nourishing Speculation: A Feminist Reading of Aristotelian Science' in Bar-On (1994), pp. 145–87.

(1998a) (ed.) *Feminist Interpretations of Aristotle* (University Park, PA: Pennsylvania State University Press).

(1998b) 'On Irigaray on Aristotle' in Freeland (1998), pp. 59–92.

Freudenthal, G. (1995) *Aristotle's Theory of Material Substances: Heat and Pneuma, Form and Soul* (Oxford: Clarendon).

Fricker, M. and J. Hornsby (2000) (eds.) *The Cambridge Companion to Feminism in Philosophy* (Cambridge: Cambridge University Press).

Furley, D. (1985) 'The Rainfall Example in Physics ii.8' in Gotthelf (1985), pp. 177–82.

Furth, M. (1987) 'Aristotle's Biological Universe: An Overview' in Gotthelf and Lennox (1987), pp. 21–52.

(1988) *Substance, Form and Psyche in Aristotle's Metaphysics* (Cambridge: Cambridge University Press).

Garry, A. and M. Pearsall (1989) (eds.) *Women, Knowledge, and Reality: Explorations in Feminist Philosophy* (Boston, MA: Unwin Hyman).

Gelber, J. (2010) 'Form and Inheritance in Aristotle's Embryology', *Oxford Studies in Ancient Philosophy* 39, 183–212.

Gerson, L. (2005) *Aristotle and Other Platonists* (Ithaca, NY: Cornell University Press).

Giallongo, A. (1981) *L'Immagine Della Donna Nella Cultura Greca* (Rimini: Maggioli).

Gill, M. L. (1989a) *Aristotle on Substance: The Paradox of Unity* (Princeton: Princeton University Press).

(1989b) 'Aristotle on Matters of Life and Death' in J.J. Cleary and D. Shartin (eds.) *Boston Area Colloquium on Ancient Philosophy* 7, pp. 187–205.

(1994) 'Individuals and Individuation in Aristotle' in Scaltsas *et al.* (1994) pp. 55–72.

Gimbutas, M. (1956) 'Prehistory of Eastern Europe, Part I. Mesolithic, Neolithic and Copper Age cultures in Russia and the Baltic area', *American School of Prehistoric Research, Peabody Museum, Harvard University, Bulletin* No. 20.

Gotthelf, A. (1976) 'Aristotle's Conception of Final Causality', *Review of Metaphysics* 30, 226–54. Repr. in Gotthelf and Lennox (1987), pp. 204–42.

(1985) (ed.) *Aristotle on Nature and Living Things* (Bristol: Bristol Classical Press).

(1998a) 'The Place of the Goal in Aristotle's Natural Teleology' in J.J. Cleary and G.M. Gurtler (eds.) *Proceedings of the Boston Area Colloquium on Ancient Philosophy* 4, 113–39.

(1998b) 'Historiae I: plantarum et animalium' in W.W. Fortenbaugh and R.W. Sharples (eds.) *Theophrastean Studies: On Natural Science,*

Philosophy and Metaphysics, Ethics, Religion and Rhetoric (Oxford: Transaction Books), pp.100–35.
(1997a) 'The Elephant's Nose: Further Reflections in the Axiomatic Structure of Biological Explanation in Aristotle' in Kullmann and Föllinger (1997), pp. 85–95.
(1997b) 'Understanding Aristotle's Teleology' in H.F. Hassings (ed.) *Final Causality in Nature and Human Affairs* (Washington DC: Catholic University of America Press), pp. 71–82.
(1997c) 'Division and Explanation in Aristotle's Parts of Animals' in Günther and Pengakos (eds.) (1997), pp. 215–30.
(2012a) *Teleology, First Principles, and Scientific Method in Aristotle's Biology* (Oxford: Oxford University Press).
(2012b) 'Teleology and Embryogenesis in Aristotle's GA II.6' in Gotthelf (2012a), pp. 90–116.
(2012c) 'Data-Organisation, Classification, and Kinds: The Place of the *History of Animals* in Aristotle's Biological Enterprise' in (2012a), pp. 261–92.
Gotthelf, A. and J.G. Lennox (1987) (eds.) *Philosophical Issues in Aristotle's Biology* (Cambridge: Cambridge University Press).
Gotthelf, A. and M. Leunissen (2010) 'What's Teleology Got to Do With It? A Reinterpretation of Aristotle's *GA* V', *Phronesis* 55, 325–56.
Gould, C.C. and M.W. Wartofsky (1976) (eds.) *Women and Philosophy: Toward a Theory of Liberation* (New York: G.P. Putnam's Sons).
Graf, F. (1996) 'Zeus' in Hornblower and Spawforth (1996), pp. 1636–8.
Granger, H. (1987) 'Deformed Kinds and the Fixity of Species', *Classical Quarterly* 37, 110–16.
(1990) 'Aristotle and the Functionalist Debate', *Apeiron* 23, 27–49.
Green, J.M. (1992) 'Aristotle on Necessary Verticality, Body Heat, and Gendered Proper Places in the Polis: A Feminist Critique', *Hypathia* 7, 70–96.
Grosz, E. (1990) 'Philosophy' in S. Gunew (ed.) *Feminist Knowledge: Critique and Construct* (London: Routledge), pp. 147–74.
(1994) *Volatile Bodies: Toward a Corporeal Feminism* (Bloomington: Indiana University Press).
Günther, H. and A. Pengakos (1997) *Beiträge zur antiken Philosophie: Festschrift für Wolfgang Kullmann* (Stuttgart: Steiner Verlag).
Guthrie, W.K.C. (1957) *In the Beginning: Some Greek Views on the Origins of Life and the Early State of Man* (London: Methuen).
(1962–81) *A History of Greek Philosophy*, 5 vols. (Cambridge: Cambridge University Press).
Hackett, A. and S. Pomeroy (1972) 'Review: Making History, "The First Sex"', *Feminist Studies* 1, 97–108.
Halperin, D.M. (1990) *One Hundred Years of Homosexuality: And Other Essays on Greek Love* (London: Routledge).
Hammond, W.A. (1895) 'Hylozoism: A Chapter in the Early History of Science', *The Philosophical Review* 4, 394–406.

Hankinson, J. (ed.) (2008) *The Cambridge Companion to Galen* (Cambridge: Cambridge University Press).

Hanson, A.E. (1991) 'Continuity and Change: Three Case Studies in Hippocratic Gynecological Therapy and Theory' in S. Pomeroy (1991), pp. 73–110.

(1990) 'The Medical Writers' Woman' in Halperin (1990), pp. 309–38.

Haraway, D. (1988) 'Situated Knowledges: The Science Question in Feminism and the Privilege of Partial Perspective', *Feminist Studies* 14, 575–99.

Harding, S. and M.B. Hintikka (1983) (eds.) *Discovering Reality: Feminist Perspectives on Epistemology, Metaphysics, Methodology, and the Philosophy of Science* (Dordrecht, Holland: D. Reidel).

Hass, M. (1998) 'Feminist Readings of Aristotelian Logic' in Freeland (1998a), pp. 19–40.

Heinaman, R. (1990) 'Aristotle and the Mind-Body Problem', *Phronesis* 35, 83–102.

Hekman, S.J. (1996) (ed.) *Feminist Interpretations of Michel Foucault* (Bloomington: Indiana University Press).

Henry, D. (2003) 'Themistius and Spontaneous Generation in Aristotle's *Metaphysics*', *Oxford Studies in Ancient Philosophy* 24, 183–208.

(2005) 'Embryological Models in Ancient Philosophy', *Phronesis* 50, 1–42.

(2006a) 'Understanding Aristotle's Reproductive Hylomorphism', *Apeiron* 39, 257–88.

(2006b) 'Aristotle on the Mechanism of Inheritance', *Journal of the History of Biology* 39, 425–55.

(2007a) 'Review of M.R. Johnson, Aristotle's on Teleology', *Ancient Philosophy* 27, 191–200.

(2007b) 'How Sexist is Aristotle's Developmental Biology?', *Phronesis* 52, 251–69.

(2008) 'Organismal Natures' in Mouracade (2008a), pp. 47–74.

Hewson, M.A. (1975) *Giles of Rome and the Medieval Theory of Conception* (London: Athlone Press).

Hirshman, L.R. (1998) 'The Book of "A"' in Freeland (1987), pp. 201–47.

Hite, S. (1976) *The Hite Report: A Nationwide Study on Female Sexuality* (New York: Macmillan).

Holland, N.J. (1997) (ed.) *Feminist Interpretations of Jacques Derrida* (Bloomington: Indiana University Press).

Homiak, M.L. (1993) 'Feminism and Aristotle's Rational Ideal' in Antony and Witt (1993), pp. 1–17.

Hornblower, S. and A. Spawforth (eds.) (1996) *The Oxford Classical Dictionary*, 3rd. edn. (Oxford: Clarendon),

Horowitz, M. C. (1976) 'Aristotle and Woman', *Journal of the History of Biology* 9, 183–213.

Howard, C (2006) 'Aristotle's Catfish' www.non-contradiction.com/ac_aristotle catfish.asp

Huby, P.M. (1990) 'Soul, Life, Sense, Intellect: Some Thirteenth Century Problems' in Dunstan, G.R. (ed.) (1990) *The Human Embryo: Aristotle and the Arabic and European Traditions* (Exeter: Exeter University Press), pp. 113–22.

Hull, D. (1968) 'The Conflict Between Spontaneous Generation and Aristotle's Metaphysics', *Proceedings of the Inter-American Congress of Philosophy* 7, 245–50.

Hursthouse, R. (1999) *On Virtue Ethics* (Oxford: Oxford University Press).

(2006) 'The Central Doctrine of the Mean' in R. Kraut (ed.) *The Blackwell Guide to Aristotle's Nicomachean Ethics*, pp. 96–115.

Irby-Massie, G.L. (1993) 'Women in Ancient Science' in M. DeForest (1993), pp. 354–72.

Irigaray, L. (1985a) *Speculum of the Other Woman*, trans. G. Gill (Ithaca, NY: Cornell University Press). Originally in French as *Speculum de l'Autre Femme* (Paris: Minuit, 1974).

(1985b) *This Sex Which Is Not One*, trans. C. Porter (Ithaca, NY: Cornell University Press). Originally in French as *Ce Sexe Qui N'en est Pas Un* (Paris: Editions de Minuit, 1977).

Jacobsen, A.J. (2000) (ed.) *Feminist Interpretations of David Hume* (Bloomington: Indiana University Press).

Jacquart, D. and C. Thomasset (1988) *Sexuality and Medicine in the Middle Ages*, trans. M. Adamson (Cambridge: Polity Press).

Jaeger, W. (1913) 'Das Pneuma im Lykeion', *Hermes* 48, 29–74.

(1923) *Aristoteles, Grundlegung einer Geschichte seiner Entwicklung* (Berlin); trans. R. Robinson (1948) *Aristotle: Fundamentals of the History of His Development* (London: Oxford University Press).

Jaggar, A.M. and S.R. Bordo (1989) *Gender/Body/Knowledge: Feminist Reconstructions of Being and Knowing* (New Brunswick, NJ: Rutgers University Press).

Jaggar, A.M. and I.M. Young (eds.) (1998) *A Companion to Feminist Philosophy* (Oxford: Blackwell).

James, S. (1998) 'Feminism' in E. Craig, *Routledge Encyclopedia of Philosophy* (London: Routledge), vol. 3, pp. 576–83.

Johansen, T.K. (1998) *Aristotle on the Sense Organs* (Cambridge: Cambridge University Press).

(2004) *Plato's Natural Philosophy* (Cambridge: Cambridge University Press).

(2012) *The Powers of Aristotle's Soul* (Oxford: Oxford University Press).

Johnson, M.R. (2005) *Aristotle's Teleology* (Oxford: Clarendon).

Jones, B. (1974) 'Aristotle's Introduction of Matter', *Philosophical Review* 83, 474–500.

Judson, L. (1991) *Aristotle's Physics: A Collection of Essays* (Oxford: Oxford University Press).

(2005) 'Aristotle's Teleology', *Oxford Studies in Ancient Philosophy* 29, 341–66.

Karbowski, J. (2009) 'Endoxa and the Bees in *GA* III.10' Unpublished manuscript.

Katayama, E.G. (1999) *Aristotle on Artifacts: A Metaphysical Puzzle* (Albany: State University of New York Press).

(2008) 'Substantial Unity and Living Things in Aristotle' in Mouracade (2008a), pp. 99–127.

Kember, O. (1972) 'Aristotle, De Gen An. 761b35', *Classical Review* 22, 172–73.

(1973) 'Anaxagoras' Theory of Sexual Difference and Heredity', *Phronesis* 18, 1–14.

Keuls, E.C. (1985) *The Reign of the Phallus: Sexual Politics in Ancient Athens* (New York: Harper and Row).

King, H. (1998) *Hippocrates' Woman: Reading the Female Body in Ancient Greece* (London: Routledge).

(2013) Helen King, *The One-Sex Body on Trial: The Classical and Early Modern Evidence. The History of Medicine in Context* (Burlington, VT: Ashgate).

King, R.A.H. (2001) *Aristotle on Life and Death* (London: Duckworth).

Kosman, I.A. (1984) 'Substance, Being and Energeia', *Oxford Studies in Ancient Philosophy* 2, 121–150.

Kullmann, W. (1974) *Wissenschaft und Methode: Interpretationen zur Aristotelischen Theorie der Naturwissenschaft* (Berlin: de Gruyter).

Kullmann, W. and S. Föllinger (1997) (eds.) *Aristotelische Biologie: Intentionen, Methoden, Ergebnisse* (Stuttgart: Franz Steiner Verlag).

Lacey, W.K. (1968) *The Family in Classical Greece* (Ithaca, NY: Cornell University Press).

Lang, H.S. (1998) *The Order of Nature in Aristotle's Physics* (Cambridge: Cambridge University Press).

Lange, L. (1983) 'Women Is Not a Rational Animal: On Aristotle's Biology of Reproduction' in Harding and Hintikka (1983), pp. 1–16.

Laqueur, T. (1990) *Making Sex: Body and Gender from the Greeks to Freud* (Cambridge, MA: Harvard University Press).

Lear, J. (1988) *Aristotle: The Desire to Understand* (Cambridge: Cambridge University Press).

Lefèvre, C. (1972) *Sur l'évolution d'Aristote en psychologie* (Louvain: Éditions de l'Institute Superieur de Philosophie).

Lefkowitz, F. and M.B. Fant (1982) (eds.) *Women's Life in Greece and Rome* (London: Duckworth).

Lennox, J.G. (1982) 'Teleology, Chance and Aristotle's Theory of Spontaneous Generation', *Journal of the History of Philosophy* 20, 219–38. Repr. in Lennox (2001) as ch. 10.

(1985a) 'Demarcating Ancient Science: A Discussion of G.E.R. Lloyd Science, Folklore and Ideology', *Oxford Studies in Ancient Philosophy* 3, 307–24.

(1985b) 'Plato's Unnatural Teleology' in D. O'Meara (ed.) *Platonic Investigations* (Washington, DC: Catholic University of America Press), pp. 195–218. Repr. in Lennox (2001) as ch.13.

(1987) 'Divide and Explain: *The Posterior Analytics* in Practice' in Gotthelf and Lennox (1987), pp. 90–119.
(1991) 'Between Data and Demonstration: *The Analytics* and the *Historia Animalium*' in Bowen (1991), 269–94. Repr. in Lennox 2001 as ch. 2.
(1996) 'Material and Formal Natures in Aristotle's De Partibus Animalium' in J. Cleary and W. Wians (eds.), *Proceedings of the Boston Area Colloquium on Ancient Philosophy* 14, pp. 217–40. Repr. in Lennox (2001) as ch. 8.
(1997) 'Nature Does Nothing in Vain' in Günther and Rengakos (1997), pp. 199–214.
(1999) 'The Place of Mankind in Aristotle's Zoology', *Philosophical Topics* 27, 1–16.
(2001) *Aristotle's Philosophy of Biology* (Cambridge: Cambridge University Press).
(2005) 'Aristotle's Biology and Aristotle's Philosophy' in M.L. Gill and P. Pellegrin (eds.) *A Companion to Ancient Philosophy* (London: Blackwell), pp. 292–315.
(2012) 'The Complexity of Aristotle's Study of Animals' in C. Sheilds (2012), pp. 287–305.

Lesky, E. (1950) 'Die Zeugungs– und Vererbungslehren der Antike und ihr Nachwirken' *Akademie der Wissenschaften und der Literatur, Abhandlungen der Geists– und sozialwissenschaftlichen Klasse*, 70/19 (Wiesbaden).

Leunissen, M. (2010) *Explanation and Teleology in Aristotle's Science of Nature* (Cambridge: Cambridge University Press).

Lewis, F.A. (1988) 'Teleology and Material/Efficient Causes in Aristotle', *Pacific Philosophical Quarterly* 69, 54–98.
(1994) 'Aristotle on the Relation Between a Thing and Its Matter' in Scaltsas *et al.* (1994), pp. 247–278.

Liatsi, M (2000) *Aristoteles. De generatione animalium, Buch V, Einleitung und Kommentar* (Trier: Wissenschaflicher Verlag Trier).

Liddell, H.G., R. Scott and H.S. Jones (1990) *A Greek English Dictionary*, 9th edn., (Oxford: Oxford University Press).

Lloyd, E. (1996) 'Pre-Theoretical Assumptions in Evolutionary Explanations of Female Sexuality' in Fox Keller and Longino (1996), pp. 91–102.
(2005) *The Case of the Female Orgasm: Bias in the Science of Evolution* (Harvard: Harvard University Press).

Lloyd, G. (1984) *The Man of Reason: Male and Female in Western Philosophy* (London: Routledge).
(1996) 'Reason, Science and the Domination of Matter' in Fox Keller and Longino (1996), pp. 41–53.
(2000) 'Feminism in History of Ideas: Appropriating the Past' in Fricker and Hornsby (2000), pp. 245–63.

Lloyd, G.E.R. (1966) *Polarity and Analogy: Two Types of Argumentation in Early Greek Thought* (Cambridge: Cambridge University Press).

(1983) *Science, Folklore and Ideology* (Cambridge: Cambridge University Press).

(1987) 'Empirical Research in Aristotle's Biology' in Gotthelf and Lennox (1987), pp. 53–63.

(1990) 'Aristotle's Zoology and His Metaphysics: The Status Quaestionis. A Critical Review of Some Recent Theories' in Pellegrin and Devereux (1990), pp. 7–35.

(1992) 'Aspects of the Relationship Between Aristotle's Psychology and his Zoology' in Nussbaum and Rorty (1992), pp. 147–68.

(1996) *Aristotelian Explorations* (Cambridge: Cambridge University Press).

(2003) *In the Grip of Disease: Studies in the Greek Imagination* (Oxford: Oxford University Press).

(2007) '*Pneuma* Between Body and Soul', *Journal of the Royal Anthropological Institute* 13, Supple., 135–46.

Longino, H.E. (1999) 'Feminist Epistemology' in J. Greco and E. Sosa (eds.), *The Blackwell Guide to Epistemology* (Malden, MA: Blackwell), pp. 327–53.

Longino, H.E. and R. Doell (1983) 'Body, Bias and Behaviour: A Comparative Analysis of Reasoning in Two Areas of Biological Science', *Signs: Journal of Women in Culture and Society* 9, 206–27. Repr. in Fox Keller and Longino (1996), pp. 73–90.

Longrigg, J. (1985) 'A Seminal Debate in the Fifth Century BC?' in Gotthelf (1985), pp. 277–283.

Loux, M. (1991) *Primary Ousia: An Essay on Aristotle's Metaphysics Z and H* (Ithaca, NY: Cornell University Press).

Lovibond, S. (2000) 'Feminism in Ancient Philosophy: The Feminist Stake in Greek Rationality' in Fricker and Hornsby (2000), pp. 10–28.

Lowe, C.R. and R.G. Record (1951) 'Risk of Stillbirth in Twin Pregnancy Related to Sex and Maternal Age', *British Journal of Social Medicine* 5, 34–50.

Łukasiewicz, J. (1971) 'On the Principle of Contradiction in Aristotle', *Review of Metaphysics* 24, 485–509.

Lynch, V. and G. Wagner (2005) 'The Birth of the Uterus: Genes That Help Determine What Goes Right or Wrong in Pregnancy Also Enabled Early Mammals to Switch from Laying Eggs to Bearing Live Young', *Natural History* 114/10, 36–41.

MacDowell, D.M. (1978) *The Law in Classical Athens* (London: Thames and Hudson).

MacIntrye, A. (1981) *After Virtue: A Study in Moral Theory* (London: Duckworth).

Mackenzie, C. (1992) 'Abortion and Embodiment', *Australasian Journal of Philosophy* 70, 136–55.

Maclaren, K. (2009) 'Emotional Metamorphosis: The Role of Others in Becoming a Subject' in Campell *et al.* (2009), pp. 25–45.

MacLean, I. (1980) *The Renaissance Notion of Woman* (Cambridge: Cambridge University Press).

Malinowski, B. (1929) *The Sexual Life of Savages in Northwestern Melanesia* (New York: Horace Liveright).

Margulis, L. and D. Sagan (1986) *Origins of Sex: Three Billion Years of Genetic Recombination* (London: Yale University Press).

Marinella, Lucrezia (1999) *The Nobility and Excellence of Women and the Defects and Vices of Men*, ed., trans. A. Dunhill, intro. L. Panizza (Chicago: University of Chicago Press).

Martin, E. (1991) 'The Egg and the Sperm: How Science Has Constructed a Romance Based on Stereotypical Male-Female Roles', *Signs: Journal of Women in Culture and Society* 16/3, 485–501.

Matthen, M. (1989) 'The Four Causes in Aristotle's Embryology', in T. Penner, and R. Kraut (eds), *Nature, Knowledge and Virtue, Apeiron* 22, 159–79.

(2009) 'Teleology in Living Things' in G. Anagnostopoulos (ed.), *A Companion to Aristotle* (Chichester: Wiley-Blackwell), pp. 335–47.

Matthews, G.B. (1986) 'Gender and Essence in Aristotle', *Australasian Journal of Philosophy* 64, Supple., 16–25.

(1990) 'Aristotelian Essentialism', *Philosophical and Phenomenological Research*, 50 Supple., 251–62.

Mayhew, R. (1999) 'King-Bees and Mother-Wasps: A Note on Ideology and Gender in Aristotle's Entymology', *Phronesis* 44, 127–34.

(2004) *The Female in Aristotle's Biology: Reason or Rationalization* (Chicago: University of Chicago Press).

McKirahan, R.D. (1994) *Philosophy Before Socrates* (Indianapolis, IN Hackett).

McLaren, A. (1985) 'The Pleasures of Procreation', in W.F. Bynum and R. Porter (eds.), *William Hunter and the Eighteenth Century Medical World* (Cambridge: Cambridge University Press), pp. 323–41.

Meek, C.K. (1937) *Law and Authority in a Nigerian Tribe* (Oxford: Oxford University Press).

Menn, S. (2002) 'Aristotle's Definition of Soul and the Programme of the de Anima', *Oxford Studies in Ancient Philosophy* 22, 83–139.

(2012) 'Aristotle's Theology' in C. Shields (2012), pp. 422–64.

Merchant, C. (1980) *The Death of Nature: Women, Ecology, and the Scientific Revolution* (San Francisco: Harper and Row).

Meynell, C. (2009) 'Minding Bodies' in S. Campbell, C. Meynell and S. Sherwin (2009) *Embodiment and Agency* (University Park, PA.: Penn State University Press), pp. 1–23.

Miller Jr., F. (1999) 'Aristotle's Philosophy of Soul', *Review of Metaphysics* 53, 309–37.

(2012) 'Aristotle on the Separability of Mind' in C. Sheilds (2012), pp. 422–64.

Miller, S. (2010) *Medieval Monstrosity and the Female Body* (London: Routledge).

Mills, P.J. (ed.) (1996) *Feminist Interpretations of G.W.F. Hegel* (Bloomington: Indiana University Press).

Modrak, D. (1998) 'Aristotle's Theory of Knowledge and Feminist Epistemology' in Freeland (1998), pp. 93–117.

Molière [Jean-Baptiste Poquelin] (1962) 'Le Malade Imaginiare' in *Oeuvres Complète* (Paris: Édition Seuil), pp. 628–61.

Möller Okin, S. (1993) 'Is Multiculturalism Bad for Women?' in S. Möller Okin (ed.) *Is Multiculturalism Bad for Women* (Princeton, NJ: Princeton University Press), pp. 7–18.

Moraux, P. (1963) 'Quinta Essentia' in A.F. Pauly and G. Wissowa, *Realencyclopädie*, vol. 24 (Stuttgart: Metzler), pp. 1171–1263.

Morsink, J. (1979) 'Was Aristotle's Biology Sexist?', *Journal of the History of Biology* 12, 83–112.

(1982) *Aristotle on the Generation of Animals: A Philosophical Study* (Washington, DC: University Press of America).

Mouracade, J. (2008a) (ed.) *Aristotle on Life* (Kelowna, BC: Academic Printing and Publishing).

(2008b) 'Aristotelian Hylomorphism and Non-Reductive Materialism' in Mouracade (2008a), pp. 153–78.

Muir, H. (2007) 'Goldilock's Planet May Be Just Right for Life', *New Scientist* 25 April 2007, www.newscientist.com/article/dn11710-goldilocks-planet-may-be-just-right-for-life.html.

Mulgan, R. (1994) 'Aristotle and the Political Role of Women', *History of Political Thought* 15/2, 179–202.

(1999) 'Debate: Aristotle, Ethical Diversity and Political Argument', *The Journal of Political Philosophy* 7/2, 191–207.

Needham, J. (1959) *A History of Embryology*, 2nd edn. (Cambridge: Cambridge University Press).

Neumann, E. (1956) *Die Grosse Mutter: Der Archetyp des Grossen Weiblichen* (Zürich: Rhein-Verlag), trans. R. Manheim (1972) *The Great Mother: An Analysis of the Archetype* (Princeton, NJ: Princeton University Press).

Nicholson, L.J. (1990) (ed.) *Feminism/Postmodernism* (London: Routledge).

Nielsen, K. (2008) 'The Private Parts of Animals: Aristotle on the Teleology of Sexual Difference', *Phronesis* 53, 373–405.

Nussbaum, M.C (1978) *Aristotle's De Motu Animalium* (Princeton, NJ: Princeton University Press).

(1982) 'Saving Aristotle's Appearances' in Schofield and Nussbaum (1982), pp. 267–93.

(1986) *The Fragility of Goodness* (Cambridge: Cambridge University Press).

(1998) 'Aristotle, Feminism, and Needs for Functioning' in Freeland (ed.) (1998), pp. 248–59.

(2000) *Women and Human Development: The Capabilities Approach* (Cambridge: Cambridge University Press).

(2007) 'Constitutions and Capabilities: "Perception" Against Lofty Formalism', *Harvard Law Review* 121, 4–97.

Nussbaum, M.C. and H. Putnam (1992) 'Changing Aristotle's Mind' in Nussbaum and Rorty (1992), pp. 27–56.

Nussbaum, M.C. and A.O. Rorty (1992) (eds.) *Essays on Aristotle's De Anima* (Oxford: Oxford University Press); 2nd edn. (1995).

Nutton, V. (2004) *Ancient Medicine* (London: Routledge).

Nuyens, F. (1948) *L'Évolution de la Psychologie d'Aristote* (Louvain: Institute Supérieur de Philosophie).

Nye, A. (1990) *Words of Power: A Feminist Reading of the History of Logic* (London: Routledge).

O'Connor, T. and H. Wong (2012) 'Emergent Properties' in E. Zalta (ed.), *The Stanford Encyclopedia of Philosophy* (Spring 2012 Edition). http://plato.stanford.edu/archives/spr2012/entries/properties-emergent/.

O'Keefe, T. (1997) 'The Ontological Status of Sensible Qualities for Democritus and Epicurus', *Ancient Philosophy* 17, 119–34.

(2005) *Epicurus on Freedom* (Cambridge: Cambridge University Press).

Oliver, K. and M. Pearsall (eds.) (1998) *Feminist Interpretations of Friedrich Nietzsche* (Bloomington: Indiana University Press).

Olshewsky, T. (1992) 'Functionalism Old and New', *History of Philosophy Quarterly* 9, 265–86.

Omitowoju, R. S. (2002) *Rape and the Politics of Consent in Classical Athens* (Cambridge: Cambridge University Press).

Ortner, S.B. (1974) 'Is Female to Male as Nature Is to Culture?' in Rosaldo and Lamphere (1974), pp. 67–88.

Owen, G.E.L. (1961) 'Tithenai Ta Phainomena' in S. Mansion, *Aristote et Les Problèmes de Méthode* (Louvain: Publications Universitaires), pp. 83–103. Repr. in Barnes *et al.* (1975), vol. 1, pp. 113–26.

Papineau, D. (ed.) (1996) *The Philosophy of Science* (Oxford: Oxford University Press).

Peck, A.L. (1953) 'The Connate *Pneuma*: An Essential Factor in Aristotle's Solutions to the Problems of Reproduction and Sensation' in E.A. Underwood (ed.) *Science, Medicine and History* (Oxford: Oxford University Press), pp. 111–21.

Pellegrin, P. (1986) *Aristotle's Classification of Animals: Biology and the Conceptual Unity of the Aristotelian Corpus*, trans. A. Preus (Berkeley: University of California Press).

(1987) 'Logical Difference and Biological Difference: The unity of Aristotle's Thought' in Gotthelf and Lennox (1987), pp. 313–38.

Pellegrin, P. and D. Devereux (1990) *Biologie, Logique et Métaphysique chez Aristote* (Paris: Editions de Centre National de la Recherche Scientifique).

Peradotto, J. and J. Sullivan (eds.) (1984) *Women in the Ancient World: The Arethusa Papers* (Albany: State University of New York Press).

Peterson, B. (2012) 'The Pregnancy Brain: How Mothers Think Differently', *Newsweek Magazine*, February 20, 2012.

Pomeroy, S.B. (1975) *Goddesses, Whores, Wives and Slaves* (New York: Shocken).

(1991) *Women's History and Ancient History* (Chapell Hill: University of North Carolina Press).

Poster, C. (1998) '(Re)positioning Pedagogy: A Feminist Historiography of Aristotle's *Rhetorica*' in Freeland (1998a), pp. 327–49.

Price, H. (1996) *Time's Arrow and Archimedes' Point: New Directions for the Physics of Time* (Oxford: Oxford University Press).

Preus, A. (1970) 'Science and Philosophy in Aristotle's Generation of Animals', *Journal of the History of Biology* 3/1, 1–52.

(1975) *Science and Philosophy in Aristotle's Biological Works* (New York: Hildesheim).

(1977) 'Galen's Criticism of Aristotle's Conception Theory', *Journal of the History of Biology* 10/1, 65–85.

(1983) 'Aristotle and Hippocratic Gynaecology' in J. Irmscher and R. Müller (eds.), *Aristoteles als Wissenschaftstheoretiker* (Berlin: Akademie Verlag), pp. 183–96.

(1990) 'Man and Cosmos in Aristotle: Metaphysics Lambda and the Biological Works' in Pellegrin and Devereux (1990), pp. 471–90.

Putnam, H. (1975) *Mind, Language and Reality, Philosophical Papers*, vol. 2 (Cambridge: Cambridge University Press).

Quinn, W. (1984) 'Abortion, Identity and Loss', *Philosophy and Public Affairs* 13/1, 24–54.

Reiche, H.A.T. (1960) *Empedocles' Mixture, Eudoxan Astronomy and Aristotle's Connate Pneuma* (Amsterdam: Hakkert).

Rees, M. (2010) 'What We'll Never Know' from the Reith Lectures 2010: Scientific Horizons. http://downloads.bbc.co.uk/rmhttp/radio4/transcripts/20100615-reith.rtf

Reeve, C.D.C. (1995) *Practices of Reason: Aristotle's Nicomachean Ethics* (Oxford: Clarendon).

(2000) *Substantial Knowledge: Aristotle's Metaphysics* (Indianapolis, IN: Hackett).

(2012) *Action, Contemplation and Happiness: An Essay on Aristotle* (Cambridge, MA: Harvard University Press).

Rich, A. (1977) *Of Woman Born* (London: Virago).

Richardson-Lear, G. (2004) *Happy Lives and the Highest Goods: An Essay on Aristotle's Ethics* (Princeton, NJ: Princeton University Press).

Rist, J. (1989) *The Mind of Aristotle: A Study in Philosophical Growth* (Toronto: University of Toronto Press).

Robinson, E. (2012) 'Todd Akin's Comment Brings "War on Women" Back to Prominence', *The Washington Post*, 20 August 2012.

Robinson, H. (1983) 'Aristotle's Dualism', *Oxford Studies in Ancient Philosophy* 1, 123–44.

Romanoff, A.L. (1960) *The Avian Embryo: Structural and Functional Development* (New York: Macmillan).

Rorty, R. (1984) 'The Historiography of Philosophy: Four Genres' in R. Rorty, J.B. Schneewind and Q. Skinner (eds.) (1984) *Philosophy in History* (Cambridge: Cambridge University Press), pp. 49–76.

Rosaldo, M.Z. and L. Lamphere (eds.) (1974) *Woman, Culture and Society* (Stanford, CA: Stanford University Press).

Rose, H. and S. Rose (2011) 'Never Mind the Bollocks', *London Review of Books* 33, 17–8.

Ross, W.D. (1949) *Aristotle*, 5th edn. (London: Methuen).

Roussel, M. (1980), 'Éther et Chaleur Dans l'Embryologie Aristotélicienne: Influences Archaïques' in *Mélanges D'Études Anciennes offerts à Maurice Lebel* (St-Jean-Chrysostôme, Quebec: Les Éditions du Sphinx), pp. 157–60.

Rousselle, A. (1988) *Porneia: On Desire and the Body in Antiquity*, trans. F. Pheasant (Oxford: Blackwell). Originally published in French as *Porneia: De la maîtrise du corps a la privation sensorielle* (Paris: Press Universitaires de France, 1983).

Ruether, R.R. (2005) *Goddesses and the Divine Feminine: A Western Religious History* (Berkeley: University of California Press).

Saïd, S. (1983) 'Feminin, Femme et Femelle dans les Grands Traites Biologiques d'Aristote' in E. Levy (ed.) *La Femme dans les Sociétés Antiques: Actes des Colloques de Strasbourg* (Strasbourg: AECR), pp. 93–123.

Sandel, M.J. (2009) *Justice: What's the Right Thing to Do?* (London: Allen Lane).

Sauvé Meyer, S. (1992) 'Aristotle, Teleology, and Reduction', *The Philosophical Review* 101/4, 791–825.

Saxonhouse, A. (1985) 'Aristotle: Defective Males, Hierarchy, and the Limits of Politics' ch.4 in her *Women in the History of Political Thought: Ancient Greece to Machiavelli* (Westport, Conn.: Praegar), Repr. in C. Pateman and M. L. Shanley (eds.) (1991) *Feminist Interpretations and Political Theory* (Cambridge: Polity Press), pp. 32–52.

Scaltsas, T.D. (1994) *Substance and Universal in Aristotle's Metaphysics* (Ithaca, NY: Cornell University Press).

Scaltsas, T., D. Charles and M.L. Gill (1994) (eds.) *Unity, Identity, and Explanation in Aristotle's Metaphysics* (Oxford: Oxford University Press).

Schaps, D. (1977) 'The Woman Least Mentioned: Etiquette and Women's Names', *Classical Quarterly* 27/2, 323–330.

(1979) *Economic Rights of Women in Ancient Greece* (Edinburgh: Edinburgh University Press)

Scharle, M. (2008) 'The Role of Material and Efficient Causes in Aristotle's Natural Teleology' in Mouracade (2008a), pp. 27–45.

Scheman, N. (1993) *Engenderings: Construction of Knowledge, Authority, and Privilege* (London: Routledge).

Schofield, M. (1990) 'Ideology and Philosophy in Aristotle's Theory of Slavery' in G. Patzig (ed.) *Aristoteles' 'Politik'. Akten des XI Symposium Aristotelicum* (Göttingen), pp. 1–27.

Schofield, M. and M.C. Nussbaum (1982) *Language and Logos: Studies in Ancient Greek Philosophy Presented to G.E.L. Owen* (Cambridge: Cambridge University Press).

Schott, R.M. (ed.) (1997) *Feminist Interpretations of Immanuel Kant* (Bloomington: Indiana University Press).

Scott, D. (2000) 'Aristotle on Posthumous Fortune', *Oxford Studies in Ancient Philosophy* 17, 211–29.

Sedley, D.N. (1991) 'Is Aristotle's Biology Anthropocentric?', *Phronesis* 36, 179–96.

(1999) 'Stoic Physics and Metaphysics' in K. Algra, J. Barnes, J. Mansfeld and M. Schofield (eds.) (1999) *The Cambridge History of Hellenistic Philosophy* (Cambridge: Cambridge University Press), pp. 382–411.

(2007) *Creationism and Its Critics in Antiquity* (Berkeley: University of California Press).

Sharples, R. (1985) 'Species, Form and Inheritance: Aristotle and After' in Gotthelf (1985) pp. 117–28.

(1995) *Theophrastus of Eresus : Sources of His Life, Writings, Thought and Influence. Commentary 5, Sources on Biology: Human Physiology, Living Creatures, Botany – texts* 328–435 (Leiden: Brill).

(2005) 'Some Thoughts on Aristotelian Form: With Special Reference to Metaphysics Z 8', *Science in Context* 18, 93–109.

Sherwin, S. (1991) 'Abortion Through a Feminist Ethics Lens', *Dialogue* 30, 327–42.

Shields, C. (1988) 'Soul and Body in Aristotle', *Oxford Studies in Ancient Philosophy* 6, 103–38.

(1990a) 'The Generation of Form in Aristotle', *History of Philosophy Quarterly* 7/4, 367–90.

(1990b) 'The First Functionalist' in J.-C. Smith (ed.) *Historical Foundations of Cognitive Science* (Dordrecht: Kluwer Academic Publishers), pp. 19–34.

(2007) *Aristotle* (London: Routledge).

(2012) (ed.) *Oxford Handbook of Aristotle* (Oxford: Oxford University Press).

Shildrick, M. (1997) *Leaky Bodies and Boundaries: Feminism, Post Modernism and (Bio) Ethics* (London: Routledge).

Sier, K. (2009) 'Weiblich und Männlich. Ihre Funktion bei der Zeugung nach Aristoteles und Platon' in C. Brockmann, W. Brunschön and O. Overwien (eds.) *Antike Medizin im Schnittpunkt von Geistes– und Naturwissenschaften* (Berlin: Walter de Gruyter).

Singer, P. and K. Dawson (1988) 'IVF Technology and the Argument from Potential', *Philosophy and Public Affairs* 17/2, 87–104.

Siraisi, N. (1990) *Medieval and Early Renaissance Medicine: An Introduction to Knowledge and Practice* (Chicago: Chicago University Press).

Sissa, G. (1990) *Greek Virginity*, trans. A. Goldhammer (Cambridge, MA: Harvard University Press).

(1992) 'The Sexual Philosophies of Plato and Aristotle' in G. Duby and M. Perrot (general eds.), P.S. Pantel (ed. vol 1); A. Goldhammer (trans.), *A History of Women in the West*, vol. 1, pp. 46–81. Originally published in Italian as *Storia delle Donne in Occidente*, vol. 1. (Rome: L'Antichita, 1990).

Sly, D. (1990) *Philo's Perception of Women* (Atlanta, GA.: Scholars Press).

Smith, J.M. (1988) *Games, Sex and Evolution* (Brighton: Harvester Wheatsheaf).

Solmsen, F. (1957) 'The Vital Heat, the Inborn Pneuma and the Aether', *Journal of Hellenic Studies* 77, 119–26.

Sorabji, R. (1974) 'Body and Soul in Aristotle', *Philosophy* 49, 63–89. Repr. in Barnes *et al.* (1975–9), pp. 42–64.

(1980) *Necessity, Cause and Blame: Perspectives on Aristotle's Theory* (London: Duckworth).

Sparshott, F. (1985) 'Aristotle on Women', *Philosophical Inquiry* 7/3–4, 177–200.

Spelman, E.V. (1982) 'Woman as Body: Ancient and Contemporary Views', *Feminist Studies* 8/1, 109–31.

(1988) *Inessential Woman: Problems of Exclusion in Feminist Thought* (Boston: MA: Beacon Books).

Sprague, R. (1991) 'Plants as Aristotelian Substances', *Illinois Classical Studies* 16, 221–9.

Steinman, G. (2006) 'Mechanisms of Twinning: VII. Effect of Diet and Heredity on Human Twinning Rate', *Journal of Reproductive Medicine* 51, 406–10.

Sussman, L.S. (1978) 'Workers and Drones: Labor, Idleness and Gender Definition in Hesiod's Beehive', *Arethusa* 11, 43–87.

Taylor, C.C.W. (1990) 'Aristotle's Epistemology' in S. Everson (ed.) *Cambridge Companion to Ancient Thought: Epistemology* (Cambridge: Cambridge University Press), pp. 116–42.

Thomasen, A. (1980) 'Historia Animalium Contra Gynaecia in der Literatur des Mittelalters', *Clio Medica* 15/1–2, 5–23.

Thompson, D'A.W. (1895) *A Glossary of Greek Birds* (Oxford: Clarendon).

(1913) *On Aristotle as a Biologist* (Oxford: Clarendon).

(1947) *A Glossary of Greek Fishes* (Oxford: Oxford University Press).

Tracy, T.J. (1969) *Physiological Theory and the Doctrine of the Mean in Plato and Aristotle* (The Hague: Moulton).

Tress, D.M. (1992) 'The Metaphysical Science of Aristotle's GA and His Feminist Critics', *Review of Metaphysics* 46, 307–41.

(1997) 'Aristotle's Child: Development Through Genesis, Oikos and Polis', *Ancient Philosophy* 17, 63–84.

(1999) 'Aristotle Against the Hippocratics on Sexual Generation: A Reply to Coles', *Phronesis* 44/3, 228–41.

Tuana, N. (1989) 'The Weaker Seed: The Sexist Bias of Reproductive Theory' in N. Tuana (ed.), (1989) *Feminism and Science* (Bloomington: Indiana University Press, 1989), pp. 147–71.

(1992) *Woman and the History of Philosophy* (New York: Paragon Press).

(1993) *The Less Noble Sex: Scientific, Religious, and Philosophical Conceptions of Woman's Nature* (Bloomington: Indiana University Press).

(1994a) 'Aristotle and the Politics of Reproduction' in B. Bar On (ed.), *Critical Feminist Readings on the History of Western Philosophy* (Albany: State University of New York Press), pp. 189–206.

(1994b) (ed.) *Feminist Interpretations of Plato* (Bloomington: Indiana University Press).

Tuana, N. and R. Tong (1995) *Feminism and Philosophy: Essential Readings in Theory, Reinterpretation and Application* (Boulder, CO: Westview).

Unger, P. (1996) *Living High and Letting Die: Our Illusion of Innocence* (Oxford: Oxford University Press).

Van der Eijk, P. (1997) 'The Matter of Mind: Aristotle on the Biology of Psychic Processes and the Bodily Aspects of Thinking' in Kullmann and Follinger (1997), pp. 231–58.

(1999) 'On Sterility ('HA X'), A Medical Work by Aristotle?', *Classical Quarterly* 49/2, 490–502.

(2005) *Medicine and Philosophy in Classical Antiquity: Doctors and Philosophers* (Cambridge: Cambridge University Press).

(2007) 'Les Mouvements de la Matière dans la Génération des Animaux selon Aristote' in V. Boudon-Millot, A. Guardasole, C. Magdelaine (eds.) *La Science Médicale Antique, Nouveaux Regards: Études Réunites en l'Honneur de Jacques Jouanna* (Paris: Beauchesne), pp. 405–24.

Verbeke, G. (1945) *L'Evolution de la Doctrine de Pneuma: Du Stoicisme à S. Augustin* (Louvain: Editions de l'Institut Supérieur de Philosophe).

(1978) 'Doctrine du Pneuma et Entéléchisme chez Aristote' in G.E.R. Lloyd and G.E.L. Owen, *Aristotle on the Mind and the Senses* (Cambridge: Cambridge University Press), pp. 191–214.

Vinci, T. and J. Scott Roberts (2005) 'Aristotle and Modern Genetics', *Journal of the History of Ideas* 66/2, 201–21.

Vlastos, G. (1947) 'Equality and Justice in Early Greek Cosmologics', *Classical Philology* 62/3, 156–78.

Von Staden, H. (1989) *Herophilus: The Art of Medicine in Early Alexandria: Edition, Translation and Essays* (Cambridge: Cambridge University Press).

Waithe, M.E. (1987) *A History of Women Philosophers* (Boston, MA: Nijhoff).

Ward, J.K. (2008) *Aristotle on Homonymy: Dialectic and Science* (Cambridge: Cambridge University Press).

Wardy, R. (1990) *The Chain of Change: A Study of Aristotle Physics VII* (Cambridge: Cambridge University Press).

(1993) 'Aristotelian Rainfall or the Lore of Averages', *Phronesis* 38/1, 18–30.

Warren, J. (2002) 'Democritus, the Epicureans, Death, and Dying', *Classical Quarterly* 52, 193–206.

Weber, B. (2003) 'Life', *Stanford Encyclopedia of Philosophy* on-line http://plato.stanford.edu/entries/life.

Weiner, C. (2010) 'Aristotle's Use of Division in *Generation of Animals* II.1' Unpublished manuscript.

Weisheipl, J. (1965) 'Matter in Fourteenth-Century Science' in E. McMullin (1965) *The Concept of Matter in Greek and Medieval Philosophy* (Notre Dame, IN: University of Notre Dame Press), pp. 147–69.

Whitbeck, C. (1973–4) 'Theories of Sexual Difference', *The Philosophical Forum* 5/1–2. Repr. in Gould and Wartofsky (1976), pp. 54–80.

(1989) 'A Different Reality: Feminist Ontology' in Garry and Pearsall (1989), pp. 51–76.

Whitford, M. (1991a) *The Irigaray Reader* (Oxford: Blackwell).

(1991b) *Luce Irigaray: Philosophy in the Feminine* (London: Routledge).

Whiting, J. (1992) 'Living Bodies' in Nussbaum and Rorty (1992), pp. 75–92.

Wians, W. (1996) (ed.) *Aristotle's Philosophical Development: Problems and Prospects* (Lanham, MD: Rowman & Littlefield).

Wieland, W. (1962) *Die Aristotelische Physik: Untersuchungen über die Grundlegung der Naturwissenschaft und die sprachlichen Bedingungen der Prinzipienforschung bei Aristoteles* (Göttingen: Vandenhoeck and Ruprecht).

(1975) 'The Problem of Teleology' in Barnes *et al.* (1975) vol. 1, pp. 141–60.

Williams, B. (1994) 'Descartes and the Historiography of Philosophy' in J. Cottingham (ed.) *Reason, Will and Sensations: Studies in Descartes' Metaphysics* (Oxford: Clarendon Press), pp. 19–27.

Wilshire, D. (1989) 'The Uses of Myth, Image, and the Female Body in Re-visioning Knowledge' in Jaggar and Bordo (1989), pp. 92–114.

Witt, C. (1985) 'Form, Reproduction, and Inherited Characteristics in Aristotle's *GA*', *Phronesis* 30/1, 46–57.

(1989) *Substance and Essence in Aristotle: An Interpretation of Metaphysics VII–IX* (Ithaca, NY: Cornell University Press).

(1998) 'Form, Normativity, and Gender in Aristotle' in Freeland (1998), pp. 118–37.

(2000) 'Feminist History of Philosophy', *The Stanford Encyclopedia of Philosophy* (Winter 2000 Edition), E.N. Zalta (ed.), http://plato.stanford.edu/archives/win2000/entries/feminism-femhist/.

Wolf, S. (1987) 'Sanity and the Metaphysics of Responsibility' in F. Shoeman (ed.) *Responsibility, Character and the Emotions: New Essays in Moral Psychology* (Cambridge: Cambridge University Press), pp. 46–62.

Woods, M.J. (1968) 'Problems in Metaphysics Z, Chapter 13' in J. Moravcsik (ed.) *Aristotle: A Collection of Critical Essays* (London: Macmillan), vol. 1, pp. 215–38.

(1994) 'The Essence of a Human Being and the Individual Soul in Metaphysics Z and H' in Scaltsas *et al.* (1994) pp. 272–90.
Zeitlin, F. I. (1996) *Playing the Other: Gender and Society in Classical Greek Literature* (Chicago: University of Chicago Press).
Zuby, G. (2000) *Origins of Life: On Earth and in the Cosmos* (London: Elsevier).

INDEX LOCORUM

GENERAL INDEX